PROBLEMS OF CONDENSED MATTER PHYSICS

Contemporary Fundamental Physics
Valeri V. Dvoeglazov (Editor)

Problems of Condensed Matter Physics
S. J. Vlaev and L. M. Gaggero-Sager
ISBN 1-56072-899-2

Symmetry and Electromagnetic Interactions of Fields with Multi-Spin
S. I. Kruglov
ISBN 1-56072-880-9

Photon and Old Problems in Light of New Ideas
Valeri V. Dvoeglazov
ISBN 1-56072-810-8

Photon and Poincaré Group
Valeri V. Dvoeglazov
ISBN 1-56072-718-7

Instantaneous Action at a Distance in Modern Physics: "Pro" and "Contra"
Andrew Chubykalo, Viv Pope and Roman Smirnov-Rueda
ISBN 1-56072-698-9

The Weyl-Dirac Theory and Our Universe
Mark Israelit
ISBN 1-56072-551-6

PROBLEMS OF CONDENSED MATTER PHYSICS

S. J. VLAEV AND L. M. GAGGERO-SAGER
EDITORS

Nova Science Publishers, Inc.
Huntington, New York

Library of Congress Cataloging-in-Publication Data

p. cm.
Includes bibliographical references and index.
ISBN 1-56072-889-2

CIP

227 Main Street, Suite 100
Huntington, New York 11743
Tele. 631-424-6682 Fax 631-425-5933
e-mail: Novascience@earthlink.net
Web Site: http://www.nexusworld.com/nova

Printed in the United States of America

CONTENTS

Editorial Introduction

The present book includes eight works on some contemporary problems of condensed matter physics. The choice of the topics reflects the scientific interest of the editors. We hope the book will find a wide circle of readers.

Chapter 1 examines critically the studies of the quasiregular (quasiperiodic) heterostructures. The minimal basis of mathematical background required for a more rigorous characterization of spectra of quasiregular sequences is summarized and then the selfsimilarity and the fractal nature of the spectrum are discussed on this basis.

Chapter 2 presents some new aspects of the empirical tight-binding model and its applications for semiconductor heterostructure calculations. The objective is to show how best to exploit the power and circumvent the limitations of the method as applied to heterostructure calculations.

An introduction to the phenomenology of type II superconductivity and sandpile physics is given in Chapter 3. Key experiments on vortex avalanches performed in the last five years are revieved and some new experiments are suggested.

Chapter 4 shows that the joint use of the surface Green function matching method and the full transfer matrix technique simplifies greatly the study of inhomogeneous media. As applications, the transverse elastic waves in finite Fibonacci superlattices are studied.

Chapter 5 presents recent progress in understanding the physics of a novel class of one-dimensional ferrimagnetic Heisenberg systems composed of different antiferromagnetically coupled quantum spins in the elementary cell.

Finite temperature properties of electron-phonon systems in semiconducting heterostructures are considered in Chapter 6.

The long standing Boltzmann-Loschmidt polemic which confronts irreversible Thermodynamics with reversible Mechanics is revived in Chapter 7. The possibility offered by NMR to invert the direction of time (quantum Loschmidt daemon) in a many-body system is explored. The whole set of experiments indicates that there is quantum chaos intrinsic to the many-body spin dynamics in the thermodynamic limit.

Chapter 8 describes a simple model based on the Thomas-Fermi-Dirac approximation for an inhomogeneous hole gas. This model gives results in good agreement with the experiment and self-consistent calculations for delta-doped quantum wells.

Finally, the editors wish to express their gratitude to the following: Dr. E. Altshuler for his very ingenious cartoons; Annamaria D'Amore for his editorial collaboration; Frank and Nadya Columbus, the President and Vice-President of Nova Science Publishers, Inc.; Dr. V. Dvoeglazov, the Editor of the series of which this book is a part, for his kind invitation to compile this book; the reviewers for their valuable advice on the selection of papers.

The Editors

PORTRAIT OF A NICE, SIMPLE-MINDED, SELF-SIMILAR FAMILY

The Spectrum of Quasiregular Heterostructures

R. Pérez-Alvarez* and F. García-Moliner†

Abstract

Quasiregular -also named quasiperiodic- sequences (e.g. Fibonacci, Thue-Morse, Period Doubling, etc.) have interesting mathematical properties. On the other hand, heterostructures following chosen quasiregular sequences can also be experimentally grown, an additional feature which increases their interest. The study of these problems is in fact attracting increasing attention and this is what is critically examined here. Early studies of idealised quasiregular sequences revealed novel and attractive features of their eigenstate spectra and this has stimulated a vast amount of work, mostly theoretical and also some experimental, searching for the same properties in actual quasiregular heterostructures. The general conclusion of this survey is that, while the field is attractive, in the present condition the claims abounding in the literature are largely based on rather vague and imprecise appreciation, basically lacking in conceptual clarity and formal rigour. On the other hand this has attracted the attention of a number of mathematicians and some interesting developments have taken place in the last decade which appear to be ignored by most physicists working on these problems. The minimal basis of mathematical background required for a more rigorous characterisation of the spectra of quasiregular sequences is first summarized and then their topological and morphological properties are discussed on this basis. Irrespective of the specific physical modes, states or quasiparticles studied, the rule by which the quasiregular heterostructure is defined determines the character of the spectrum. While attention is most often paid to the supposed selfsimilarity -with the usual imprecision pointed out- it is stressed that the fractal nature of the spectrum should be studied more carefully. The spectrum in the most commonly studied cases is always fractal, but not in the simple manner usually claimed or implied. In fact in the three instances mentioned above it is multifractal and this implies that there is no possible simple selfsimilarity. The study of quasiregular heterostructures is an open and attractive field but a critical survey suggests that it is still vastly unknown in its most basic aspects, with substantial unanswered questions which so far remain an interesting open task.

1. Introduction

For close to thirty years research on various types of quantum nanostructures and, in particular, on periodic superlattices has been very intensive [1, 2], including the development of epitaxial growth techniques with which the control of samples with prescribed design parameters has reached high quality levels. These techniques have also been successfully used to grow non periodic *digital heterostructures* [3] which have interesting properties of

*Permanent address: Departamento de Física Teórica, Universidad de La Habana, Ciudad Habana 10400, Cuba. Email: rpa@ff.oc.uh.cu

†Corresponding author. Email: garmol@sg.uji.es; Cátedra de Ciencia Contemporánea, Universitat Jaume I, 12071 Castellón, Spain

their own [4]. On the other hand, starting from the early eighties a number of theoretical studies [5-15] originated a great deal of interest in another type of structures. Suppose one takes a linear -1D- chain of atoms and studies, for instance, electronic states in a tight binding model in which there is one state per atom with an energy which can take one of two values, E_A or E_B, according to a prescribed sequence. When this is *quasiregular* or *quasiperiodic*, that is to say, when it follows a selfreplicating rule -see Section 2- then the spectrum has novel and intriguing properties which ressemble those of a simple Cantor set [16]. The most often studied case is that of the Fibonacci sequence, but other cases have been studied and will be here discussed. The details of the model could change -for instance, the same atomic level but different values of the hopping parameter- or one could study a different physical model like, say, longitudinal vibrational modes in a 1D chain with atomic masses m_A, m_B varying according to the prescribed sequence. The systems studied in these investigations constituted essentially academic problems but the intriguing features found in the resulting spectra strongly suggested that there are some interesting basic features inherent to the properties of nonperiodic but quasiregular sequences and this soon stimulated new research aimed at studying the same issues for "real" systems, either in terms of simple models [17-20] or else by actually growing quasiregular heterostructures [21-23], a development made possible by the progress in the type of epitaxial growth techniques which also facilitated the studies of digital heterostructures [3]. Research on various aspects of the spectra of quasiregular heterostructures has since then continued very active, in a few cases delving on further aspects of the academic problems and in most cases trying to approximate actual physical systems, with different degrees of elaboration in the models employed [24 - 48]. By contrast with the comparatively abundant theoretical studies, experimental work in this field has been altogether rather scarce [29,49-51]. In parallel with this, several interesting mathematical studies of the properties of the quasiregular sequences -henceforth QS- have been published in the nineties which appear to be largely ignored in most studies of the physical properties of QS or models of actual quasiregular heterostructures -henceforth QH. After describing in Section 2 the most common examples of QS we summarise the main aspects of their mathematical properties, namely the *trace maps* and their use to establish the *nature of the spectrum* (Section 3), and the *multifractal analysis* (Section 4). Section 5 discusses the application of this to the various types of physical problems -based on a QS or a QH- and Section 6 summarises the current situation. Final comments are made in Section 7 where outstanding open questions are stressed.

2. Quasiregular sequences and heterostructures

The first historical example of a QS appears to have been introduced by Fibonacci in the XIII Century. An example, studied by this author, is the growth of rabbit population described in Fig. 1. It is assumed that an adult pair (A) takes one month to produce a baby pair (B) and a baby pair takes one month to reach fertile age, after which it can produce another baby pair in one month. The figure describes the population of A and B rabbit pairs for as long as all stay alive. If F_n is the number sequence of A pairs, then the number of B pairs follows the same sequence one step behind (F_{n-1}) while the total number of pairs is one step ahead (F_{n+1}) in the same sequence and obviously

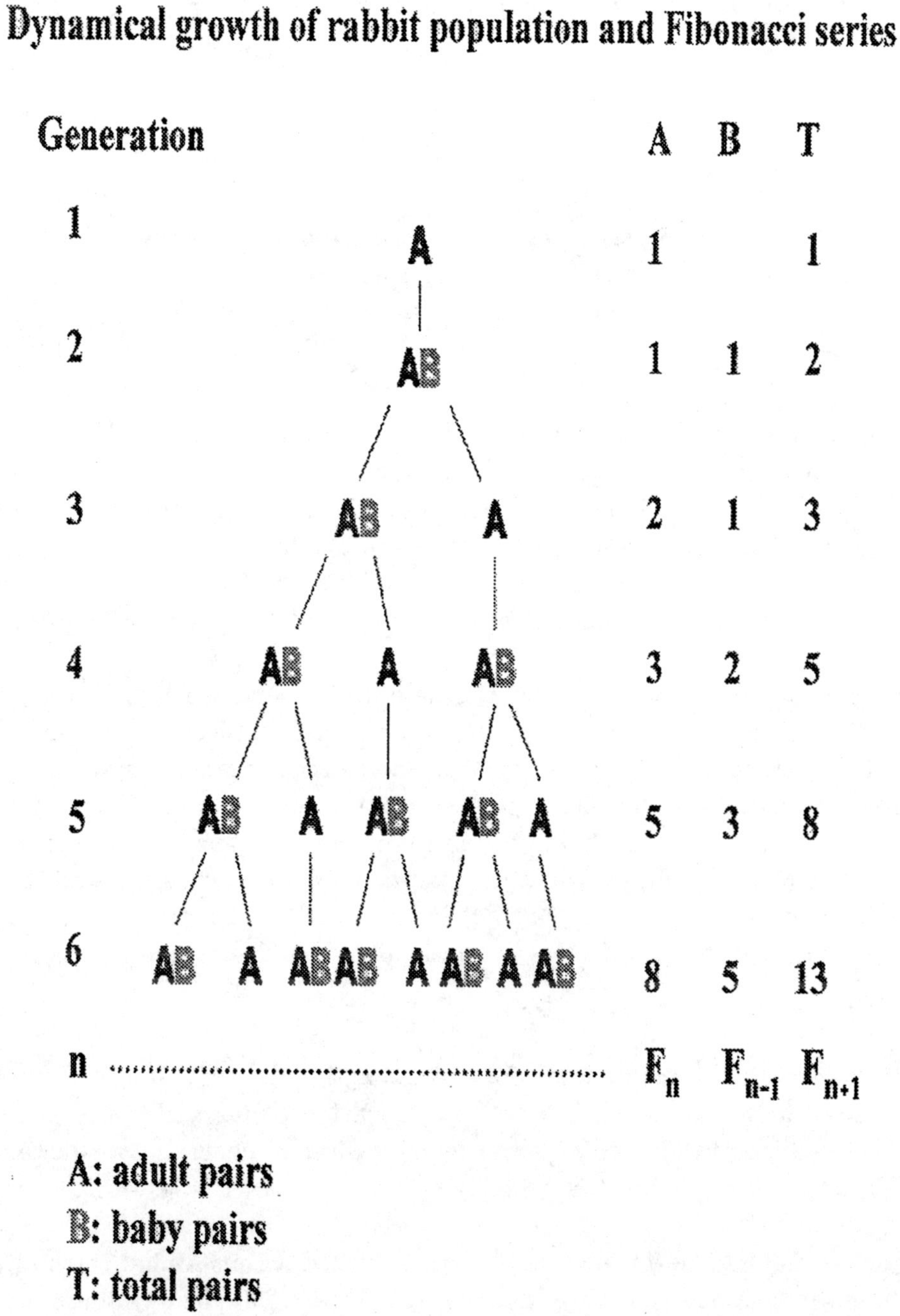

Figure 1: Growth of rabbit population and Fibonacci sequence. A: Adult pairs, B: Baby pairs, T: Total pairs.

$$F_{n+1} \quad = \quad F_n + F_{n-1} \,. \tag{1}$$

The F_n are the *Fibonacci numbers.* Given the first two numbers Eq. (1) determines the value of the generic number F_n for the n-th generation in the sequence and it is found that

$$\lim_{n\to\infty} \frac{F_n}{F_{n-1}} = \tau = \frac{1+\sqrt{5}}{2}. \tag{2}$$

τ is the *golden ratio.* We can also look at Fig. 1 and Eq. (1) in a different way. Let A denote a layer of a given material and B a layer of another material. Then the successive rows of Fig. 1 describe successive heterostructures, which we may indicate S_n, formed with A and B layers. The rule is given by (1), which we now express as

$$S_{n+1} \quad = \quad S_n \bigoplus S_{n-1}, \tag{3}$$

i.e. the juxtaposition of S_{n-1} and S_n gives the resulting structure for the generation $n+1$. We then have a *quasiregular heterostructure* (QH) built according to a prescribed *quasi-regular sequence* (QS). This is not a regular periodic heterostructure, but the prescription for successive buildup contains a selfreplicating rule. The layers A and/or B can consist of one or more materials, just as in the case of polytype superlattices. Fig. 2 shows a simple example of a polytype heterostructure built according to the Fibonacci sequence, which was studied in [46] and [47].

The mathematical structure of the QS determines some basic properties, even some theorems, which provide a formal basis to study the key features of the spectra of QH and it is interesting to take note of some key features in the developments which have taken place in this direction in the last decade. It is customary to define an *alphabet* [52] $\mathcal{A}$ consisting of a number of *letters* with which one forms *words*, out of which the infinite QS is formed. One then defines just the first term and a *substitution rule* ξ which maps the alphabet $\mathcal{A}$ onto a set of words of finite length [52]. For instance, for the Fibonacci sequence [53,54] the alphabet $\mathcal{A}$ consists of the two letters A and B and the substitution rule is

$$\xi(A) = AB \qquad , \qquad \xi(B) = A, \tag{4}$$

which maps $\mathcal{A}$ onto the set of words of finite length $\{AB, A\}$. It then suffices to define just the first term $S_1 = A$ and successive application of (4) yields the successive terms -generations- depicted in Fig. 1. This can be defined more generally, for instance, with AB in (4) replaced by AB$^\ell$, where ℓ is an integer > 1, but $\ell = 1$ is the most common case in physics.

Many other QS can be formed in this way, with different alphabets and substitution rules. The most common cases are Thue-Morse [15, 53, 54], Rudin-Shapiro [15, 52 - 54], Period Doubling [15, 54, 55], Circular [15, 52 - 54], Binary non-Pisot [52, 54, 56] and Ternary non-Pisot [52, 54, 56]. Table 1 summarises for the most common of these cases the substitution rules and gives typical examples of a low order S_n for each one. By convention we take $S_1 = A$ and then apply successively the rule $S_n = \xi(S_{n-1})$ according to the details given in the table for each case. This displays the different degrees of complexity of the varios QSs. Henceforth the different sequences will be denoted by the initials appearing in the lower part of this table.

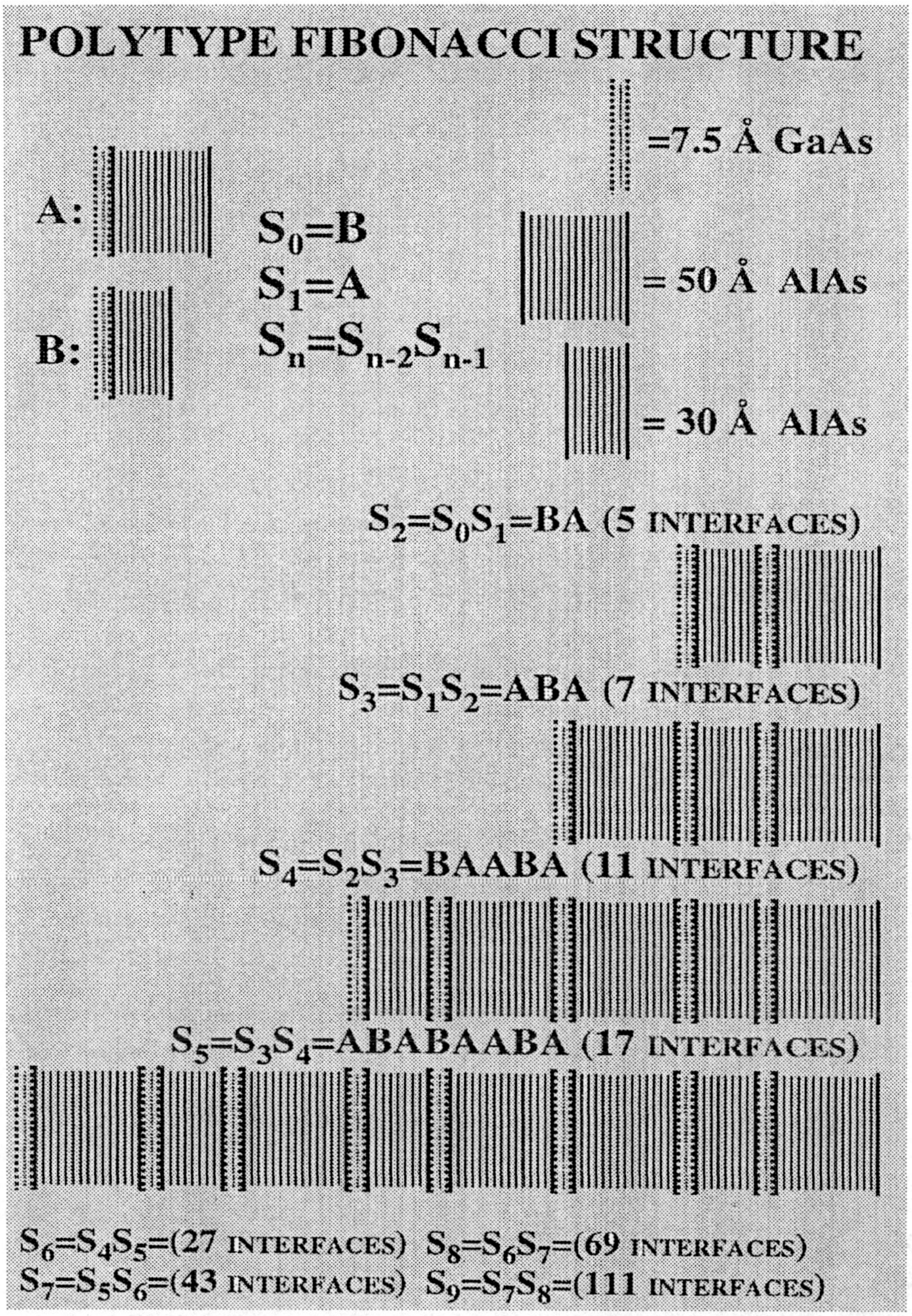

Figure 2: An example of a Fibonacci heterostructure in which the building blocks, A and B consist themselves of two layers of different materials. This QH was studied in Refs. [46] and [47] .

SEQUENCE	$\xi(A)$	$\xi(B)$	$\xi(C)$	$\xi(D)$
FIBONACCI	AB	A		
THUE-MORSE	AB	BA		
PERIOD DOUBLING	AB	AA		
CIRCULAR	CAC	ACCAC	ABCAC	
RUDIN-SHAPIRO	AC	DC	AB	DB

F:	$S_6 = ABBABBABABBAB$
T-M:	$S_4 = ABBABAABBAABABBA$
PD:	$S_4 = ABAAABABABAAABAA$
C:	$S_2 = ABCACCACABCAC$
R-S:	$S_4 = ACABACDCACABDBAB$

Table 1: Substitution rules for the most common quasiregular sequences (QS). The first term is assumed to be A.

In order to study the spectra of physical problems defined on QSs it is convenient to consider some of the basic findings resulting from many detailed mathematical studies. In this the concept of *transfer matrix* plays an important role. Transfer matrix -henceforth TM- methods have been abundantly used in the study of periodic superlattices and are reviewed in this volume [57] . We consider that definition of TM [58] which for continuos systems described by a differential problem, like a Schrödinger equation, transfers both amplitude and derivative. Note that this differs from the often used definition of **M** which transfers only the coefficients of the amplitude when expressed as a linear combination of basis functions [57]. If ψ and ψ' are known at some initial point z_0, then for any z:

$$\left| \begin{array}{c} \psi(z) \\ \psi'(z) \end{array} \right| = \mathbf{M}(z, z_0) \cdot \left| \begin{array}{c} \psi(z_0) \\ \psi'(z_0) \end{array} \right|. \tag{5}$$

This defines **M** as the TM. For discrete systems described in terms of finite difference equations, like a vibrating linear chain, the relationship defining **M** involves the amplitude at two different consecutive points. Thus

$$\left| \begin{array}{c} u(m) \\ u(m-1) \end{array} \right| = \mathbf{M}(m, m_0) \cdot \left| \begin{array}{c} u(m_0) \\ u(m_0 - 1) \end{array} \right|. \tag{6}$$

We note for further reference that all the fundamental mathematical studies of the properties of QS apply to cases in which the TM is *unimodular and is a* 2×2 *matrix.* From the physics point of view this is restricted to a class of *simple problems* in which there is only one amplitude. Such are, for instance, electron states in the one band tight binding model or purely longitudinal or transverse vibrational modes of linear atomic chains following a QS. For continuos systems modelling "real life" QHs the main examples are: The one band effective mass Schrödinger equation, the Poisson equation for a simple dielectric model -of, say, plasmons in a phenomenological model of the electron gas or polar optical modes in a simple model for ionic crystals- or pure transverse elastic modes. This proviso must be kept in mind from here on.

3. Simple problems: Trace maps and nature of the spectrum

As mentioned above, the mathematical properties of many QSs have been studied in detail [52, 54, 55, 59 - 73]. In particular, Refs. [52, 54, 55] are interesting as they contain an effort to translate the results of the mathematical analysis into a language closer to the physicists with the purpose of trying to help close the gap which at present seems to exist between the two communities.

We stress that the problem has two ingredients, namely (i)the nature of the QS, embodied in the initial alphabet and the substitution rules and (ii)the *simple* nature of the problem defined on the given QS, which means that we are dealing with unimodular 2×2 TMs. In this case application of the *Cayley-Hamilton* theorem

$$\mathrm{M}^2 - (tr\mathrm{M})\mathrm{M} + \mathbf{I} \;=\; \mathbf{0} \tag{7}$$

yields the following relations [74]:

$$\begin{aligned}
tr(\mathrm{M}_1\mathrm{M}_2) &= tr(\mathrm{M}_1)tr(\mathrm{M}_2) - tr(\mathrm{M}_1^{-1}\mathrm{M}_2), && (8)\\
tr\mathrm{M}^2 &= (tr\mathrm{M})^2 - 2, && (9)\\
tr(\mathrm{M}_1\mathrm{M}_2\mathrm{M}_1\mathrm{M}_3) &= (tr\mathrm{M}_1\mathrm{M}_2)(tr\mathrm{M}_1\mathrm{M}_3) - (tr\mathrm{M}_2)(tr\mathrm{M}_3) + \\
&\quad tr(\mathrm{M}_2\mathrm{M}_3), && (10)\\
tr(\mathrm{M}_1^2\mathrm{M}_2) &= (tr\mathrm{M}_1)(tr\mathrm{M}_1\mathrm{M}_2) - tr\mathrm{M}_2, && (11)\\
\mathrm{M}^3 &= (tr\mathrm{M})^2\mathrm{M} - \mathrm{M} - (tr\mathrm{M})\mathrm{I}, && (12)\\
tr\mathrm{M}^3 &= (tr\mathrm{M})^3 - 3(tr\mathrm{M})\;. && (13)
\end{aligned}$$

. These equalities prove useful in constructing an object which plays a central role in the said mathematical analysis, namely the so called *trace map* [52, 54, 55, 59 - 73], which is constructed as follows.

Denote the S_n of a given QS by A_n/B_n when by convention the starting generation is $S_1 = A/B$ and associate a TM denoted $\mathcal{A}_n/\mathcal{B}_n$ to the n-th generation in each case. Then the substitution rules of the sequence induce recurrence relations between the TMs for the successive generations. For instance, in the F sequence, from (4),

$$\begin{aligned}
\mathcal{A}_n &= \mathcal{B}_{n-1}\mathcal{A}_{n-1}\;, && (14)\\
\mathcal{B}_n &= \mathcal{A}_{n-1}\;. && (15)
\end{aligned}$$

These recurrence relations can be very practical for doing actual computations but they also prove useful to induce something similar between their traces. We define, in all cases,

$$x_n = tr\mathcal{A}_n \qquad , \qquad y_n = tr\mathcal{B}_n\;. \tag{16}$$

In general, application of the substitution rules makes it necessary to introduce new traces. In the F case we must introduce another trace, namely,

$$z_n = tr(\mathcal{A}_n\mathcal{B}_n) \tag{17}$$

and since, from (14)-(15),

$$\begin{aligned}
x_{n+1} &= z_n && (18)\\
y_{n+1} &= x_n && (19)
\end{aligned}$$

we need another recurrence relation for z_n. In general this process brings into the analysis further traces, but the equalities (7) through (13) can be used to break the chain. In fact it is proved [53] that for simple problems it is always possible to limit the chain to a finite number of matrices and traces. For instance, in the F case, from (17) and (11), making use of the cyclic invariance of the trace, one obtains readily that

$$z_{n+1} = x_n z_n - y_n . \tag{20}$$

Equations (18)-(20) constitute a limited trace map for the F sequence and involve a finite number of traces corresponding to a finite number -three in this case- of matrices, namely $\mathcal{A}_n$, $\mathcal{B}_n$ and $\mathcal{B}_n\mathcal{A}_n$. The corresponding finite set of words $\{A, B, AB\}$ is said to constitute the *lexicon* of the sequence [54]. The form of the trace map is not unique and can be cast in different ways. What matters is that simple problems always have an associated trace map [53]. It is also possible -and customary- to cast the trace map in a more concise form which we shall call *compact*.

Returning to the F case, by eliminating y and z terms, (18) through (20) can be condensed into a recurrence formula for x_n, namely

$$x_n = x_{n-1}x_{n-2} - x_{n-3} . \tag{21}$$

This is the *compact* trace map for the F sequence. A more general F-type sequence can be defined, for instance,

$$\xi(A) = AB^{\ell} \quad , \quad \xi(B) = A \quad , \quad \ell > 1 . \tag{22}$$

By using, for $\ell = 2$, the method to be explained presently, we obtain the compact trace map given in the second row of Table 2. These results illustrate a general principle. It is proved [52] that if L is the cardinality of the lexicon, then the recurrence relation for x_n can involve at most down to x_{n-L}. For the two cases the alphabet is the same and for $\ell = 2$ the lexicon is $\{A, B, ABB\}$, which also has cardinality $L = 3$. In both cases the recurrence relation for x_n involves down to x_{n-3}, but $\ell = 2$ is more complicated and has a more complicated compact trace map.

For the regular periodic sequence

$$\xi(A) = AB \quad , \quad \xi(B) = AB , \tag{23}$$

and it is very easy to obtain the compact trace map given in the last row of Table 2, which also lists the compact trace maps for another sequence, denoted LWWSM, which was introduced in Ref. [51] and will be discused later.

In all cases so far mentioned, and also in T-M and PD the compact trace map has been cast as a recurrence formula in terms of the x only, but this is not a general rule. The C and R-S sequences, based on alphabets with more letters, are more complicated.

For the C sequence [53] the compact trace map given in the table involves six traces, namely (16) and

$$\begin{aligned} z_n &= tr\mathcal{C}_n , \\ s_n &= tr(\mathcal{C}_n\mathcal{A}_n) , \\ u_n &= tr(\mathcal{B}_n\mathcal{A}_n) , \\ w_n &= tr(\mathcal{A}_n\mathcal{C}_n\mathcal{B}_n) \end{aligned} \tag{24}$$

Sequence	Compact trace map	Ref.
F	$x_n = x_{n-1}x_{n-2} - x_{n-3}$	[5]
F ($\ell = 2$)	$x_n = x_{n-1}x_{n-2}^2 - x_{n-2}x_{n-3}^2 + 2x_{n-2} - x_{n-1}$	
T-M	$x_n = 4x_{n-2}^2(x_{n-1} - 2) + 2 \; ; \; n \geq 3$	[15]
PD	$x_n = (x_{n-2}^2 - 1)x_{n-1} - x_{n-2}(x_{n-3}^2 - 2)$	[15]
C	$x_n = z_{n-1}s_{n-1} - x_{n-1}$	[53]
	$y_n = s_{n-1}x_n - z_{n-1}$	
	$z_n = s_{n-1}w_{n-1} - y_{n-1}$	
	$s_n = x_n z_n - u_{n-1}$	
	$u_n = x_n y_n - s_{n-1}$	
	$w_n = y_n s_n - w_{n-1}$	
R-S	$x_n = x_{n-1}z_{n-1} + s_{n-1}$	[53]
	$y_n = z_{n-1}w_{n-1} + t_{n-1}$	
	$z_n = x_{n-1}y_{n-1} + t_{n-1}$	
	$w_n = y_{n-1}w_{n-1} + s_{n-1}$	
	$s_n = r_{n-1} - y_{n-1}z_{n-1}$	
	$t_n = q_{n-1} - x_{n-1}w_{n-1}$	
	$q_n = x_{n-1}w_{n-1}r_{n-1} - x_{n-1}^2 - w_{n-1}^2 + 2$	
	$r_n = y_{n-1}z_{n-1}q_{n-1} - y_{n-1}^2 - z_{n-1}^2 + 2$	
LWWSM	$x_n = 2x_{n-2}x_{n-3} - 2x_{n-3}x_{n-5} + x_{n-7}$	
Periodic	$x_n = 2x_{n-1}^2 - 1$	

Table 2: Compact trace maps for the QSs of Table 1 and some more commented in the text.

and for the R-S the eight traces, namely (16) and

$$\begin{aligned} z_n &= trC_n \,, \\ s_n &= tr(C_nA_n) - trC_ntrA_n \,, \\ t_n &= tr(C_nD_n) - trC_ntrD_n \,, \\ q_n &= tr(A_nD_n) \,, \\ w_n &= trD_n \,, \\ r_n &= tr(B_nC_n) \,. \end{aligned} \quad (25)$$

The LWWSM sequence was introduced in [51] and will be discussed later in physical terms. The substitution rule is

$$\xi(A) = AC \quad , \quad \xi(B) = A \quad , \quad \xi(C) = B \,. \quad (26)$$

Hence the matrix relations

$$A_n = C_{n-1}A_{n-1} \quad , \quad B_n = A_{n-1} \quad , \quad C_n = B_{n-1} \,. \quad (27)$$

We define the traces

$x_n = trA_n$, $y_n = trB_n$, $z_n = trC_n$,
$p_n = tr(C_nA_n)$, $q_n = tr(A_nB_nC_n)$, $r_n = tr(B_nC_n)$,
$s_n = tr(A_nB_n)$.

The relations

$$x_n = p_{n-1} \quad , \quad y_n = x_{n-1} \quad , \quad z_n = y_{n-1} \tag{28}$$

are immediately obvious and

$$p_{j+1} = tr(\mathcal{C}_{j+1}\mathcal{A}_{j+1}) = tr(\mathcal{B}_j\mathcal{C}_j\mathcal{A}_j)$$

whence

$$p_n = q_{n-1} \tag{29}$$

For the rest we use (7) through (13) and the cyclic invariance of the trace, besides resorting to (27) whenever it is convenient. Thus, for instance,

$$\begin{aligned} q_{j+1} &= tr(\mathcal{A}_{j+1}\mathcal{B}_{j+1}\mathcal{C}_{j+1}) = tr(\mathcal{C}_j\mathcal{A}_j\mathcal{A}_j\mathcal{B}_j) = tr(\mathcal{A}_j^2\mathcal{B}_j\mathcal{C}_j) \\ &= tr(\mathcal{A}_j)tr(\mathcal{A}_j\mathcal{B}_j\mathcal{C}_j) - tr(\mathcal{B}_j\mathcal{C}_j) \end{aligned} \tag{30}$$

i.e.

$$q_n = x_{n-1}q_{n-1} - r_{n-1} \tag{31}$$

Likewise

$$r_n = s_{n-1} \quad , \quad s_n = x_{n-1}p_{n-1} - z_{n-1} \, . \tag{32}$$

With (28), (29), (31) and (32) we have seven relations giving x_n through s_n in terms of x_{n-1} through x_{n-1}. The Jacobian is

$$J\left(\frac{x_n, y_n, z_n, p_n, q_n, r_n, s_n}{x_{n-1}, y_{n-1}, z_{n-1}, p_{n-1}, q_{n-1}, r_{n-1}, s_{n-1}}\right) =$$

$$\begin{vmatrix} 0 & 0 & 0 & 1 & 0 & 0 & 0 \\ 1 & 0 & 0 & 0 & 0 & 0 & 0 \\ 0 & 1 & 0 & 0 & 0 & 0 & 0 \\ 0 & 0 & 0 & 0 & 1 & 0 & 0 \\ r_{n-1} & 0 & 0 & 0 & x_{n-1} & -1 & 0 \\ 0 & 0 & 0 & 0 & 0 & 0 & 1 \\ p_{n-1} & 0 & -1 & x_{n-1} & 0 & 0 & 0 \end{vmatrix} = -1 \, . \tag{33}$$

Since the Jacobian does not vanish, the reduced map involves seven equalities for seven traces. The lexicon has cardinality $L = 7$ and is $\{A, B, C, AC, ABC, BC, AB\}$. The least repeated traces in the seven relations are y, z, q, s. Thus we seek to eliminate these. Then from (28), (31) and (32) we obtain

$$q_n = x_{n-1}q_{n-1} - x_{n-3}p_{n-3} + x_{n-5} \tag{34}$$

which, by using (28) and (29), can be cast in terms of x variables only and we obtain the compact trace map given in the last but one row of Table 2. Note that x_n involves down to x_{n-7}. In the same way the compact trace map can be obtained for any given sequence.

In the first studies of F sequences it was customary to define x_n as $(1/2)tr\mathcal{A}_n$, probably because the dispersion relation for simple problems on F sequences can always be obtained in terms of one half of a trace. The form of the compact trace map is then the same with some trivial changes in numerical factors. Other authors have defined x_n without the factor $(1/2)$. This is the criterion we have followed here in all cases for the sake of uniformity.

At this stage it is convenient to pause and consider a basic property of the eigenvalue spectrum of a given problem -which, we recall, is assumed to be *simple* in the sense explained above. Consider, for instance, the electronic state eigenvalues of a regular -periodic- superlattice. The spectrum consists of minibands and is piecewise continuous. More specifically, it is ***absolutely continuous***, which means that given any point of the spectrum and any small interval about this point, all the points of the interval belong to the spectrum. The obviuous exceptions are the miniband edges, for which the interval must be one sided, just as in the free electron spectrum for the lower bound -say, $E = 0$- of the one band of states. The result is that on the real line the points of the spectrum form a onedimensional set. This specification for the band edges will always be understood and need not be explicitly stated henceforth. Consider a different situation: Given any point of a spectrum in any arbitrarily small interval about it there are always points of the spectrum, but not all points of the interval belong to it. Then the spectrum is *singular continuous*; its points no longer cover the intervals of the real line densely as in the absolutely continuous case.

The characterization of this topological property will be discussed in the next section. For the moment it suffices to note that it is a fundamental property which makes the spectrum of some QHs distinctly different from that of a regular superlattice. Now, given a QS, the *Bovier-Ghez theorem* [52] establishes that if some conditions are satisfied, then the spectrum of any simple problem defined on it is singular continuous. The structure of the trace map plays a central role in this analysis. When the sequences of Table 2 are studied in this way, then one finds that in the periodic and R-S cases the conditions of the theorem are not satisfied, hence it cannot be proved whether the spectrum is or not singular continuous. In fact for the superlattice it is known to be absolutely continuous. The R-S case will be discussed in the next section. From this analysis it can be proved that the spectrum is singular continuous for the cases F, T-M, PD, C, and also for the LWWSM sequence.

4. Fractal analysis

Having started from linear equations, the mathematical analysis is based on the study of the trace maps, which consist of *nonlinear* equations. Thus there are some affinities with the studies of chaotic phenomena described in terms of nonlinear dynamics. The similarity manifests itself in a particularly clear way with respect to the topological properties of singular continuous spectra. In an abstract way the term *event* may mean either finding an eigenvalue in an interval of the real line or having a dynamical trajectory pass through an elementary domain in a phase space. In the nomenclature employed by some authors [5] the trace map is defined as the dynamical system under study. The fact, mentioned above, that singular continuous spectra do not cover the intervals of the real axis densely means that such spectra are fractal, with a dimension less than unity. The literature on nonlinear dynamics contains abundant examples of fractal analysis [75 - 78] and we now summarise a brief reminder of some basic definitions and relations together with some observations related to the systems of concern here.

Let $\{E_i\}$, with $i = 1, 2, \ldots, N_E$ be a set of points on the real axis which we shall refer to as *the spectrum*, although it could be any geometrically defined set of points. Consider the interval -or set of intervals- occupied by the spectrum, define elementary intervals -1D boxes- of size ϵ and cover the entire interval -or set of intervals- with $N(\epsilon)$ boxes of size ϵ. Let q be a continuous parameter ranging form $-\infty$ to $+\infty$ and p_j the fraction of points in box j. The *generalised information of order* q is defined as ([78], p. 199)

$$I(\epsilon, q) = \frac{1}{1-q} \, log \sum_{j=1}^{N(\epsilon)} p_j^q \,, \tag{35}$$

and from this the *generalised box-counting dimension*

$$D_q = -\lim_{\epsilon \to 0} \frac{I(\epsilon, q)}{log\,[\epsilon]} \,. \tag{36}$$

Different concepts of dimension are defined in the literature, but for our purposes it suffices to consider this and the *simple box-counting dimension* D_{bc} defined as follows. We consider again the spectrum and cover it with $M(\epsilon)$ boxes of size ϵ. Then ([77], p.394 and [78], p. 72)

$$D_{bc} = -\lim_{\epsilon \to 0} \frac{log\,[M(\epsilon)]}{log\,[\epsilon]} \,. \tag{37}$$

The difference between $N(\epsilon)$ and $M(\epsilon)$ is that the latter, by definition, does not include any empty box, i.e. with no point of the spectrum in it. It is proved that ([77], p. 444 and [78], p. 199)

$$D_{q=0} \equiv D_0 = D_{bc} \tag{38}$$

and this, under rather general conditions, is also equal to the *simple Hausdorff dimension* [75 - 78].

In the simplest cases D_q does not depend on q and we have just one fractal dimension D_0. A case in point is the simple Cantor set, which we briefly consider here for future reference. The process of construction of this set is wellknown [75 - 78]. One starts from a finite interval of the real line, which we can take as $(0, 1)$, and divides it in three equal parts. The central part is discarded, so one is left with the intervals $(0, 1/3)$ and $(2/3, 1)$ and applies to both the same procedure. This is repeated for the four resulting intervals and so on. It is easily seen that in this case there is only one fractal dimension

$$D_0 = \frac{log2}{log3} = 0.6309\ldots \tag{39}$$

The simple Cantor set has another interesting property which is demonstrated in Fig. 3

In order to make the points of the set -the "spectrum" under study in this case- more visible we draw vertical lines through them (Fig. 3.a). We then take a fragment of the spectrum and make a zoom magnification (Fig. 3.b). The result reproduces the entire spectrum and the same happens when we take the same numerical fraction of Fig. 3.b and make a further zoom magnification into Fig. 3.c with the same magnification factor. This is the property of *selfsimilarity*. For future reference we stress that the standard literature

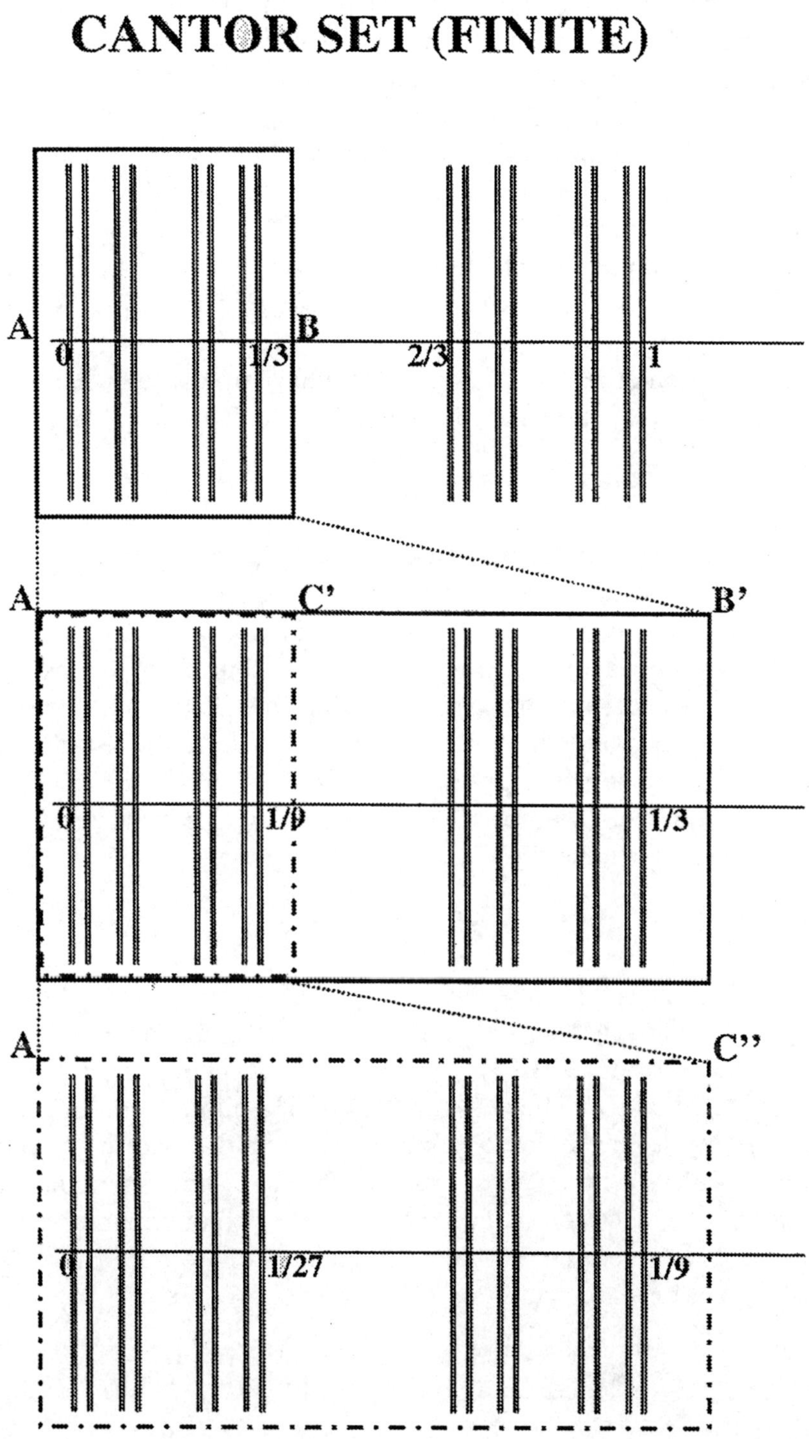

Figure 3: Selfsimilarity in a simple Cantor set. One third of the set, after magnification by a factor 3, reproduces the entire set. Zooming into one third of the result thus obtained and doing the same yields again the same result. The process can be repeated with always the same outcome. For a finite set there is obviously a limit to the number of times one can go on zooming in, determined by the finite order of the generation.

on fractal analysis is very specific on this point [75]. In a fractal -i.e. simple fractal- object, if one looks at some fraction of it at some level of magnification, it is identical to any other fraction at some other level of magnification. Thus there is no inherent length scale; with some suitable magnification factor it is impossible to tell which length scale one is seeing because there are no differences. It is proved [77] that selfsimilar sets have a fractal dimension D_q independent of q and in the field of nonlinear dynamics, where experience with fractal analysis is very abundant, it is stressed [76] that finding a selfsimilar fractal set arising from an actual physical problem is rather excepcional. The more general situation is adequately described in terms of a *multifractal* analysis [75 - 77], of which we outline here the basic elements.

Let p_j be the fraction of points of a given set which are contained in the j-th cell of size ϵ_j. It is assumed that $p_j(\epsilon_j)$ satisfies a scaling relation of the form

$$p_j(\epsilon_j) \;=\; k\,\epsilon_j^{\alpha_j} \tag{40}$$

where k is some unimportant proportionality constant and α_j is the *scaling index* for cell j. The different cells, or boxes, are now labelled to indicate that one is focusing on different regions of the spectrum, but the problem is analysed in the limit in which all $\epsilon_j \to \epsilon \to 0$. Then the number of cells $N(\epsilon)$ increases and the distribution of values of the α_j becomes continuous, so $n(\alpha)d\alpha$ is the number of cells having a scaling index between α and $\alpha + d\alpha$. Note that this approach, where we start from the fact that there are different scaling indices already implies that we no longer contemplate the idea of a selfsimilarity as discussed above.

At this stage the crucial assumption is made that $n(\alpha)$ scales with the size of the cells according to a power law of the form

$$n(\alpha) = K\,\epsilon^{-f(\alpha)} \tag{41}$$

The proportionality constant K is again irrelevant and $f(\alpha)$ can be interpreted as the fractal dimension of the -partial- set of points with scaling index α. The characteristic exponent $f(\alpha)$, first introduced in Ref. [7], is of central importance in the multifractal analysis, and is related to D_q. The basic properties of a multifractal object are: (i) D_q does vary with q. (ii) The scaling index α takes different values and, correspondingly, so does $f(\alpha)$ and (iii) The object is no longer selfsimilar although some particular choice in the way of effecting a zoom magnification might give an impression of apparent selfsimilarity which, however, does not hold when more careful stringent tests are performed [76].

The following relations are proved among D_q, α and $f(\alpha)$:

$$f(\alpha) \;=\; q\,\frac{d}{dq}\,[(q-1)D_q] - (q-1)D(q) \tag{42}$$

$$\alpha(q) \;=\; \frac{d}{dq}\,[(q-1)D_q] \tag{43}$$

and

$$D(q) \;=\; \frac{1}{q-1}\,[q\alpha(q) - f(\alpha(q))] \tag{44}$$

In practice one starts from a given set of points -in our case a spectrum-, which can result from an experiment or a calculation, so what one directly evaluates numerically is D_q and then, in principle, from the above relations one ought to be able to evaluate $f(\alpha)$. An alternative method for a direct evaluation of $f(\alpha)$ from the given spectrum has been proposed in Ref. [79].

It is in order to remark that a multifractal analysis starts formally from a given set of points, which need not constitute an eigenvalue spectrum. If they do, the problem need not be simple in the sense defined in Sections 2 and 3. Nevertheless the two formalisms constitute complementary and related aspects of the formal mathematical analysis in the current state of the art. A case in point, which is trivial but instructive, is the observation one can make that the existence of a density of states function has a relevant, direct consequence in the fractal analysis. Let $\rho(E)$ be this function. Then we can evaluate analytically

$$\begin{aligned} D_q &= \lim_{\epsilon \to 0} \frac{1}{q-1} \frac{log \sum_j \rho(E_j)^q \epsilon^q}{log \epsilon} \\ &= \lim_{\epsilon \to 0} \frac{1}{q-1} \frac{log[\epsilon^{q-1} \sum \rho(E_j)^q \epsilon]}{log \epsilon} \\ &= \lim_{\epsilon \to 0} \frac{1}{q-1} \frac{log \epsilon^{q-1}}{\log \epsilon} \\ &= 1 \end{aligned} \tag{45}$$

Thus the mere existence of a density of states function implies that the spectrum is absolutely continuous with a dimension equal to unity. This is the case of the regular periodic superlattice, in accordance with the fact, stressed in Section 3, that the periodic sequence does not satisfy the conditions of the Bovier-Ghez theorem [52].

It was stressed in Section 3 that the R-S sequence also fails to satisfy the conditions of the Bovier-Ghez theorem, but no density of states appears to be known for this case, so nothing can be stated from first principles. However, it is possible to perform a numerical experiment. We take the Schrödinger equation for two different materials and define a QH of the Kronig-Penney type on a R-S sequence. We obtain the eigenvalues and then evaluate D_q. The parameters used in the calculation were $d_A = 50$ a.u., $d_B = 60$ a.u., $d_C = 70$ a.u., $d_D = 80$ a.u., $V_A = 1/100$ a.u., $V_B = 2/100$ a.u., $V_C = 3/100$ a.u., $V_D = 4/100$ a.u., $m_A = m_B = m_C = m_D = 7/100$ a.u. The result, within the attainable numerical accuracy, is that D_q *is a constant, independent of* q *and equal to unity*, which would indicate an absolutely continuous spectrum. This is only a numerical experiment, but it strongly suggests that the spectrum need not be singular for all QSs, a fact which does not appear to be mentioned in the literature.

Before leaving this topic it is convenient to comment on some practical aspects. Firstly, the theory of all the mathematical and formal properties of QSs and/or QHs holds for infinitely large sequences, something never actually ocurring in experiments or calculations. In practice we always deal with some finite realisation, something which is not new. It is the same with regular superlattices: experimentally one grows a sample containing a finite number of periods and for the calculations one puts, say, periodic Born von Karman boundary conditions on a ring with a given number of periods. The spectrum of the infinite superlattice is absolutely continuous while the actually obtained eigenvalues form, strictly speaking, a discrete set. The point is that for a sufficiently large number of periods the spacing between eigenvalues is sufficiently small that for some purposes it may be a good approximation to treat the spectrum as continuous. Except for the different nature of the spectrum, the same situation arises with QSs. A finite realisation is the n-th generation which results from applying the substitution rule n times and this is what one grows experimentally and what one calculates. So one may think that for QSs what one obtains, if n is sufficiently large, is a good approximation to a singular continuous spectrum. While this may be so for some purposes, it may create practical difficulties on trying to perform a correct fractal analysis following the principles outlined in Section 4, something which in principle should be done in any study of the spectrum of a QH. This will be illustrated by the example discussed in Section 6, where more will be said about the practical aspects of the numerical work needed for this purpose.

The setup of the calculations raises the question of boundary conditions and in this context it is in order to note that a term like *Fibonacci superlattice*, often used in the literature, may be a misnomer and cause some ambiguity. Sometimes this means that one takes a finite generation F_n of the QH under study, imposes periodic boundary conditions and does the theory for a superlattice in which F_n is the period. Other times it is used instead of Fibonacci heterostructure, which means simply F_n, without being too explicit about boundary conditions. In fact one can put infinite barriers at the extremes, or adequate semiinfinte media -say, GaAs for $Al_xGa_{1-x}As$ heterostructures- and study the problem in terms of scattering states solutions. The same remarks apply to any other QS.

In order to avoid ambiguities we shall define a finite generation QH -say, F_n- together with the boundary conditions as an *approximant.* The superlattice in which F_n is the period will thus be denoted as the periodic approximant to F_∞ and other non periodic approximants may also be considered. The point is that in practice, for some purposes, with values of n of order 10 one has a sufficient number of eigenvalues to characterise the spectrum. These considerations should always be kept in mind.

A case in point is the calculation of the fractal dimension. Suppose we take a simple Cantor set as "the spectrum" and calculate its fractal dimension D_0 by the box counting formula (37). Figure 4 gives the log-log plot for the 5th generation. D_0 is obtained from the slope of the "linear" stretch, which in this plot is a staircase of regular steps with given average ascent slope. The picture is very crude and it soon levels off to a horizontal plateau, when a box size is reached which is smaller than the smallest difference between any two contiguous points. But on going to higher generations -$n = 7$ and $n = 9$ are shown in the figure- the only difference is that the horizontal plateaus are increasingly higher while the average slope of the stepwise linear stretch is the same. In fact from a very low generation one can already obtain an excellent numerical approximation to the correct value (37). This is, of course, because the simple Cantor set is ideal in every sense, which is why in the literature on QHs it is always taken as a reference. But the situation is not drastically different in actual physical models with considerable complexity. Figure 5 gives the log-log plot needed to calculate D_0 -which, we recall, in this case is only D_q for $q = 0$- for polar optical modes in a GaAs-AlAs F heterostructure. The calculations were made [47] for a rather elaborate model involving two mechanical vibration amplitudes and an electrostatic potential, all coupled, which proves fairly realistic in practice [80]. It is seen that the 6-th generation is practically as good as the 8-th. A least squares fit to the linear stretch gives $D_0 = 0.54 \pm 0.03$, which is already a fairly accurate result.

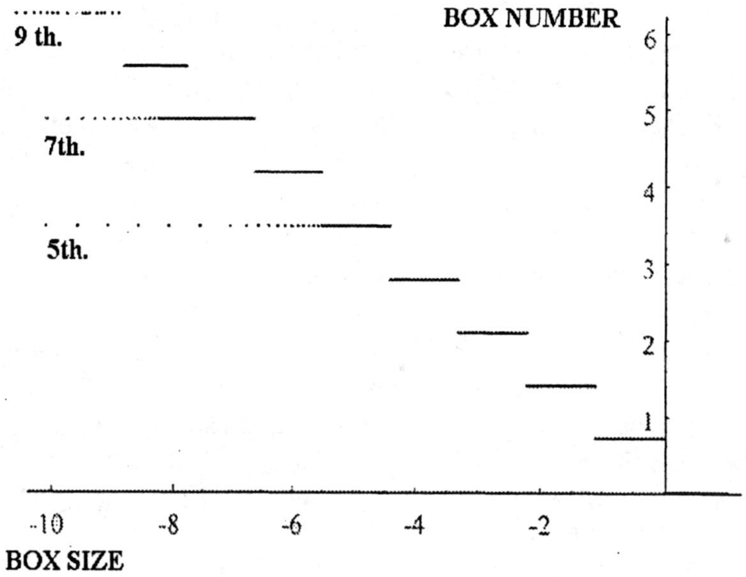

Figure 4: Log-log box counting plots for the 5-th, 7-th and 9-th simple Cantor sets.

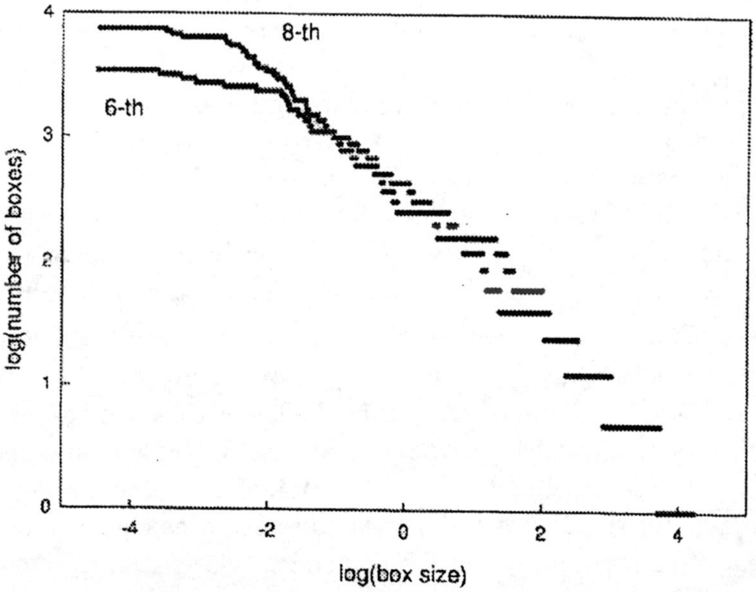

Figure 5: Log-log counting plot -corresponding to $q = 0$- for the spectrum of polar optical modes in a Fibonacci heterostructure and $n = 6, 8$. For higher generations the horizontal plateau appears higher above as the on the average linear stretch grows longer without appreciably changing the mean slope. The calculation was based on a rather elaborate model, involving two mechanical amplitudes and the electrostatic potential mutually coupled [47] .

Some practical aspects of this type of calculations deserve comment. In general, as seen in the examples considered, the scaled -on the average linear- stretch of the log-log plot for comparatively low generations can yield fairly accurate results. On going to a higher generation the "linear" stretch is obviously longer but in the part in which it overlapps one usually finds that there are only minor changes in some points while some more intermediate points appear without significantly altering the general trend. This is why, although we deal with a finite discrete set, to a good approximation this may carry significant information concerning the distribution of eigenvalues in the infinite set. However, there may be subtleties which require some care. Thus, for instance, if the set contains some finite subset which by itself would yield a generalised information $I(\epsilon, q)$ which gives a different average slope, then we may obtain an initial finite linear stretch having a fictitious slope. After a kink in the log-log plot, another linear stretch will start with the right slope and the actual D_q will not be obtained until we go well into box sizes ϵ smaller than the minimum difference between contiguous points in the said subset. This is known from experience with dynamical systems [77]. More will be said in Section 6. Otherwise, finite size effects have been studied in [5, 6, 11, 14,].

5. Characterisation of the spectrum

While the ultimate goal is to understand the physical properties of QHs, the first step is to study their spectra. In periodic superlattices there is little to say in this respect. One must, for instance, calculate an electronic structure but it is known that the spectrum is absolutely continuous. The situation is different with QHs, where a characterisation of the spectrum is a necessary first step for a basic understanding. Everything else follows from it and as long as some basic issues remain misunderstood, it is necessary to clarify these before proceeding to the analysis of the physical properties in search of new significant features supposedly characteristic of the new type of heterostructures. In order to assess the current situation it suffices to concentrate on just this question. Also, since the F sequence has been by far the most often studied, we shall discuss problems predicated on an F sequence, but the general conclusions are equally applicable to any other type of QHs.

While in a regular superlattice the result for, say, the electronic structure yield dispersion relations, i.e. eigenvalues against the momentum, or wavevector, in the growth direction, the situation is essentially different in a QH. In the latter case one also ends up with graphs plotting the bands of eigenvalues but the independent variable cannot be interpreted as a momentum. It is just a label which takes successive integral values and serves to identify the eigenvalues. Sometimes this is presented in the literature as eigenvalues versus eigenvalue number, but very often it is simply called a dispersion relation which, having explained this, is the term we shall henceforth use for brevity.

An apparently different plot is that of the integrated Density of States (IDOS) versus the eigenvalue variable. However, both pictures contain identically the same information. For a given eigenvalue, its label counts precisely the number of solutions up to it. Figure 6 shows this for the results obtained for the electronic structure of a model F heterostructure of a Kronig-Penney type -i.e. one band, effective mass models. By simply interchanging and relabelling the two axes one changes the "dispersion relation" E_i vs i into the IDOS vs E graph. A vertical gap in the upper part of the figure is a horizontal stretch in the IDOS

q	-5	0	5
D_q	1.0	0.6	0.4

Table 3: Random Cantor set. D_q for a few illustrative values of q.

-lower part.

One may study other aspects of the spectrum, like the localisation of the states [81 - 85], the supposed selfsimilarity of the amplitudes [81 - 85] or the level statistics [84, 86], but with just the dispersion relation or the IDOS it suffices to discuss the basic issues.

6. The current situation

As mentioned in the introduction, the attractive novel features found in the spectra of linear atomic chains where some parameters follow a QS stimulated numerous studies of QHs. The fact that there is a selfreplicating pattern in the rule for the buildup of the successive generations suggests that one may expect to find a manifestation of it in the properties of the heterostructures following the growth rule of the QS. This is very suggestive indeed but unfortunately a critical scrutiny of the literature reveals that there has been a notorious lack of rigour in the analysis of the results. Whether in the dispersion relations, the IDOS, the spatial dependence of the amplitudes or some other physical property, the claim to having found selfsimilarity is usually a matter of essentially visual appreciation with no formal justification based on rigoruos arguments and precise definitions or even explicit verification. A magnification of a fragment of a dispersion relation plot showing again the trifurcation of a band, for instance, is taken to mean that the system is selfsimilar, but according to the precise and unambiguos definition discussed in Section 5 this zoom magnification must reproduce the entire spectrum and this is nowhere shown. Moreover, this should hold not only for the chosen segment of the spectrum, but for any other segment as well [76]. The Cantor set is usually taken as a reference in the literature on QHs and the spectra obtained are said to be Cantor-like and thus selfsimilar.

The basic issue is that it is taken for granted that it follows from the fractal nature of the spectrum that it is selfsimilar, but a topological property -fractality- and a morphological property -selfsimilarity- are of a different fundamental rank; they may appear together, but they also may not. A very sharp evidence is provided by the Cantor set. We take this as a "spectrum" and plot its IDOS for the 10-th generation in Fig. 7, upper part. This is a perfect example of a so called *devil's staircase*: by zooming into a fraction and magnifying it we can reproduce the entire spectrum. The situation is the same as in Fig. 3. But the lower part of Fig. 7 shows the IDOS for a different Cantor set which is topologically equivalent but differently constructed: Given two points belonging to one generation and the segment defined by them, the two intermediate points of this segment for the next generation are obtained through a random algorithm. Obviously this set is also fractal, but in no way is the IDOS of Fig. 7 selfsimilar.

In fact there is a formal theory available. We can calculate D_q according to Eq. (36). Just a few values of q evaluated to the first decimal figure -Table 3- suffice to see that this set is actually multifractal, which according to general theory -Section 4- means that it cannot be selfsimilar since there are different scale factors.

Actually a fractal analysis of the results is truly exceptional in the literature on QHs

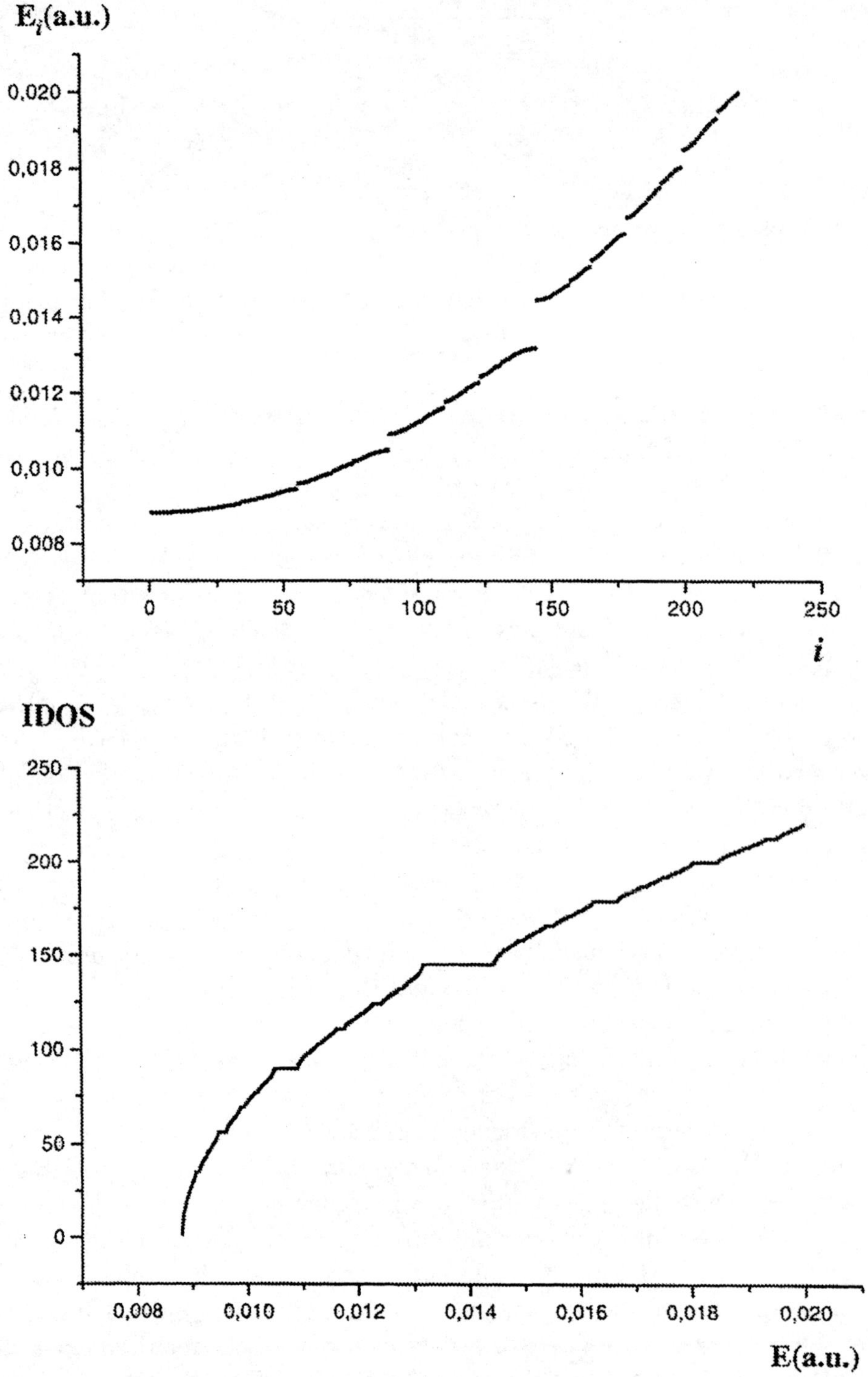

Figure 6: Top: "Dispersion relation" -see text. Bottom: Integrated density of states. Both figures contain the same information with simply the coordinates axes interchanged. The calculations were made for a periodic approximant with a Kronig Penney 13-th generation Fibonacci heterostructure with $d_A = 10$ a.u., $d_B = 50$ a.u., $V_A = 0$, $V_B = 0.01$ a.u., $m_A = 0.01$ a.u., $m_B = 0.05$ a.u. and 2D wavevector $\kappa = 0$.

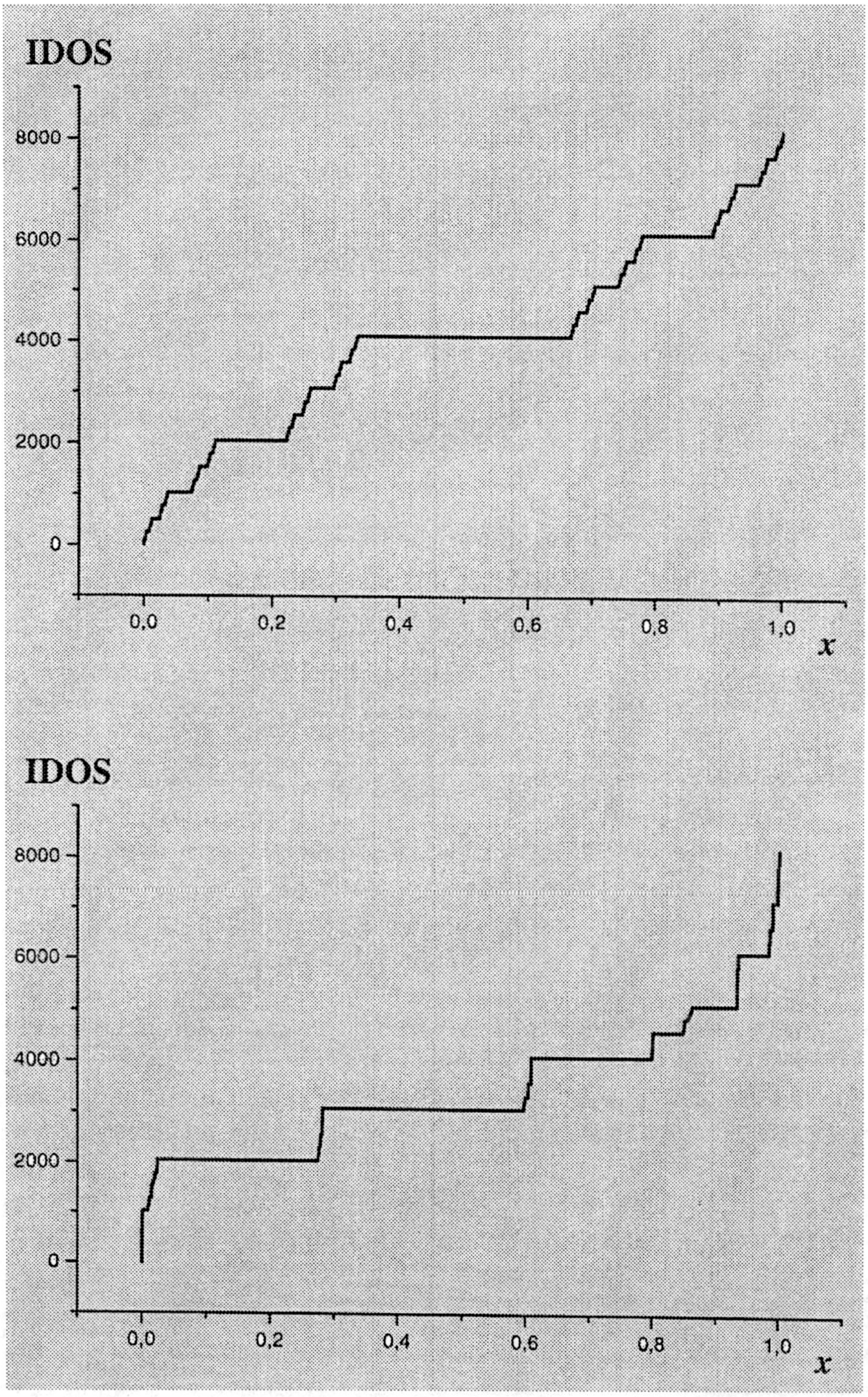

Figure 7: Top: IDOS for the regular simple Cantor set. Bottom: IDOS for a random Cantor set. 10th generation in both cases.

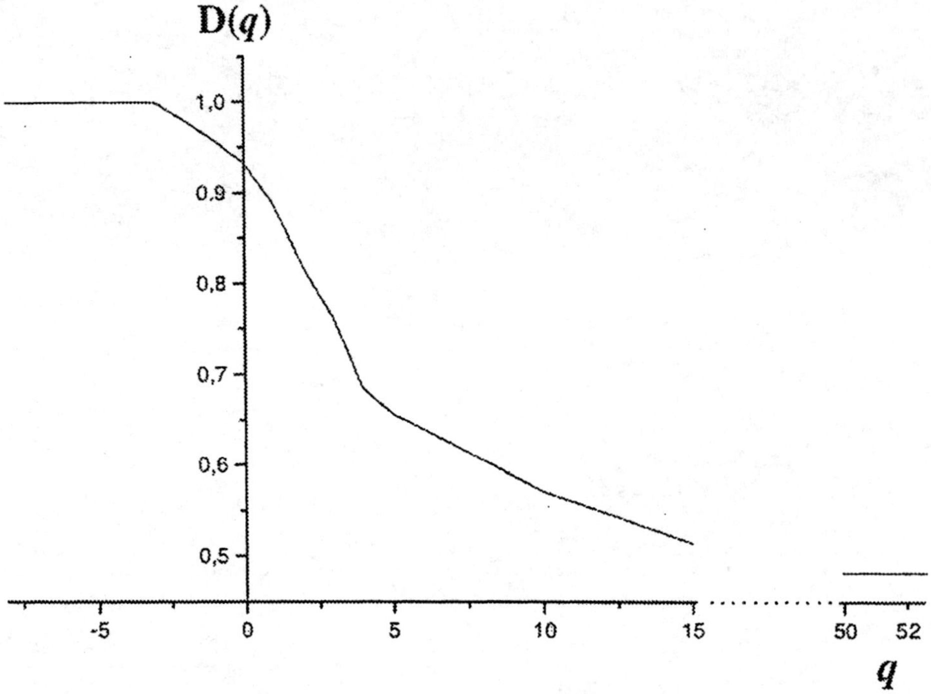

Figure 8: D_q versus q for the system of Figure 6.

[29] and it will be instructive to pause and delve into this question in some detail. For this we take the Kronig-Penney QH of Fig. 6. The problem is as simple as can be and the dispersion relation shows the customary trifurcation which is usually identified with selfsimilarity. Both, for the dispersion relation and for the IDOS these results, obtained for the 13-th generation, are as persuasive by virtual inspection as one can find in the open literature. Yet the spectrum is decidedly multifractal, as is seen in Figure 8, which gives $D(q)$ for q ranging from -10 to $\sim +50$. At this stage it is convenient to consider again the practical aspects of this type of calculations, taking up the discussion where it was left in Section 4. The extremes -$|q|$ large- are always inherently more difficult as regards numerical accuracy, but the situation is always better for $q > 0$. In this case, with generation $n = 13$ the calculation can be carried out with no problems up to q at least > 50 and the asymptotic value of $D_{+\infty}$ can be established with comfortable accuracy. The situation is worse for q large and negative, for the reasons explained in Section 4. In this case for $q < -10$ the errors are already too large. We are reaching the limits allowed in practice by the smallest meaningful boxes.

In practice, in order to go beyond this point we would have to repeat the entire calculation for a higher generation $n > 13$. It is also significant that as the errors in the value of D_q become large, they are even larger when one tries to extract from this, numerically, the derivative needed -(42), (43) to evaluate $f(\alpha)$, which is what one would like to obtain in order to complete the multifractal analysis. The general theory is quite clear on three facts [77], namely: the lower bound for the range of values covered by α is $D_{+\infty}$, the maximum in the f *vs* α is $f_{max} = D_0$ and the upper bound for the range of values covered by α is $D_{-\infty}$.

If one can obtain as many points of the spectrum as desired with no difficulty -for instance, if an analytical treatment is possible, or feasable up to a very advanced stage- then one can cover comfortably te range of very small boxes, extend the D_q down to very large negative values of q and carry out the calculation for α very close to $D_{-\infty}$. The practical situation is described in Figure 9.

Of the three factors established by general theory the first two are fairly well borne out by the calculation. However, for the reason just explained, the quality of D_q for $-10 < q < 0$ is not sufficiently good in practice for a reliable estimate of $f(\alpha)$ and with the generation employed in this calculation -$n = 13$- this stops in practice at about the apex of the f *vs* α plot, which corresponds to $q = 0$. On the other hand it is fairly easy to guess what the rest of the curve should look like.

The point is that the spectrum of electron states has no upper bound. Thus the fact that one is dealing with a finite number of points, as commented in Section 4, may then be due to two reasons: one is that the calculation is done for some finite generation; the other is that one looks in practice only a finite region of the spectrum, thus truncating it at some cut. This eliminates the range of high energies, where, for periodic approximants or for non-periodic approximants with propagating media at both extrema, the electrons tend to be essentially free. Thus if one looked only at that range of the spectrum, one would find scaling exponents tending to unity, which is the fractal dimension for a free electron spectrum. Hence one should expect to find $D_{-\infty} = 1$ and thus the span of α, whatever the starting value $D_{+\infty}$, should reach up to $\alpha = 1$. If one chooses a sufficiently high upper cut, then the range of values of $f(\alpha)$ for $\alpha \simeq 1$ will be fairly well estimated but, as the upper cut decreases, so this range will deteriorate until eventually the domain of $f(\alpha)$ may stop short of $\alpha = 1$. This may be an additional reason why $f(\alpha)$ stops short of α. One way or another, these facts should be kept in mind when doing practical calculations and the outcome for $f(\alpha)$ should be observed to assess the situation.

Now, returning to the heterostructure here discussed -Figures 6, 8, 9- the results attainable in practice for $n = 13$ suffice to establish unambiguously that the spectrum is multifractal and, if it is, then it has different scaling factors and it cannot be truly selfsimilar. It is tempting to just look at the better looking part of it and thence to conclude that one has obtained selfsimilarity, but this superficial visual appreciation may be misleading, as experience with nonlinear dynamics abundantly shows [76]. We shall now illustrate these ideas with a detailed study of the IDOS of Figure 6.

This is reproduced in Fig. 10 -top- where some ranges are indicated by numerical figures and different types of lines. A one digit figure denotes a region selected for first magnification and is framed with dash-dotted lines. Likewise with a two digit figure for second zoom, with dashed lines and a three figure for third zoom with dotted lines.

The prescription for zoom magnification requires some definition. Let x/y denote the abscissae/ordinates axis. Given the region of the spectrum selected for magnification we choose a fixed point (x_0, y_0) which will remain invariant and then define the zoom transformation

$$x \longrightarrow x_0 + C_{xx}(x - x_0) + C_{xy}(y - y_0) \tag{46}$$

$$y \longrightarrow y_0 + C_{yx}(x - x_0) + C_{yy}(y - y_0) \tag{47}$$

For Figure 10 we choose A as the fixed point. Then -bottom- after a first magnification D goes into D′ and I into I′. The original spectrum is also shown -full line- in the Figure

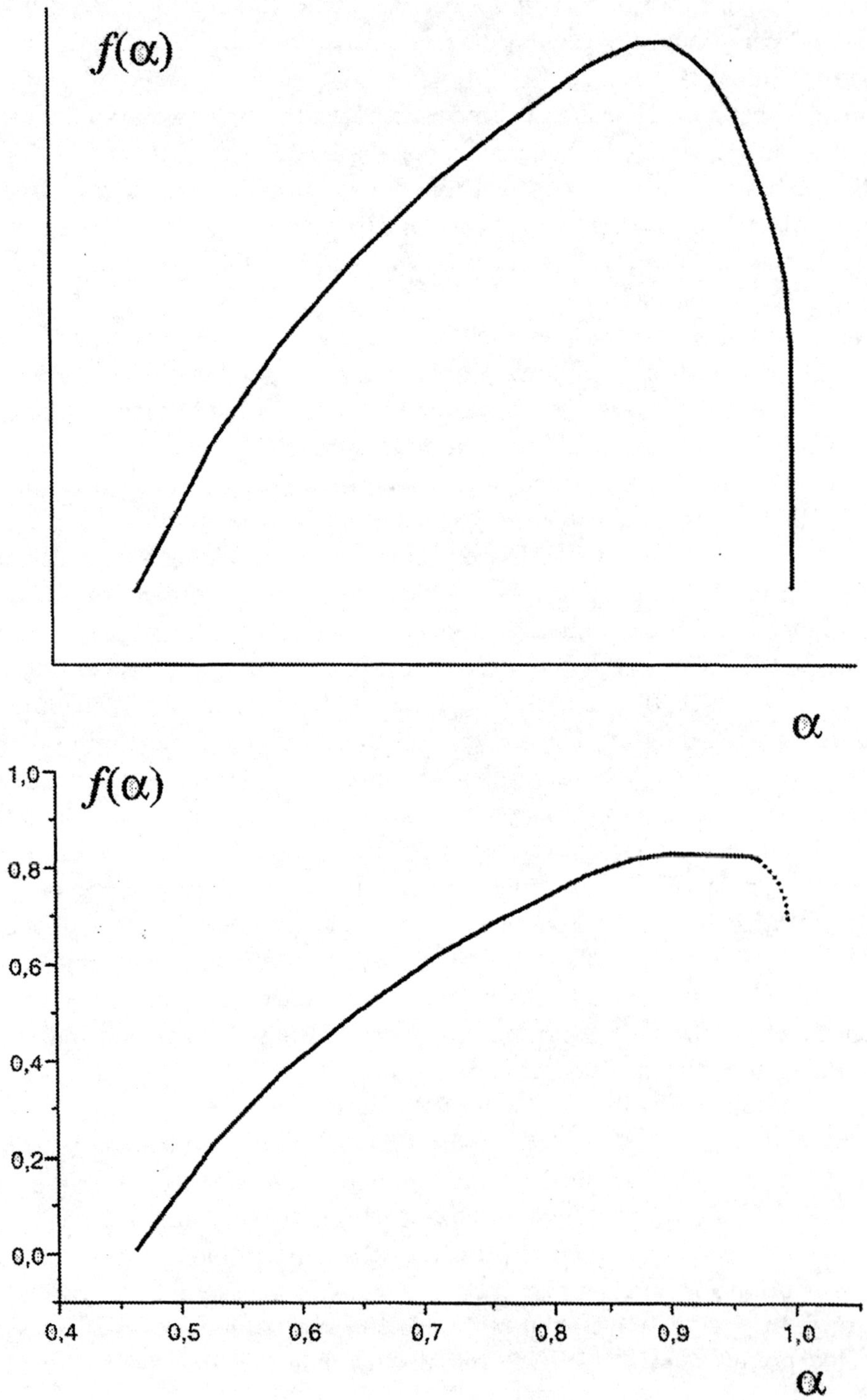

Figure 9: Top: $f(\alpha)$ for an ideal analytical construct. Bottom: The part of $f(\alpha)$ which can reliably calculate for the F heterostructure of Figures 6 and 8 with the 13 t generation.

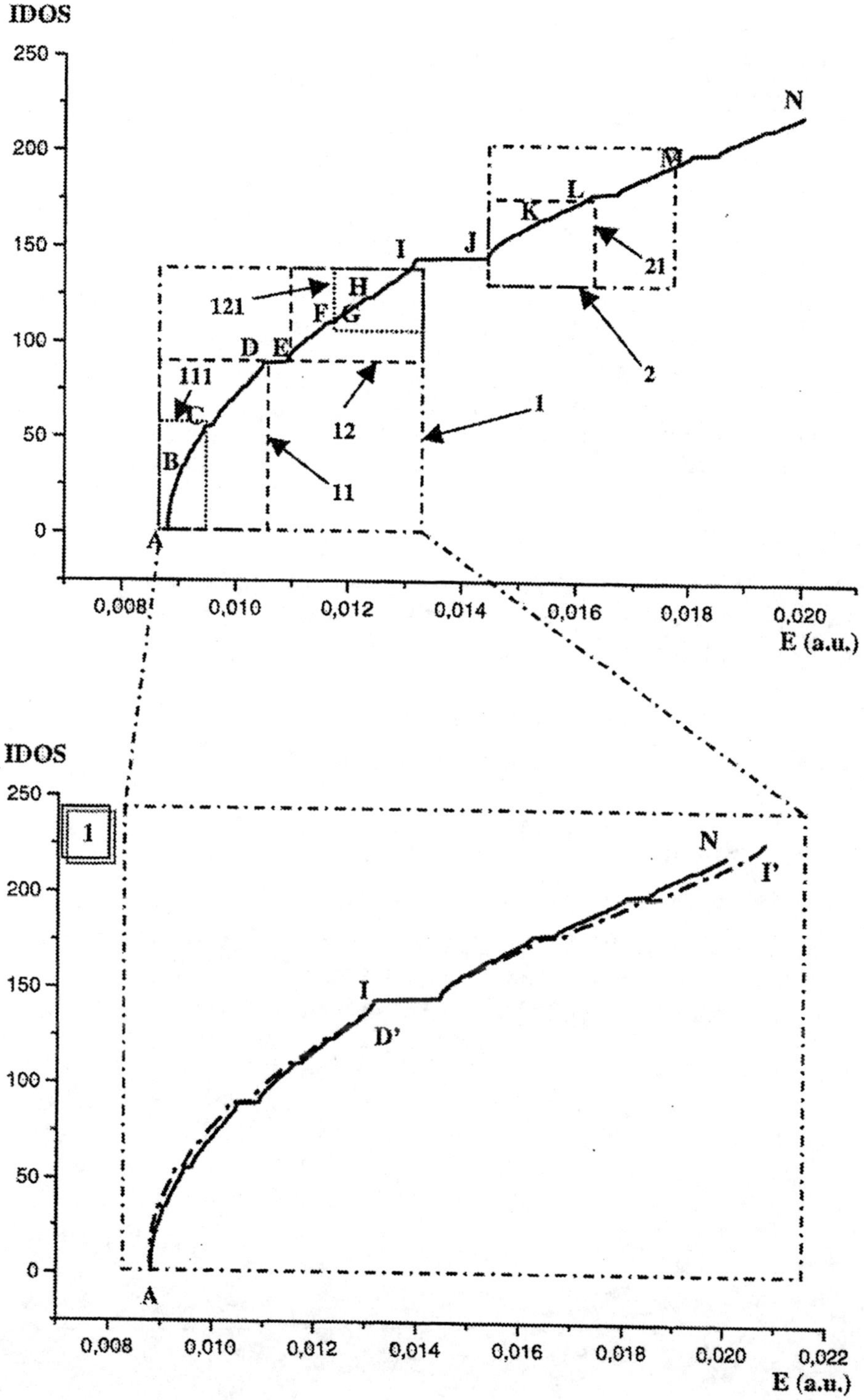

Figure 10: Top: The IDOS of Figure 6 with fractions of it marked for zoom amplification Bottom: First amplification of fraction 1. See text.

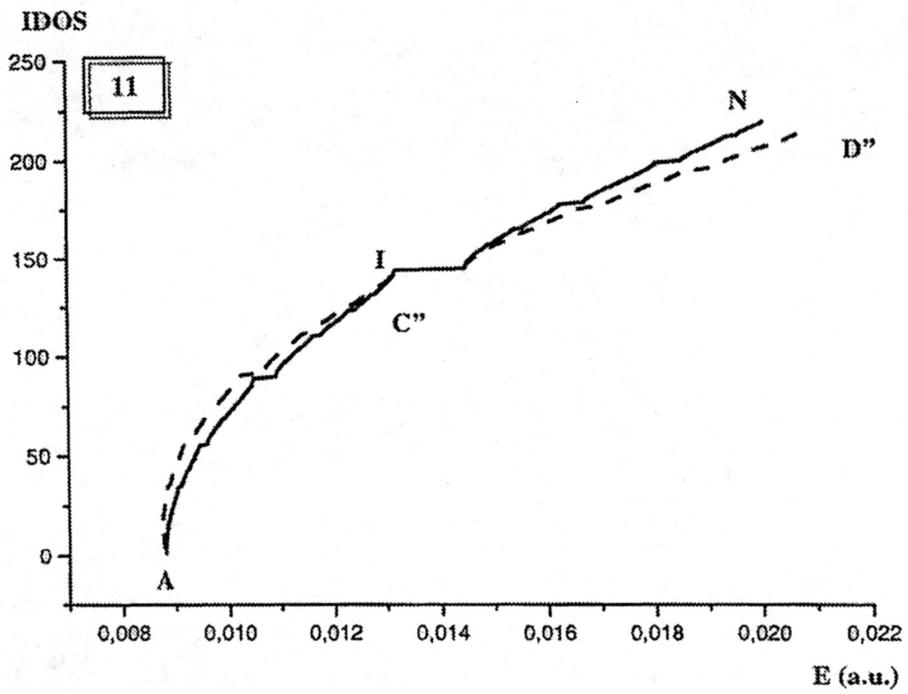

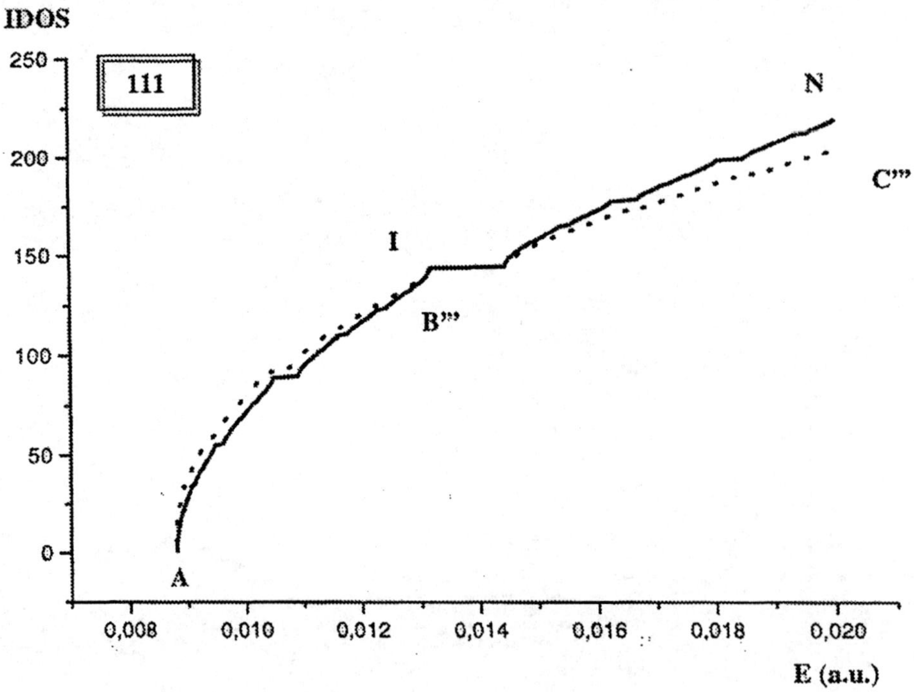

Figure 11: Top: Second zoom. A fraction of 1 is magnified into 11. Bottom: Third zoom. A fraction of 11 is magnified into 111.

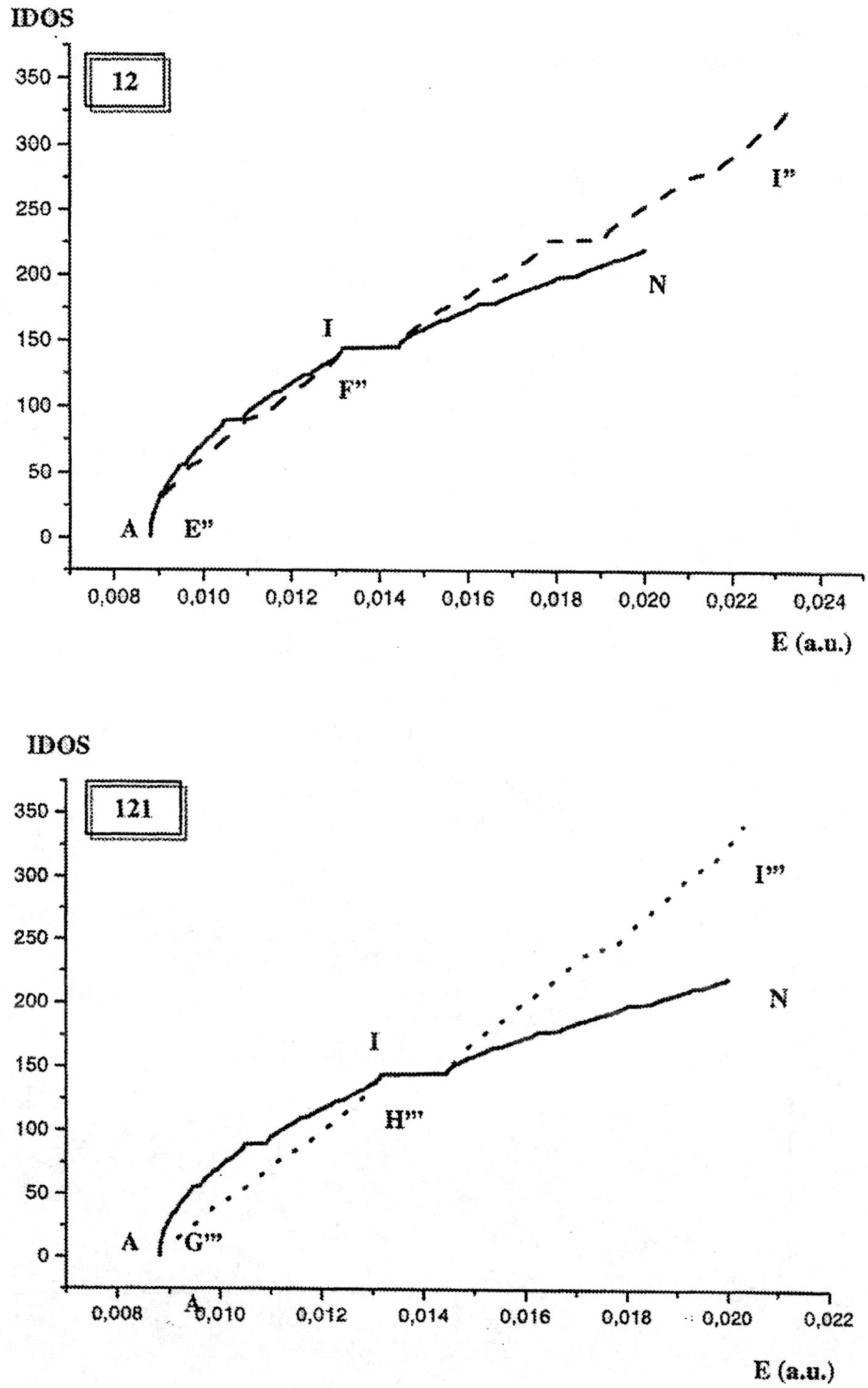

Figure 12: Top: Second zoom. A fraction of 1 is magnified into 12. Bottom: Third zoom A fraction of 12 is magnified into 121

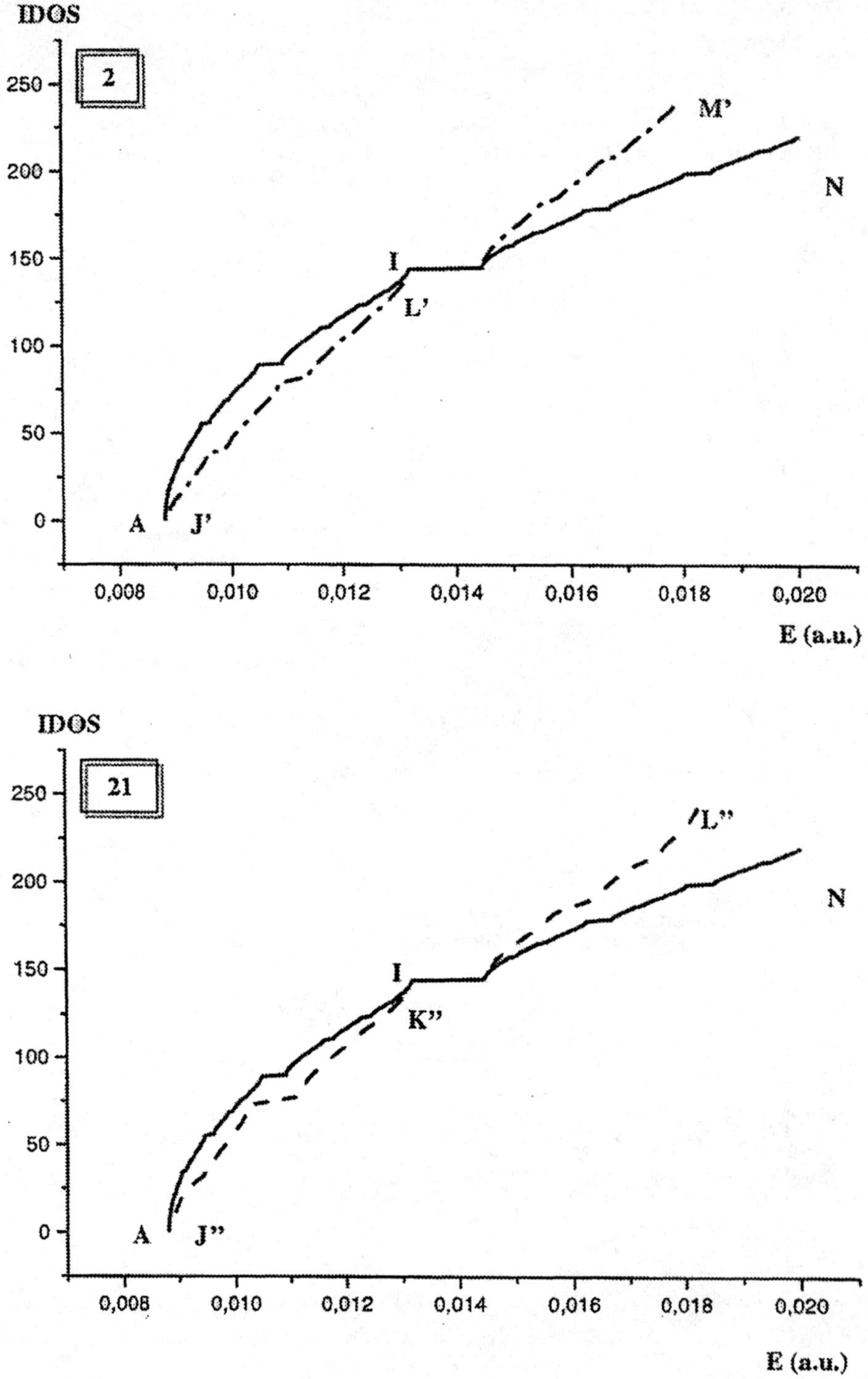

Figure 13: Top: First zoom. A fraction of the IDOS of Figure 10 is magnified into 2 Bottom: Second zoom. A fraction of 2 is magnified into 21.

-bottom. The matrix **C** of numerical coefficients is chosen so that the principal gap is reproduced. The same criterion is followed for the second and third zooms for fractions 11 and 111 -Figure 11. After the first zoom, fraction 1 reproduces the original spectrum rather well. Within acceptable numerical accuracy, by just looking at this one would say that the spectrum is selfsimilar. After this, the magnified fractions 11 and 111 do not quite show the same "quality" and, what is more significant, the **C** matrices of zoom factors differ substantially for the three levels of magnification, so the grounds for claiming selfsimilarity become somewhat weaker.

Now, consider another fraction within region 1. Figure 12 shows the results of second -fraction 12- and third -121- zoom magnifications. The deterioration of the selfsimilarity is evident and growing worse and the same happens for the results in Figure 13. After just a first zoom of fraction 2 -top- it would be totally unjustified to claim any selfsimilarity. And, of course, there is a wide range of **C** matrices with no relation among them, which corroborates again that the spectrum, in spite of its simple appearance, contains many different scale factors, as it had to in view of the result of the fractal analysis. Thus we must conclude that in spite of the promising look of the spectrum of Figure 6, there are no sound grounds for claiming selfsimilarity, as is to be expected from general theory given that it is multifractal. Just one particular choice and one zoom magnification -Figure 10, bottom- might show apparent selfsimilarity, but this does not stand further scrutiny, in line with general experience in nonlinear dynamics [76].

Similar tests performed for dispersion relations in a simple dielectric model of polar optical modes lead to similar conclusions and, since the spectrum of eigenvalues is at the root of all the physical properties of the system, a word of caution on the many claims to selfsimilarity based on just the look of diagrams for some physical property seems to be in order. It would seem reasonable to regard all this as provisional and rather uncertain until more detailed studies clarify the situation.

The result of a survey of all standard simple problems, wether for models of QH or for problems defined on a linear atomic chain following a QS is that we find no evidence of any simple problem which is simply fractal and strictly selfsimilar. In all cases the spectrum turns out to be multifractal.

Furthermore, on dealing with QHs as physical systems it is also necessary to be aware of the complications of real life. One is that the actual heterostructures are 3D; we Fourier transform the 2D in plane position vector into a 2D wavevector $\boldsymbol{\kappa}$ in the plane of the interfaces and are left with a $\boldsymbol{\kappa}$-dependent 1D problems. It is natural to ask how the situation changes with $\boldsymbol{\kappa}$ and, since an isotropic model suffices to study the basic issues, just to see the changes with the modulus κ is sufficient.

Figure 14, again for the system of Figures 6 and 10 through 13 gives -top- the IDOS for $\kappa = 0.1$ a.u., which is still a fairly long wave. The lower part of the Figure reproduces the case $\kappa = 0$ for comparison. The change is drastic and shows that any ressemblance of selfsimilarity is lost on going to $\kappa \neq 0$, a fact which stands out conspicuously upon zoom magnifications in the way of Figures 10 - 13.

Now, so far the discussion has assumed that the problem is simple in the mathematical sense defined in Section 3, but the class of acceptable models involving only one amplitude is rather limited. It suffices, for instance, to go from pure shear elastic waves to saggital modes involving two amplitudes and there is not the slightest hint of selfsimilarity in the calculated

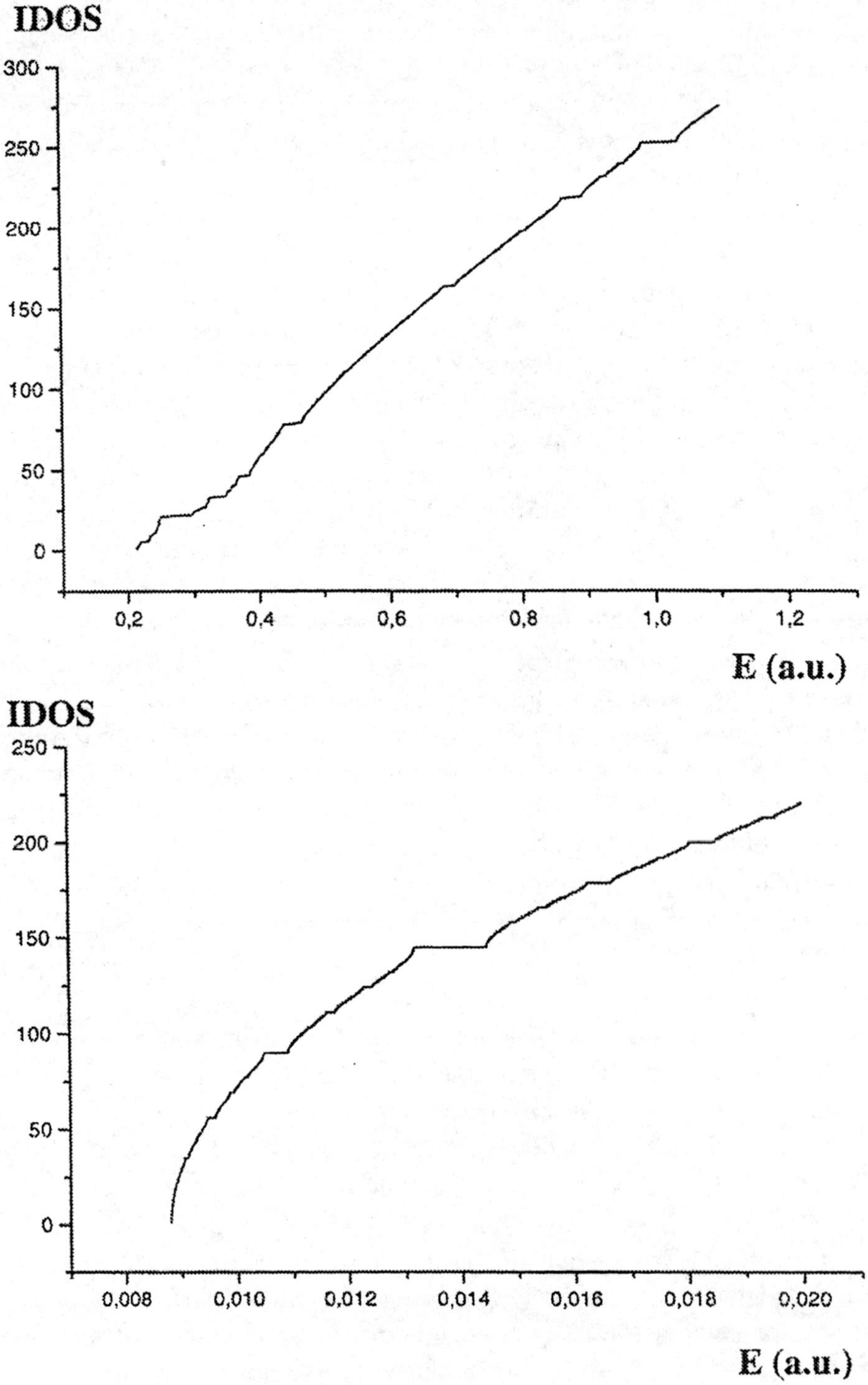

Figure 14: IDOS for the system of Figure 6. Top: $\kappa = 0.1$ a.u. Bottom: $\kappa = 0$.

dispersion relation [45]. We recall that trace maps and associated mathematical analysis, Bovier-Ghez theorem included, hold only for problems involving just one amplitude, so the TM is 2×2. For more than one amplitude the TM is of rank > 2 and there is no trace map so far for these problems. However, there is no limitation to the fractal analysis, for which one can take any spectrum as a given set of points on the real line and apply to it the multifractal analysis, which holds quite generally. Even more simply, one can just calculate the spectrum, wether in the form of IDOS or in the form of dispersion relations, and verify that there is no trace of selfsimilarity. This was also borne out by a study of the polar optical modes of Fibonacci heterostructures [47] in a model involving three coupled amplitudes.

The situation admits a physical interpretation in terms of wave diffraction theory. When a simple amplitude propagates it undergoes multiple reflections and transmissions across a heterostructure which has a built in selfreplicating pattern and in some way the diffraction of this amplitude reflects this pattern. Thus, for example, one finds repeated trifurcations of the bands of eigenvalues. As has been abundantly discussed here, this is only a qualitative feature. It does not mean that one can conclude from this that the spectrum is strictly speaking selfsimilar according to the a precise definition of selsimilarity, but at least in a qualitative sense it can be claimed that one sees a selfreplicating feature which reflects the selfreplicating structure of the system. When the heterostructure is periodic, then the spectrum becomes absolutely continuous and has all the features of a regular band structure. This holds even when the object undergoing multiple diffraction is more complicated because the diffractions of the different coupled amplitudes are coherent on account of the regularity of the structure. However, when two or more coupled amplitudes propagate across a QH, each amplitude sees the structure in a different way, the diffractions are no longer coherent and the pattern becomes more disorganised as more coupled components are involved. A similar situation arises when a four band tight binding model is used to study the propagation through inhomogeneous digital heterostructures [4]. The same idea is also essentially put forward in studies based on eight band tight binding models where the electronic spectrum of F heterostructures is calculated and it is stressed that all trace of selfsimilarity is lost in the regions of the spectrum with significant band hybridisation [81,83].

It is noteworthy that the comparative scarcity of experimental studies of QHs is in sharp contrast with the abundance of theoretical calculations. Besides the calculations of a spectrum in search of the expected selfsimilarity, when some physical property is studied the situation is basically the same. By and large, existing theories of QHs appear to be of an essentially interpretative nature in contrast with many calculations of other heterostructures, like the various types of quantum wells or regular superlattices where one can not only interpret what has been observed experimentally, but also make calculations of a predictive type in which one anticipates significantly new physical properties to be expected from a given design. The mutual stimulation between theoretical and experimental work is then very intense, but this is not the case with heterostructures, where so much theoretical work by and large has had comparatively little effect in stimulating experimental work. Indeed, if theories do not appear so far to be able to predict significant new features and even on a fundamental level the situation requires a great deal of critical caution, it seems reasonable to ask where are we and what have we really learnt so far.

Some studies of harmonic generation in QHs are in this respect different and deserve separate comment. Consider, for instance, second harmonic generation. A quasiphase matching condition of the form

$$\mathbf{k}_2 - 2\mathbf{k}_1 - \mathbf{G} = \mathbf{0} \tag{48}$$

in a periodic structure relies on reciprocal vectors $\mathbf{G}$ of the form

$$\mathbf{G}_n = n\,\frac{\pi}{d} \quad , \quad d = d_A + d_B\ . \tag{49}$$

Suppose one takes a Fibonacci heterostructure. Then, if the generation is sufficiently high, the reciprocal vectors are of the form [50]

$$\mathbf{G}_{n,m} = (m + n\tau)\,\frac{2\pi}{D} \quad , \quad D = \tau d_A + d_B\ , \tag{50}$$

where instead of the period d of (49) D acts as an "average structure parameter". So now there are two numbers -n and m- to play with, and this increases the posibilities of meeting the quasi phase matching condition (48), thus opening possibly new perspectives for devices [50]. Further steps can be taken in this direction . The FWWSM sequence was described and studied in Section 2. This sequence, which was introduced in Ref. [51], is not actually a Fibonacci sequence. From the point of view of harmonic generation it is even more interesting because in a FWWSM heterostructure the reciprocal vectors are given by a formula involving three numerical coefficients instead of two and the implications on harmonic generation are even more interesting [51].

These two appear to be cases in which theory has some predictive character. Indeed in the first case the problem was also studied experimentally [50], so one can expect these studies to open new perspectives for optical devices while hardly anything of the sort appears to stem so far from other trheoretical calculations. The point deserves further analysis. There may be other QS where the situation is similar. We do not know what happens, for instance, with the R-S sequence, which at least for simple problems seems to have a non fractal spectrum, yet it remains to be seen whether a similar situation is possible. It is even conceivable that other types of heterostructures, not necessarily quasiregular, may also offer more flexible choices of reciprocal vectors. It is in any case an interesting question to pursue. A feature which appears to characterise the current situation is that so far it is not easy to find other questions of a similar nature.

7. Conclusions. Open questions

Quasiregular sequences are intriguing and attractive, but the real situation when it comes to studying actual physical heterostructures designed to follow such a sequence suggests that, while being very suggestive, this remains largely an open and vastly unknown field. A great deal of literature has been generated but the real question is what have we learnt so far.

There is a mathematical analysis available which can be very useful to provide a formal guidance, but: (i) It appears to be largely ignored by those studying QHs and (ii) It is in any case restricted to simple problems with a 2×2 TM.

So far there is no trace map for a higher rank TM, which is necessary to cover many physical models involving more than one amplitude. Further research on this would be timely.

The weakest aspect of this field seems to be related to what is, on the other hand, its most attractive feature, namely, the idea of selfsimilarity. Trifurcation is simply a qualitative feature, but a property needs precise adequate concepts for its definition and well defined parameters for its quantitative description. All that is known from general theory is (i) that a selfsimilar object must be fractal and (ii) that in *multifractal* spectra which, as has been stressed, so far appears to be a fact in all cases studied in physics, even ths simplest ones - there are different inherent scales, but this has not been so far translated into useful concepts to analyse the situation in morphological terms. Is it possible to define something like "multisimilarity" and characterise it going beyond the simple qualitative stage? If so, what would be the significance of it?

Nevertheless, even for simple problems, there is room for numerical experiments from which one could clarify some relevant issues. This is what in a few cases has been attempted here. The Fibonacci sequence is the most studied one, but it might be useful to perform numerical experiments on several other sequences. If done with the formal guidance available from general theory, this might help to reveal some interesting features. In particular, we have seen strong indications that the R-S sequence may be in a distinct class of QSs with a non fractal spectrum. We have no ironclad proof that the spectrum of the R-S sequence is absolutely continuos, but the numerical experiment mentioned in Section 3 strongly suggests that this is so and this raises some interesting questions. One would not expect the properties of a QS to be in all respects like those of a regular periodic sequence but, in physical terms, if it has a fundamental property -the topology of the spectrum- in which it is not like the other QSs, then what are the distinctive properties of such a sequence -or heterostructure? Are there more QSs in this class? This is again a question worth investigating.

Another question is the following. Consider, for instance, a full phenomenological model of longwave polar optical phonon modes which involves two mechanical amplitudes and an electrostatic potential φ mutually coupled [80]. These modes are sometimes described in a much simpler model, called the dielectric model, which is reduced to a Poisson equation for φ. It is known that the eigenvalues of the dielectric model describe only the interface matching modes, while the solutions of the full phenomenological model include other modes as well. Now, if one studied plasmons in a phenomenological model for the dielectric function of an electron gas -say, a hydrodynamical model in a heterostructure following a QS, then one would have a well defined simple problem and one would expect to find a situation similar to that of the electronic spectrum of the Kronig-Penney QH studied here in Figures 6 and 10 to 13. That is to say, we would expect to find at least a qualitative type of selfreplicating feature, namely, trifurcation. So if one studied the interface optical modes separately by means of a dielectric model, then this what one would expect to find. As a matter of fact such a study exists [35] and this is indeed what one finds.

So the question is the following. We know that the spectrum of a more complete model contains, together with others, the interface modes and we know that these by themselves constitute a subset with distinctive features of their own. Is there any way of detecting and characterising this subset in the more complete study of the problem? Does this have any consequences of interest for the physical properties of the QH under study? More generally, are there, in more complicated spectra, partial subsets with distinct features of their own and can this be in some way characterised?

A wild guess is that something siimilar to a kind of group theoretical analysis might be a way of characterising something like "invariant subspaces". A first attempt has been made at introducing a group theoretical view in the study of QSs [87] and this might be perhaps as interesting line to pursue.

There is a clear need for more studies of physical properties of QHs having more predictive power. We do not seem to have learnt a great deal from simply looking at results and saying by inspection that they are selfsimilar. What is needed is more efforts trying to bring this field into a situation similar to that of standard heterostructures, which one can understand to the point of being able to monitor desired physical properties by design. This should stimulate more experimental work, of which there is at present great scarcity. In this respect the discussion of the work on harmonic generation [50, 51] seems to bear out the relevance of the Fourier components of the structure, a fact pointed out in Ref. [15]. This could be a promising line to pursue. It would be interesting to see whether something general can be ascertained which characterises the Fourier components of this type of spectra and has some predictable consequences on the physical problems of the system. As stressed in Section 6, on superficial inspection there appears to be no reason why a special case like that of the R-S should be excluded but, would the result of such an analysis be as in the case of the other QSs?

And, of course, more studies are needed of systems involving more elaborate physical models and other aspects of real life, such as 3D nature of the actual heterostructures, which follow some QS in one dimension but are in reality threedimensional.

Acknowledgments

We have benefitted from many useful exchanges with H. Rodríguez-Coppola, C. Trallero-Giner and V.R. Velasco, to whom we express our gratitude. The most efficient help generously given by Pilar Jiménez has been instrumental for the graphical work. One of us (R. P-A) is greattly indebted fore the hospitality of the Jaume I University of Castellón (Spain) where nearly all this work was carried out.

References

[1] Esaki L. 1984. *J. Physique*, **45**, C5-3.

[2] Vinter B. and C. Weisbuch. 1991. *Quantum Semiconductor Structures*, Academic Press, San Diego.

[3] Capasso F. *et al.* 1984. *Appl.Phys. Lett.*, **45**, 1193; Gossard A.C., R.C. Miller and W. Wiegmann. 1986. *Surf. Sci.*, **174**, 131; Sundaram M. *et al.* 1991. *J. Vac. Sci. Technol.*, **B9**, 1524.

[4] Vlaev S., F. García-Moliner and V.R. Velasco. 1995. *Phys. Rev.*, **B52**, 13784; Aldea A. *et al.* 1996. *J. Phys. Condens. Matter*, **8**, 7733.

[5] Kohmoto M., L.P. Kadanoff and C. Tang. 1983. *Phys. Rev. Lett.*, **50**, 1870.

[6] Kohmoto M. and Y. Oono. 1984. *Phys. Lett.*, **102A**, 145.

[7] Halsey T.C. *et al.* 1986. *Phys. Rev.*, **A33**, 1141.

[8] Lu J.P., T. Odagaki and J.L. Birman. 1986. *Phys. Rev.*, **B33**, 4809.

[9] Kohmoto M. and J.R. Banavar. 1986. *Phys. Rev.*, **B34**, 563.

[10] Nori F. and J.P. Rodriguez. 1986. *Phys. Rev.*, **B34**, 2207.

[11] Kohmoto M. 1986. *Phys. Rev.*, **B34**, 5043.

[12] Niu Q. and F. Nori. 1986. *Phys. Rev. Lett.*, **57**, 2057.

[13] Fujita M. and K. Machida. 1986. *Solid State Commun.*, **59**, 61.

[14] Kohmoto M., B. Sutherland and C. Tang. 1987. *Phys. Rev.*, **B35**, 1020.

[15] Luck J.M. 1989. *Phys. Rev.*, **B39**, 5834.

[16] Kaplan D.E. and L. Glass. 1995. *Understanding nonlinear dynamics*, Springer Verlag, NY.

[17] Ostlund S. *et al.* 1983. *Phys. Rev. Lett.*, **50**, 1873.

[18] Ninomiya T. 1996. *J. Phys. Soc. Japan*, **55**, 3709.

[19] Lin Y. and R. Riklund. 1987. *Phys. Rev.*, **B35**, 6034.

[20] Tamura S. and J.P. Wolfe. 1987. *Phys. Rev.*, **B36**, 3491.

[21] Merlin R. *et al.* 1985. *Phys. Rev. Lett.*, **55**, 1768.

[22] Todd R.J. *et al.* 1986. *Phys. Rev. Lett.*, **57**, 1157.

[23] Hurley D.C., S. Tamura and J.P.Wolfe. 1988. *Phys. Rev.*, **B37**, 8829.

[24] Niu Q. and F. Nori. 1990. *Phys. Rev.*, **B42**, 10329.

[25] Tamura S. and F. Nori. 1990. *Phys. Rev.*, **B41**, 7941.

[26] Cruz H., F. Piazza and L. Pavesi. 1992. http://xxx.unizar.es/cond-mat/9210020.

[27] Huang D., G. Gumbs and M. Kolár. 1992. *Phys. Rev.*, **B46**, 11479.

[28] Nishiguchi N., S-i Tamura and F. Nori. 1994. http://xxx.unizar.es/cond-mat/9402079.

[29] Munzar D. *et al.* 1994. *J. Phys.: Condens. Mattter*, **6**, 4107.

[30] Maciá E., F. Domínguez-Adame and A. Sánchez. 1994. *Phys. Rev.*, **E 50**, R679.

[31] Carpena P., V. Gasparian and M. Ortuño. 1995. *Phys. Rev.*, **B51**(18), 12813.

[32] Domínguez-Adame F. *et al.* 1995. *Semiconductor Science and Technology*, **10**, 797.

[33] Domínguez-Adame F. 1996. http://xxx.unizar.es/cond-mat/9604177.

[34] Hiramoto H. and M. Kohmoto. 1996. http://xxx.unizar.es/cond-mat/9607185.

[35] Jin G.J. *et al.* 1996. *Phys. Rev.*, **B54**, 11883.

[36] You J.Q., L.Zhang and Q.B. Yang. 1997. *Phys. Rev.*, **B55**, 1314.

[37] Ghosh A. and S.N. Karmakar. 1998. http://xxx.unizar.es/cond-mat/9801043.

[38] Ghosh A. and S.N. Karmakar. 1998. http://xxx.unizar.es/cond-mat/9801047.

[39] Thakur P.K. and P. Biswas. 1998. http://xxx.unizar.es/cond-mat/9811350.

[40] Oh G.Y. 1998. , *J.of the Korean Phys. Soc.*, **33**, 617.

[41] Tong P.Q. 1998. *Physica B*, **254**, 212.

[42] Sibilia C., F. Tropea and M. Bertolotti. 1998. *J. of Modern Optics*, **45**, 2255.

[43] Cuesta I.G. and I.I. Satija. 1999. http://xxx.unizar.es/cond-mat/9904022.

[44] Fernández-Alvarez L. and V.R. Velasco. 1998. *Phys. Rev.*, **B57**, 14141.

[45] Zárate J.E., L. Fernández-Alvarez and V.R. Velasco. 1999. *Superlatt. and Microstr.*, **25**, 519.

[46] Huang D. *et al.* 1995. *Phys. Lett.*, **A200**, 459.

[47] Pérez-Alvarez R. *et al.* 1999. *Journal of Raman Spectroscopy.* In the press.

[48] Velinho M.T. and I.R. Pimentel. 2000. *Phys. Rev.*, **B**. In the press.

[49] Peyriere J., E. Cockayne and F. Axel. 1995. *J. Phys. I France*, **5**, 111.

[50] Zhu S-n, Y-y Zhu and N-b Ming. 1997. *Science*, **278**, 843.

[51] Liu X. *et al.* 1998. *Phys. Rev.*, **B58**, 12782.

[52] Bovier A. and J.M. Ghez. 1995. *J. Phys.. A: Math. Gen.*, **28**(8), 2313.

[53] Kolár M. and F. Nori. 1990. *Phys. Rev.*, **B42**, 1062.

[54] Bovier A. and J.M. Ghez. 1993. *Commun. Math. Phys.*, **158**(1), 45.

[55] Béllissard J., A. Bovier and J.M. Ghez. 1991. *Comm. Math. Phys.*, **135**(2), 379.

[56] Hof A., O. Knill and B. Simon. 1995. *Comm. Math. Phys.*, **174**(1), 149.

[57] V.R. Velasco, *Problems of Condensed Matter Physics*, Nova Science Publishers 2001 127-144.

[58] Mora E., R. Pérez Alvarez and Ch. B. Sommers. 1985. *J. Physique*, **46** (7), 1021.

[59] Béllissard J., A. Bovier and J.M. Ghez. 1992. *Rev. Math. Phys.*, **4**(1), 1.

[60] Baake M., U. Grimm and D. Joseph. 1993. *Int. J. of Mod. Phys.*, **B7**, 1527.

[61] Béllissard J., A. Bovier and J.M. Ghez. 1993. *Math. Sci. Engrg.*, **192**, 13.

[62] Hof A. 1993. *J. Statist. Phys.*, **72**(5-6), 1353.

[63] Roberts J.A.G. and M. Baake. 1994. *NATO Adv. Sci. Inst. Ser. B Phys.*, **331**, Plenum, New York, 275.

[64] Jitomirskaya S.Ya. and Y. Last. 1996. *Phys. Rev. Lett.*, **76**, 1765.

[65] Rudinger A. and C. Sire. 1996. http://xxx.unizar.es/cond-mat/9607072.

[66] Rudinger A. and C. Sire. 1996. *J. Phys. A: Math. Gen.*, **29**, 3537.

[67] Damanik D. 1997. *Univ. Iagel. Acta Math.*, **34**, 45.

[68] Hof A. 1997. *Comm. Math. Phys.*, **184**(3), 567.

[69] Damanik D. 1998. *Lett. Math. Phys.*, **46**(4), 303.

[70] Damanik D. 1998. *Helv. Phys. Acta*, **71**(6), 667.

[71] Damanik D. 1998. *Comm. Math. Phys.*, **196**(2), 477.

[72] Gazeau J.P. and J. Miekisz. 1998. *J. Phys. A*, **31**(23), L435.

[73] Poisson O. 1999. *Math. Methods in the Applied Sciences*, **22**(9), 773.

[74] Barnett S. 1990. *Matrices. Methods and Applications*, Oxford Applied Mathematics and Computing Sciences Series, Clarendon Press, Oxford.

[75] Moon F.C. 1992. *Chaotic and fractal dynamics*, Wiley-Interscience, NY.

[76] Ott E.O. 1993. *Chaos in dynamical systems*, Cambridge University Press.

[77] Hilborn R.C. 1994. *Chaos and nonlinear dynamics*, Oxford University Press.

[78] Rasband S.N. 1997. *Chaos dynamics of nonlinear systems*, Wiley Professional Paperback Series.

[79] Chhabra A. and R.V. Jensen. 1989. *Phys. Rev. Lett.*, **62**, 1327.

[80] Trallero-Giner C. , R. Pérez-Alvarez and F. García-Moliner.1998. *Long wave polar modes in semiconductor heterostructures*, Pergamon/Elsevier Science, London.

[81] Hirose K., D.Y.K. Ko and H. Kamimura. 1992. *J. Phys.: Condens. Matter*, **4**, 5947.

[82] Macia E. and F. Domínguez-Adame. 1996. *Phys. Rev. Lett.*, **76**, 2957.

[83] Arriaga J. and V.R. Velasco. 1997. *J. Phys.: Condens. Matter*, **9**, 8031.

[84] Carpena P., V. Gasparian and M. Ortuño. 1997. *Z. Phys.*, **B102**, 425.

[85] Carpena P. 1997. *Phys. Lett.*, **A231**, 439.

[86] Zhong J.X. 1997. *J. Phys. Condens. Matter*, **9**, L105.

[87] Gesztesy F. and H. Holden. X. http://www.ma.utexas.edu/mpu_arc-bin/mpa?yn=95-432.

THE TOPIC OF TODAY'S LECTURE IS THE NONSTANDARD PHYSICAL WORLD MODEL...

Tailoring Empirical Tight-Binding Models for Semiconductor Heterostructure Calculations

Timothy B. Boykin *

Abstract

Using the Nonstandard Physical World (NSP-world) model, a model that cannot be eliminated from scientific discourse, it was predicted that specific processes exist such that information between objects within our natural environment would appear to be transmitted instantaneously. In this article, certain aspects of this model are discussed as well as experimental evidence that establishes as fact the model's prediction that a type of nonlocal instantaneous action-at-a-distance does occur in objective reality.

1. Introduction

The premise of the empirical tight-binding method is that the electronic properties of a solid may be described by relatively short-range couplings between atomic-like orbitals centered on neighboring atoms. The empirical character of the method arises from the fact that the interaction integrals, or parameters, are chosen so that the resulting energy bands reproduce as closely as possible those determined by experiment or more complete calculations. Although empirical in nature, the model has real predictive power since it can be used to calculate bulk properties other than electronic structure (eg., optical absorption) or applied to heterointerface systems. The attraction and widespread popularity of the method are doubtless due to its foundation upon familiar atomic-like orbitals.

Although tight-binding models are quite old, most modern research descends directly from the pioneering work of Slater and Koster[1], who consider the sp^3 and sp^3d^5 models for a variety of bulk crystalline systems. More recently, much progress on bulk (and also surface and interface) systems has been made by Harrison; his books *ElectronicStructure* and *Elementary*
ElectronicStructure[2] conveniently bring together many of the most important results. Semiconductor modeling especially has benefited from the work of Vogl, Hjalmarson, and Dow[3] who add an excited s-like orbital (s^*) to the usual sp^3 basis set in order to more accurately fit the conduction bands.

Because of the short-range nature of its interactions the tight-binding method is a natural choice for treating surface and interface systems as well. It has been applied in studies of surfaces[4], single heterointerfaces[5], quantum wells and superlattices[6, 7], complex band-structures[8, 9], and various types of resonant tunneling diodes (RTDs)[10-22]. Success in

*Department of Electrical and Computer Engineering. The University of Alabama in Huntsville, Huntsville. MD 21402-5002 USA

dealing with these systems, however, involves many subtleties often overlooked in treating bulk systems, with transport calculations proving to be the most demanding. In the first place, space-charge regions, which can be quite long, must be included in transport calculations; the numerical instabilities present in the early applications of the method precluded this, rendering realistic studies impossible. The instability remained a problem for well over a decade after the first applications of the method to single heterointerfaces: Independently derived, but related, remedies only appeared in 1991[11] and 1992[12]. Secondly, transport work requires careful fitting of *both* effective masses and energy gaps, early recognized by some[6, 11-13]. More recent systematic studies of effective masses within the tight-binding method[23-27] have improved these fits. When these and other issues are properly addressed, tight-binding transport calculations can achieve remarkable accuracy, comparing very well to room-temperature experiments[18, 20].

The empirical tight-binding method is therefore a potentially very powerful tool for quantum heterostructure calculations, although simply using *any* tight-binding model does not guarantee accuracy: An improperly parametrized or inappropriately chosen tight-binding model will produce *worse* results than a proper effective-mass calculation. Indeed, the tight-binding method *is* more comprehensive than simpler approaches (envelope-function and **kp**), but it is *not* perfectly complete, so that first and foremost one must understand which bands and extrema are most important for the structure to be modeled. This often entails examination of the bandstructures of the materials involved as calculated with more first-principles (i.e., less empirical) methods as well as experimental evidence. The calculation, too, must be performed with relatively greater care, properly addressing issues such as numerical instability[11, 12] and band non-parabolicity[14-16] which are not as prevalent in less-comprehensive calculations. Conversely, the incompleteness of the method must be properly addressed in not only the mathematics to be implemented but also the overall process. Specifically, not all tight-binding models are appropriate for all calculations (recall the improved conduction bands of the sp^3s^* model[3]) and the best one (in terms of the tradeoff of complexity vs. accuracy) must be chosen and then parametrized to reproduce the *relevant* features of the bulk materials involved as closely as possible. The full accuracy possible within the tight-binding framework can only be realized when attention is paid to these very important details.

Our objective here is to show how best to exploit the power and circumvent the limitations of the empirical tight-binding method as applied to heterostructure calculations. Because successful heterostructure simulations are built upon correct reproduction of bulk bandstructures we begin with the application of the method to bulk systems. This study will show how to properly choose and parametrize tight-binding models. We then discuss some of the issues involved when applying the method to heterostructure calculations, involving both additional physics brought out by the method as well as numerical and mathematical issues which must be addressed. Finally we conclude with a brief discussion of some of the recent successes of the method.

2. BULK SYSTEMS

We consider a perfect crystal in the absence of applied fields, described by the Hamiltonian:

$$H = H_0 + H_{so} \tag{1}$$

$$H_0 = \frac{\mathbf{p}^2}{2m} + V(\mathbf{r}) \tag{2}$$

$$H_{so} = \frac{\hbar}{4m^2c^2}(\sigma \times \nabla V) \cdot \mathbf{p} \tag{3}$$

where m is the free electron mass, σ the Pauli spin matrices, and V, the full crystal potential (within the one-electron approximation), satisfies $V(\mathbf{r}) = V(\mathbf{r}+\mathbf{R})$ for any direct lattice vector $\mathbf{R}$. We consider a total volume $\mathcal{V}$ consisting of N_i primitive unit cells in the $\mathbf{a}_i$ direction, $i = 1, 2, 3$, where $\mathbf{a}_i$ is a primitive (direct) lattice translation vector and apply cyclic (Born-von Karman) boundary conditions in all three directions.

Two useful bases satisfying Bloch's theorem are the Bloch basis and the band basis. The Bloch basis is constructed of localized atomic-like orbitals centered on each atom of the crystal:

$$|\alpha;\, \mu;\, \mathbf{k}\rangle = \frac{1}{\sqrt{N}}\sum_{j=1}^{N} \exp\left[i\mathbf{k}\cdot(\mathbf{R}_j + \mathbf{v}_\mu)\right] |\alpha;\, \mu;\, \mathbf{R}_j + \mathbf{v}_\mu\rangle \tag{4}$$

In (4) $\mathbf{k}$ is the three-dimensional wavevector, $\mathbf{R}_j$ is the location of the j-th unit cell, $N = N_1N_2N_3$ the total number of cells, μ the atom within the cell, offset by $\mathbf{v}_\mu$ from the cell location, and α the orbital type (including spin). The states of the band basis are simply linear combinations of (4) which diagonalize H:

$$|n;\, \mathbf{k}\rangle = \sum_{\alpha,\, \mu} b_{n,(\alpha,\, \mu)}(\mathbf{k})\, |\alpha;\, \mu;\, \mathbf{k}\rangle \tag{5}$$

$$H\,|n;\, \mathbf{k}\rangle = E_n(\mathbf{k})\,|n;\, \mathbf{k}\rangle \tag{6}$$

where n is the band index and the b are expansion coefficients. Both sets of states are orthonormal:

$$\langle \alpha';\, \mu';\, \mathbf{k}' \mid \alpha;\, \mu;\, \mathbf{k}\rangle = \delta_{\alpha',\alpha}\delta_{\mu',\mu}\delta_{\mathbf{k}',\mathbf{k}} \tag{7}$$

$$\langle n';\, \mathbf{k}' \mid n;\, \mathbf{k}\rangle = \delta_{n',n}\delta_{\mathbf{k}',\mathbf{k}} \tag{8}$$

For all cell-periodic operators $\Omega(\mathbf{r}) = \Omega(\mathbf{r}+\mathbf{R})$ we employ the crystal momentum representation, following Blount[28]:

$$\Omega(\mathbf{k}) = e^{-i\mathbf{k}\cdot r}\,\Omega\, e^{i\mathbf{k}\cdot r} \tag{9}$$

All wavevectors are of course taken to fall in the first Brillouin zone.

3. Band Curvatures and Incompleteness

In even the simplest models of elementary quantum mechanics the effective mass of the particle is of critical importance in determining such properties as transmission resonance energies and bound states of quantum wells. Naturally the same is true in full bandstructure models, where the effective mass of an electron in a band is proportional to the inverse curvature along a given direction (the mass may not be isotropic). Hence it is essential that a tight-binding model used for heterostructure calculations be able and parametrized to reproduce the *relevant* effective masses of the materials from which the heterostructure is fabricated. We emphasize "relevant", since tight-binding models cannot reproduce *all* bandstructure features; the focus must therefore be upon those extrema important for the heterostructure studied. Fitting the masses entails at a minimum an iterative process in which the curvatures reproduced by the model are compared with the desired values and the model parameters are adjusted accordingly; starting this process requires calculating the curvatures themselves.

There are two different methods for obtaining the curvatures of a band $E_n(\mathbf{k})$. The simpler, though less numerically accurate and physically illuminating, is simply to evaluate $E_n(\mathbf{k})$ at several closely spaced $\mathbf{k}$-values along a given direction and compute the approximate curvature from Taylor's theorem. More complicated, but far more accurate (essentially limited by the diagonalization routine employed) is the perturbation approach, in which second-order theory is applied to the Hamiltonian matrix in the basis (4), $H(\mathbf{k})$. This method is more desirable than the simpler, purely numerical approach for two additional reasons: first, it explicitly shows how the incompleteness of the model manifests itself and second, for some important models at certain points in $\mathbf{k}$-space it yields *analytic* effective mass expressions[24-27]. Indeed, the connection of the issues of incompleteness and effective masses helps to explain the relatively recent appreciation of the former[23, 29, 30], since accurate reproduction of band curvatures became important only with the widespread application of the tight-binding method to quantum heterostructures. Here we limit ourselves to the perturbation procedure for nondegenerate bands; the degenerate case is discussed elsewhere[23].

The most direct application of the perturbation approach considers the purely mathematical problem of finding the eigenvalues of the matrix $H(\mathbf{k}+\mathbf{q})$ in the basis (4)[30]. Using the Taylor expansion for this matrix,

$$H\left(\mathbf{k}+\mathbf{q}\right)=H\left(\mathbf{k}\right)+\mathbf{q}\cdot\nabla_{\mathbf{k}}H\left(\mathbf{k}\right)+\frac{1}{2!}\sum_{\alpha,\beta=\{x,y,z\}}q^{(\alpha)}q^{(\beta)}\frac{\partial^{2}H\left(\mathbf{k}\right)}{\partial k^{(\alpha)}\partial k^{(\beta)}}+O\left(q^{3}\right) \tag{10}$$

direct application of second-order nondegenerate perturbation theory for the band state $|n;\,\mathbf{k}\rangle$ yields:

$$\begin{aligned} E_n(\mathbf{k}+\mathbf{q}) \approx\ & E_n(\mathbf{k})+\mathbf{q}\cdot\langle n;\,\mathbf{k}|\,\nabla_{\mathbf{k}}H(\mathbf{k})\,|n;\,\mathbf{k}\rangle+\frac{1}{2!}\sum_{\alpha,\beta=\{x,y,z\}}q^{(\alpha)}q^{(\beta)}\left\langle n;\,\mathbf{k}\left|\frac{\partial^{2}H(\mathbf{k})}{\partial k^{(\alpha)}\partial k^{(\beta)}}\right|n;\,\mathbf{k}\right\rangle \\ & +\sum_{j\neq n}\frac{|\mathbf{q}\cdot\langle j;\,\mathbf{k}|\,\nabla_{\mathbf{k}}H(\mathbf{k})\,|n;\,\mathbf{k}\rangle|^{2}}{E_n(\mathbf{k})-E_j(\mathbf{k})} \end{aligned} \tag{11}$$

where the inverse effective mass is proportional to the second-order coefficient. While eq. (11) gives a highly accurate method for calculating the band curvature it does not shed much light on the nature of the model employed.

In order to better understand the physics behind (11) we rewrite $H(\mathbf{k}+\mathbf{q})$ in the $\mathbf{k}$-basis (4), using the Baker-Hausdorff identity, as in ref. 23:

$$H(\mathbf{k}+\mathbf{q}) = \exp(-i\mathbf{k}\cdot\mathbf{r})\left\{\exp(-i\mathbf{q}\cdot\mathbf{r})\,H\exp(i\mathbf{q}\cdot\mathbf{r})\right\}\exp(i\mathbf{k}\cdot\mathbf{r}) \tag{12}$$

$$H(\mathbf{k}+\mathbf{q}) = \exp(-i\mathbf{k}\cdot\mathbf{r})\left\{H - i[\mathbf{q}\cdot\mathbf{r},H] - \frac{1}{2!}[\mathbf{q}\cdot\mathbf{r},[\mathbf{q}\cdot\mathbf{r},H]] + \ldots\right\}\exp(i\mathbf{k}\cdot\mathbf{r}) \tag{13}$$

$$H(\mathbf{k}+\mathbf{q}) = H(\mathbf{k}) - i[\mathbf{q}\cdot\mathbf{r},H](\mathbf{k}) - \frac{1}{2!}[\mathbf{q}\cdot\mathbf{r},[\mathbf{q}\cdot\mathbf{r},H]](\mathbf{k}) + O(q^3) \tag{14}$$

Introducing the velocity operator,

$$\mathbf{v} = \frac{1}{i\hbar}[\mathbf{r},H] \tag{15}$$

eq. (14) is recast into a more familiar form:

$$H(\mathbf{k}+\mathbf{q}) = H(\mathbf{k}) + \mathbf{q}\cdot\mathbf{v}(\mathbf{k}) - \frac{i\hbar}{2}\sum_{\alpha,\beta=\{x,y,z\}} q^{(\alpha)}q^{(\beta)}\left[r^{(\alpha)},v^{(\beta)}\right](\mathbf{k}) + O\left(q^3\right) \tag{16}$$

Since coefficients of like powers of q must be the same in (10) and (16) we make the identifications:

$$\mathbf{v}(\mathbf{k}) = \frac{1}{\hbar}\nabla_{\mathbf{k}}H(\mathbf{k}) \tag{17}$$

$$\left[r^{(\alpha)},v^{(\beta)}\right](\mathbf{k}) = \frac{i}{\hbar}\frac{\partial^2 H(\mathbf{k})}{\partial k^{(\alpha)}\partial k^{(\beta)}},\ \alpha,\beta=\{x,y,z\} \tag{18}$$

Eq. (17) appears exactly as expected, but this is not the case with (18), for (15) along with (1)-(3) (and the *usual* commutation relations) gives

$$\left[r^{(\alpha)},v^{(\beta)}\right] = \frac{i\hbar}{m}\mathbf{1}\delta_{\alpha,\beta},\ \alpha,\beta=\{x,y,z\} \tag{19}$$

More troubling still is that eq. (19) implies that higher order commutators ought to vanish whereas it is clear from (10) and (16) that they generally will not. Indeed, the second-order formula familiar from solid-state texts[31, 32],

$$E_n(\mathbf{k}+\mathbf{q}) \approx E_n(\mathbf{k}) + \hbar\mathbf{q}\cdot\langle n;\mathbf{k}|\,\mathbf{v}\,|n;\mathbf{k}\rangle + \frac{\hbar^2 q^2}{2m} + \hbar^2\sum_{j\neq n}\frac{|\mathbf{q}\cdot\langle j;\mathbf{k}|\,\mathbf{v}\,|n;\mathbf{k}\rangle|^2}{E_n(\mathbf{k})-E_j(\mathbf{k})} \tag{20}$$

differs markedly from (11), so that if one substitutes the band states (5) into (20) one will generally obtain an *incorrect* value for the second-order correction and hence the effective mass.

The reason for these discrepancies is quite simply that the model is (of necessity) incomplete. Lew Yan Voon and Ram-Mohan[29] attribute the disagreement between (11) and (20) to the incompleteness of the basis and inadequately fit bands and Graf and Vogl[30] note that (19) cannot hold in any finite basis. Boykin[23] shows explicitly how the incompleteness appears by partitioning the states in (20) into those included in the model and those excluded from it. By recognizing that in (20) the term $\hbar^2q^2/2m$ actually arises from the diagonal matrix element $\langle n;\, \mathbf{k}|\, [\mathbf{q}\cdot\mathbf{r}, \mathbf{q}\cdot\mathbf{v}]\, |n;\, \mathbf{k}\rangle$, he then demonstrates that the terms involving states excluded from the model in this commutator (due to the matrix multiplications) exactly cancel those in the last term (the perturbation sum). The result, following these cancellations, is (11), in which all sums cover *only* those states included in the model, thus explicitly highlighting the incompleteness.

4. Analytic Effective Mass Expressions and Capabilities of Models

As mentioned in the preceding subsection, it is occasionally possible to obtain *analytic* formulas for the band curvatures (inverse effective masses). While such expressions are of great practical utility, as, for example, parts of either manual or automated fitting procedures, their real value lies in what they say about the capabilities of the underlying tight-binding model itself. For example, as the number of common terms in the expressions for two different inverse effective masses increases, it becomes relatively more difficult to independently select the masses. It may in fact become impossible for a simpler model to simultaneously reproduce correctly certain pairs of masses for some materials, necessitating the use of a more complete model. Analytic mass expressions can thus help guide the selection of the best tight-binding model for the materials of interest.

We are quite fortunate that for the widely employed sp^3s^* model[3], with or without the spin-orbit coupling, and including interactions up to second-near neighbor, useful analytic effective mass formulas are obtainable at $\mathbf{k} = 0$[25-27]; Loehr and Talwar give analytic formulas for the no-spin-orbit, second-near-neighbor sp^3 model[24]. Useful analytic formulas exist for these models because, as long as all first- and second-near neighbor $s - s^*$ and $s* - s*$ parameters are set to zero and Chadi's[33] treatment of the spin-orbit interaction (i.e., as a same-site only effect) is used, the Hamiltonian matrix decouples into matrices of dimension 1 or 2, the former by definition diagonal and the latter easily diagonailzable. (In principle, it is possible to analytically diagonalize an Hermitian matrix of size up to 4, but the utility of analytic solutions to the cubic or quartic equations is debatable.) In the no-spin-orbit model the initially 10-dimensional Hamiltonian matrix decomposes into one- and two-dimensional subspaces: $\{|s*a\rangle\}$, $\{|s*c\rangle\}$, $\{|sa\rangle, |sc\rangle\}$, $\{|xa\rangle, |xc\rangle\}$, $\{|ya\rangle, |yc\rangle\}$, $\{|za\rangle, |zc\rangle\}$, the last three of which are degenerate. In the spin-orbit model[25], the initially 20-dimensional Hamiltonian first factorizes into two 10-dimensional subspaces, one in the basis $\{|s\mu;\uparrow\rangle, |s*\mu;\uparrow\rangle, |z\mu;\uparrow\rangle, |x\mu;\downarrow\rangle, |y\mu;\downarrow\rangle\}$, $\mu = a$, c (anion and cation), the other with the spins reversed. Each 10-dimensional subspace may be further decomposed into one- and two-dimensional subspaces by diagonalizing H_{SO} and using the resulting states as the new basis. For the subspace above these new states and their energies are:

$$|h\mu;\,2\rangle = \frac{1}{\sqrt{2}}\left[|x\mu;\downarrow\rangle - i\,|y\mu;\downarrow\rangle\right], \lambda_\mu \tag{21}$$

$$|l\mu;\,2\rangle = \frac{1}{\sqrt{6}}\left[|x\mu;\downarrow\rangle + i\,|y\mu;\downarrow\rangle - 2\,|z\mu;\uparrow\rangle\right], \lambda_\mu \tag{22}$$

$$|so\mu;\,2\rangle = \frac{1}{\sqrt{3}}\left[|x\mu;\downarrow\rangle + i\,|y\mu;\downarrow\rangle + |z\mu;\uparrow\rangle\right], -2\lambda_\mu \tag{23}$$

where λ_μ is the spin-orbit parameter in Chadi's notation[33]. (Similar expressions obtain for the spin-reversed block, called the "1" states.) The resulting subspaces now become: $\{|s*a;\uparrow\rangle\}$, $\{|s*c;\uparrow\rangle\}$, $\{|sa;\uparrow\rangle, |sc;\uparrow\rangle\}$, $\{|la;\,2\rangle, |lc;\,2\rangle\}$, $\{|ha;\,2\rangle, |hc;\,2\rangle\}$, $\{|soa;\,2\rangle, |soc;\,2\rangle\}$.

With or without the spin-orbit coupling, the two-dimensional blocks have a common form in the $\{|na\rangle, |nc\rangle\}$ subspace, n being the orbital type:

$$H_2 = \begin{bmatrix} E_a & V \\ V & E_c \end{bmatrix}, V \in \Re \tag{24}$$

with eigenvalues,

$$E_{\pm} \doteq \bar{E} \pm \Delta, \quad \bar{E} \doteq \frac{E_a + E_c}{2}, \quad E_{\Delta} \doteq \frac{E_a - E_c}{2}, \quad \Delta \doteq \sqrt{E_{\Delta}^2 + V^2} \tag{25}$$

and corresponding eigenvectors,

$$|n\pm\rangle \doteq \nu_a^{\pm} |na\rangle + \nu_c^{\pm} |nc\rangle \tag{26}$$

$$\nu_a^+ = \nu_c^- = \frac{\Delta + E_{\Delta}}{\sqrt{2}\sqrt{\Delta^2 + \Delta E_{\Delta}}}, \quad \nu_c^+ = -\nu_a^- = \frac{V}{\sqrt{2}\sqrt{\Delta^2 + \Delta E_{\Delta}}} \tag{27}$$

The quantities in (24)-(27) of course take on different values depending upon the tight-binding model and are listed in Table I. We employ the notation of ref. 3 for the same-site and nearest-neighbor parameters, while for the second-near neighbor parameters our abbreviations in terms of the Slater-Koster[1] notation are:

$$V_{\alpha\mu,\beta\mu} \doteq 4E^{(110)}_{\alpha\mu,\beta\mu}, \quad U_{\alpha\mu,\beta\mu} \doteq 4E^{(011)}_{\alpha\mu,\beta\mu}, \quad \alpha, \beta \in \{s, x, y, z\}, \quad \mu \in \{a, c\} \tag{28}$$

Table I: Parameters for the two-dimensional subspaces in both spin-orbit (SO) and no-spin-orbit ($no-SO$) models for electrons (e), no-spin-orbit holes (h), and light- (lh), heavy- (hh), and split-off holes (soh) in the spin-orbit model. Abbreviations for same-site terms are: $E^{eff}_{s\mu} \doteq E_{s\mu} + 3V_{s\mu,s\mu}$ and $E^{eff}_{p\mu} \doteq E_{p\mu} + 2V_{x\mu,x\mu} + U_{x\mu,x\mu}$. For the nearest-neighbor spin-orbit expressions, simply set to zero all second-near-neighbor parameters in the electron and spin-orbit hole expressions.

H_2 *parameter*	*e both*	*h no* $-$ *SO*	*lh SO*	*hh SO*	*soh SO*
E_μ	$E^{eff}_{s\mu}$	$E_{p\mu}$	$E^{eff}_{p\mu} + \lambda_\mu$	$E^{eff}_{p\mu} + \lambda_\mu$	$E^{eff}_{p\mu} - 2\lambda_\mu$
V	$V_{s,s}$	$V_{x,x}$	$V_{x,x}$	$V_{x,x}$	$V_{x,x}$
$E_\pm$	$E^{(e)}_\pm$	$E^{(h)}_\pm$	$E^{(lh)}_\pm$	$E^{(lh)}_\pm$	$E^{(soh)}_\pm$
ν^+_μ	σ^+_μ	ρ^+_μ	$\rho^{l,+}_\mu$	$\rho^{l,+}_\mu$	$\rho^{so,+}_\mu$
Δ	$\Delta^{(e)}$	$\Delta^{(h)}$	$\Delta^{(lh)}$	$\Delta^{(lh)}$	$\Delta^{(soh)}$

Following the procedures given above and in [23] we calculate the curvatures along [001] at Γ, (m_0/m^*_{zz}), for the nearest-neighbor sp^3s^* model with and without spin-orbit as well as the spin-orbit second-near neighbor sp^3s^* model, expressing our results in terms of the quantities in Table I. After a little algebra, we find the electron (e), light- (lh), heavy- (hh), and split-off-hole (soh) masses[25-26]. Our convention gives negative hole masses and in the expressions below a is the conventional unit cell cube edge and m_0 the free electron mass. For the nearest-neighbor, no-spin-orbit model there is of course no split-off hole, the heavy-hole being doubly degenerate:

$$\frac{m_0}{m^*_e} = \frac{2m_0}{\hbar^2}\left(\frac{a}{4}\right)^2 \left\{ -\sigma_a^+\sigma_c^+ V_{s,s} + \frac{\left[\sigma_a^+\rho_a^+ V_{sa,pc} - \sigma_c^+\rho_c^+ V_{pa,sc}\right]^2}{E_+^{(e)} - E_-^{(h)}} + \frac{\left[\sigma_a^+\rho_c^+ V_{sa,pc} + \sigma_c^+\rho_a^+ V_{pa,sc}\right]^2}{E_+^{(e)} - E_+^{(h)}} \right\} \tag{29}$$

$$\frac{m_0}{m_{lh}^*} = \frac{2m_0}{\hbar^2}\left(\frac{a}{4}\right)^2 \left\{ \rho_a^+\rho_c^+ V_{x,x} + \frac{\left[\sigma_a^+\rho_c^+ V_{pa,sc} + \sigma_c^+\rho_a^+ V_{sa,pc}\right]^2}{E_-^{(h)} - E_-^{(e)}} + \frac{\left[\sigma_a^+\rho_a^+ V_{sa,pc} - \sigma_c^+\rho_c^+ V_{pa,sc}\right]^2}{E_-^{(h)} - E_+^{(e)}} \right.$$
$$\left. + \frac{\left(\rho_a^+ V_{s*a,pc}\right)^2}{E_-^{(h)} - E_{s*a}} + \frac{\left(\rho_c^+ V_{pa,s*c}\right)^2}{E_-^{(h)} - E_{s*c}} \right\} \tag{31}$$

$$\frac{m_0}{m_{hh}^*} = \frac{2m_0}{\hbar^2}\left(\frac{a}{4}\right)^2 \left\{ \rho_a^+\rho_c^+ V_{x,x} + \frac{V_{x,y}^2}{E_-^{(h)} - E_+^{(h)}} \right\} \tag{32}$$

The addition of the spin-orbit coupling to the nearest-neighbor model results in somewhat more complicated expressions:

$$\frac{m_0}{m_e^*} = \frac{2m_0}{3\hbar^2}\left(\frac{a}{4}\right)^2$$
$$\left\{ -3\sigma_a^+\sigma_c^+ V_{s,s} + \frac{2\left[\sigma_a^+\rho_a^{l,+} V_{sa,pc} - \sigma_c^+\rho_c^{l,+} V_{pa,sc}\right]^2}{E_+^{(e)} - E_-^{(lh)}} + \frac{\left[\sigma_a^+\rho_a^{so,+} V_{sa,pc} - \sigma_c^+\rho_c^{so,+} V_{pa,sc}\right]^2}{E_+^{(e)} - E_-^{(soh)}} \right.$$
$$\left. + \frac{2\left[\sigma_a^+\rho_c^{l,+} V_{sa,pc} + \sigma_c^+\rho_a^{l,+} V_{pa,sc}\right]^2}{E_+^{(e)} - E_+^{(lh)}} + \frac{\left[\sigma_a^+\rho_c^{so,+} V_{sa,pc} + \sigma_c^+\rho_a^{so,+} V_{pa,sc}\right]^2}{E_+^{(e)} - E_+^{(soh)}} \right\} \tag{33}$$

$$\frac{m_0}{m_{lh}^*} = \frac{2m_0}{\hbar^2}\left(\frac{a}{4}\right)^2$$
$$\left\{ \rho_a^{l,+}\rho_c^{l,+} V_{x,x} + \left(\frac{2}{3}\right)\frac{\left[\sigma_a^+\rho_c^{l,+} V_{pa,sc} + \sigma_c^+\rho_a^{l,+} V_{sa,pc}\right]^2}{E_-^{(lh)} - E_-^{(e)}} + \left(\frac{2}{3}\right)\frac{\left[\sigma_a^+\rho_a^{l,+} V_{sa,pc} - \sigma_c^+\rho_c^{l,+} V_{pa,sc}\right]^2}{E_-^{(lh)} - E_+^{(e)}} \right.$$
$$\left. + \left(\frac{2}{3}\right)\frac{\left(\rho_a^{l,+} V_{s*a,pc}\right)^2}{E_-^{(lh)} - E_{s*a}} + \left(\frac{2}{3}\right)\frac{\left(\rho_c^{l,+} V_{pa,s*c}\right)^2}{E_-^{(lh)} - E_{s*c}} + \left(\frac{1}{3}\right)\frac{V_{x,y}^2}{E_-^{(lh)} - E_+^{(lh)}} \right\} \tag{34}$$

$$\frac{m_0}{m_{hh}^*} = \frac{2m_0}{\hbar^2}\left(\frac{a}{4}\right)^2$$
$$\left\{ \rho_a^{l,+}\rho_c^{l,+} V_{x,x} + \left(\frac{1}{3}\right)\frac{V_{x,y}^2}{E_-^{(lh)} - E_+^{(lh)}} + \left(\frac{2}{3}\right)\frac{V_{x,y}^2\left[\rho_a^{so,+}\rho_a^{l,+} + \rho_c^{so,+}\rho_c^{l,+}\right]^2}{E_-^{(lh)} - E_+^{(soh)}} \right.$$
$$\left. + \left(\frac{2}{3}\right)\frac{V_{x,y}^2\left[\rho_a^{so,+}\rho_c^{l,+} - \rho_c^{so,+}\rho_a^{l,+}\right]^2}{E_-^{(lh)} - E_-^{(soh)}} \right\} \tag{35}$$

$$\frac{m_0}{m_{soh}^*} = \frac{2m_0}{3\hbar^2}\left(\frac{a}{4}\right)^2 \left\{ 3\rho_a^{so,+}\rho_c^{so,+} V_{x,x} + \frac{\left[\sigma_a^+\rho_a^{so,+} V_{sa,pc} - \sigma_c^+\rho_c^{so,+} V_{pa,sc}\right]^2}{E_-^{(soh)} - E_+^{(e)}} \right.$$
$$+ \frac{\left[\sigma_a^+\rho_c^{so,+} V_{pa,sc} + \sigma_c^+\rho_a^{so,+} V_{sa,pc}\right]^2}{E_-^{(soh)} - E_-^{(e)}} + \frac{2V_{x,y}^2\left[\rho_a^{so,+}\rho_c^{l,+} - \rho_c^{so,+}\rho_a^{l,+}\right]^2}{E_-^{(soh)} - E_-^{(lh)}}$$
$$\left. + \frac{2V_{x,y}^2\left[\rho_a^{so,+}\rho_a^{l,+} + \rho_c^{so,+}\rho_c^{l,+}\right]^2}{E_-^{(soh)} - E_+^{(lh)}} + \frac{\left(\rho_a^{so,+} V_{s*a,pc}\right)^2}{E_-^{(soh)} - E_{s*a}} + \frac{\left(\rho_c^{so,+} V_{pa,s*c}\right)^2}{E_-^{(soh)} - E_{s*c}} \right\} \tag{36}$$

Adding second-near neighbor interactions does not change the form of the above expressions, but it does lengthen the numerators of most terms:

$$\begin{aligned}\frac{m_0}{m_e^*} = \frac{2m_0}{3\hbar^2}\left(\frac{a}{4}\right)^2 \Bigg\{ & -3\sigma_a^+\sigma_c^+V_{s,s} + 12\left[\left(\sigma_a^+\right)^2 V_{sa,sa} + \left(\sigma_c^+\right)^2 V_{sc,sc}\right] \\ & + \frac{2\left[4\left(\sigma_c^+\rho_a^{l,+}V_{sc,xc} - \sigma_a^+\rho_c^{l,+}V_{sa,xa}\right) + \left(\sigma_a^+\rho_a^{l,+}V_{sa,pc} - \sigma_c^+\rho_c^{l,+}V_{pa,sc}\right)\right]^2}{E_+^{(e)} - E_-^{(lh)}} \\ & + \frac{\left[4\left(\sigma_c^+\rho_a^{so,+}V_{sc,xc} - \sigma_a^+\rho_c^{so,+}V_{sa,xa}\right) + \left(\sigma_a^+\rho_a^{so,+}V_{sa,pc} - \sigma_c^+\rho_c^{so,+}V_{pa,sc}\right)\right]^2}{E_+^{(e)} - E_-^{(soh)}} \\ & + \frac{2\left[4\left(\sigma_c^+\rho_c^{l,+}V_{sc,xc} + \sigma_a^+\rho_a^{l,+}V_{sa,xa}\right) + \left(\sigma_a^+\rho_c^{l,+}V_{sa,pc} + \sigma_c^+\rho_a^{l,+}V_{pa,sc}\right)\right]^2}{E_+^{(e)} - E_+^{(lh)}} \\ & + \frac{\left[4\left(\sigma_c^+\rho_c^{so,+}V_{sc,xc} + \sigma_a^+\rho_a^{so,+}V_{sa,xa}\right) + \left(\sigma_a^+\rho_c^{so,+}V_{sa,pc} + \sigma_c^+\rho_a^{so,+}V_{pa,sc}\right)\right]^2}{E_+^{(e)} - E_+^{(soh)}} \Bigg\} \end{aligned} \tag{37}$$

$$\begin{aligned}\frac{m_0}{m_{lh}^*} = \frac{2m_0}{\hbar^2}\left(\frac{a}{4}\right)^2 \Bigg\{ & \rho_a^{l,+}\rho_c^{l,+}V_{x,x} - \frac{2}{3}\left[\left(\rho_c^{l,+}\right)^2\left(5V_{xa,xa} + U_{xa,xa}\right) + \left(\rho_a^{l,+}\right)^2\left(5V_{xc,xc} + U_{xc,xc}\right)\right] \\ & + \left(\frac{2}{3}\right)\frac{\left[4\left(\sigma_a^+\rho_a^{l,+}V_{sc,xc} + \sigma_c^+\rho_c^{l,+}V_{sa,xa}\right) - \left(\sigma_a^+\rho_c^{l,+}V_{pa,sc} + \sigma_c^+\rho_a^{l,+}V_{sa,pc}\right)\right]^2}{E_-^{(lh)} - E_-^{(e)}} \\ & + \left(\frac{2}{3}\right)\frac{\left[4\left(\sigma_c^+\rho_a^{l,+}V_{sc,xc} - \sigma_a^+\rho_c^{l,+}V_{sa,xa}\right) + \left(\sigma_a^+\rho_a^{l,+}V_{sa,pc} - \sigma_c^+\rho_c^{l,+}V_{pa,sc}\right)\right]^2}{E_-^{(lh)} - E_+^{(e)}} \\ & + \left(\frac{2}{3}\right)\frac{\left(4\rho_c^{l,+}V_{s*a,xa} - \rho_a^{l,+}V_{s*a,pc}\right)^2}{E_-^{(lh)} - E_{s*a}} + \left(\frac{2}{3}\right)\frac{\left(4\rho_a^{l,+}V_{s*c,xc} - \rho_c^{l,+}V_{pa,s*c}\right)^2}{E_-^{(lh)} - E_{s*c}} + \left(\frac{1}{3}\right)\frac{V_{x,y}^2}{E_-^{(lh)} - E_+^{(lh)}} \Bigg\} \end{aligned} \tag{38}$$

$$\begin{aligned}\frac{m_0}{m_{hh}^*} = \frac{2m_0}{\hbar^2}\left(\frac{a}{4}\right)^2 \Bigg\{ & \rho_a^{l,+}\rho_c^{l,+}V_{x,x} - 2\left[\left(\rho_c^{l,+}\right)^2\left(V_{xa,xa} + U_{xa,xa}\right) + \left(\rho_a^{l,+}\right)^2\left(V_{xc,xc} + U_{xc,xc}\right)\right] \\ & + \left(\frac{1}{3}\right)\frac{V_{x,y}^2}{E_-^{(lh)} - E_+^{(lh)}} + \left(\frac{2}{3}\right)\frac{V_{x,y}^2\left[\rho_a^{so,+}\rho_a^{l,+} + \rho_c^{so,+}\rho_c^{l,+}\right]^2}{E_-^{(lh)} - E_+^{(soh)}} \\ & + \left(\frac{2}{3}\right)\frac{V_{x,y}^2\left[\rho_a^{so,+}\rho_c^{l,+} - \rho_c^{so,+}\rho_a^{l,+}\right]^2}{E_-^{(lh)} - E_-^{(soh)}} \Bigg\} \end{aligned} \tag{39}$$

$$\begin{aligned}\frac{m_0}{m_{soh}^*} = \frac{2m_0}{\hbar^2}\left(\frac{a}{4}\right)^2 \Bigg\{ & \rho_a^{so,+}\rho_c^{so,+}V_{x,x} - \frac{4}{3}\left[\left(\rho_c^{so,+}\right)^2\left(2V_{xa,xa} + U_{xa,xa}\right) + \left(\rho_a^{so,+}\right)^2\left(2V_{xc,xc} + U_{xc,xc}\right)\right] \\ & + \left(\frac{1}{3}\right)\frac{\left[4\left(\sigma_c^+\rho_a^{so,+}V_{sc,xc} - \sigma_a^+\rho_c^{so,+}V_{sa,xa}\right) + \left(\sigma_a^+\rho_a^{so,+}V_{sa,pc} - \sigma_c^+\rho_c^{so,+}V_{pa,sc}\right)\right]^2}{E_-^{(soh)} - E_+^{(e)}} \\ & + \left(\frac{1}{3}\right)\frac{\left[4\left(\sigma_a^+\rho_a^{so,+}V_{sc,xc} + \sigma_c^+\rho_c^{so,+}V_{sa,xa}\right) - \left(\sigma_a^+\rho_c^{so,+}V_{pa,sc} + \sigma_c^+\rho_a^{so,+}V_{sa,pc}\right)\right]^2}{E_-^{(soh)} - E_-^{(e)}} \\ & + \left(\frac{2}{3}\right)\frac{V_{x,y}^2\left[\rho_a^{so,+}\rho_c^{l,+} - \rho_c^{so,+}\rho_a^{l,+}\right]^2}{E_-^{(soh)} - E_-^{(lh)}} + \left(\frac{2}{3}\right)\frac{V_{x,y}^2\left[\rho_a^{so,+}\rho_a^{l,+} + \rho_c^{so,+}\rho_c^{l,+}\right]^2}{E_-^{(soh)} - E_+^{(lh)}} \\ & + \left(\frac{1}{3}\right)\frac{\left(4\rho_c^{so,+}V_{s*a,xa} - \rho_a^{so,+}V_{s*a,pc}\right)^2}{E_-^{(soh)} - E_{s*a}} + \left(\frac{1}{3}\right)\frac{\left(4\rho_a^{so,+}V_{s*c,xc} - \rho_c^{so,+}V_{pa,s*c}\right)^2}{E_-^{(soh)} - E_{s*c}} \Bigg\} \end{aligned} \tag{40}$$

Before discussing the expressions in detail, we examine the heavy-hole-split-off-hole coupling term, which on first inspection appears to diverge as the spin-orbit coupling is turned off. Actually it is often quite small and indeed vanishes as the spin-orbit interaction is reduced[25]. To see this for the nearest-neighbor model rewrite the last term of (34) using the quantities from Table I:

$$\frac{\left[\rho_a^{so,+}\rho_c^{l,+} - \rho_c^{so,+}\rho_a^{l,+}\right]^2}{E_-^{(lh)} - E_-^{(soh)}} = \frac{V_{x,x}^2}{D^{(lh)}D^{(soh)}} \frac{\left[\Delta^{(soh)} - \Delta^{(lh)} + (3/2)\left(\lambda_c - \lambda_a\right)\right]^2}{\Delta^{(soh)} - \Delta^{(lh)} + (3/2)\left(\lambda_c + \lambda_a\right)} \tag{41}$$

where $D \doteq 2\left[\Delta^2 + \Delta E_\Delta\right]$ is well-behaved. Now, uniformly turn off the spin-orbit coupling by setting $\lambda_c = \alpha\lambda_a$ and take the limit $\lambda_a \to 0$, so that

$$\frac{\left[\rho_a^{so,+}\rho_c^{l,+} - \rho_c^{so,+}\rho_a^{l,+}\right]^2}{E_-^{(lh)} - E_-^{(soh)}} \approx \frac{3\lambda_a}{2D^{(h)2}} \left[\frac{V_{x,x}^2\left(\alpha - 1\right)^2\left(1 + E_\Pi\right)^2}{\left(1 - E_\Pi\right) + \alpha\left(1 + E_\Pi\right)}\right], \quad E_\Pi \doteq \frac{E_{pa} - E_{pc}}{\sqrt{\left(E_{pa} - E_{pc}\right)^2 + 4V_{x,x}^2}} \tag{42}$$

which becomes smaller (as opposed to divergent) as the spin-orbit interaction is reduced.

We now turn to what (29)-(39) tell us about the capabilities of the various tight-binding models; in all cases we write the incompleteness correction term (i.e., that due to d^2H/dk^2) first. Note the similarity of the electron mass in all three models; it is particularly instructive to compare the two nearest-neighbor results, (29) and (32). In the limit of small spin-orbit coupling the sum of the second and third terms of (32) becomes approximately equal to the second term of (29); the same is true of the last two terms of (32) and the last term of (29). A second characteristic common to all three models is the relative lack of freedom available for selecting the heavy-hole mass. Remarkably, the only nearest-neighbor parameter, other than $V_{x,x}$ which enters via the expansion coefficients and the d^2H/dk^2 term is $V_{x,y}$. Moreover, since nearest-neighbor parameters are generally significantly larger than second-near neighbor parameters, it follows that even in the second-near neighbor model $V_{x,y}$ is still the most important parameter for fitting the heavy-hole mass and since $V_{x,y}$ also appears in the light- and split-off hole expressions it may not always be possible to satisfactorily fit all three hole masses.

As important as these similarities are, they are not as significant as the differences between the models with respect to the light-hole mass and its relationship to the electron mass: consider first the no-spin-orbit nearest-neighbor model. The curvature of both the electron (29) and light-hole (30) masses is largely due to the common electron-light-hole coupling term, the second in (29) and the third in (30). Were this the only term we would have $m_{lh}^* = -m_e^*$, precluding a good simultaneous fit of both since for almost all III-V semiconductors experiments yield $|m_{lh}^*| > |m_e^*|$. Unfortunately, the situation is actually much worse, for in the no-spin-orbit model it is typically the case that $|m_e^*| > |m_{lh}^*|$. To see why this is so, observe that since $V_{s,s} < 0$ and $V_{x,x} > 0$, all of the ρ_μ^+ (or $\rho_\mu^{n,+}$) are positive, as is σ_a^+, whereas σ_c^+ is negative. It is also generally the case that $|V_{s,s}| > |V_{x,x}|$, so that the d^2H/dk^2 term is typically large and negative for the electron and rather smaller and positive for the light-hole, reducing the curvature of the conduction-band much more than that of the light-hole band.

The remaining terms of each expression merely exacerbate the problem. The last term of (29) is negative, due to the negative denominator energy gap, though small (generally the denominator gap is large and the numerator small since $V_{pa,sc}$ and $V_{sa,pc} > 0$) and further reduces the conduction-band curvature, *increasing* the electron mass. In contrast, the light-hole couplings to the high-lying s^* states have *negative* gaps, which *increase* the curvature whereas the positive coupling to the lowest s-band is small due to the large positive denominator gap and the small numerator. Thus in the no-spin-orbit model the conduction-band curvature is often significantly *less* than that due to the electron-light-hole coupling whereas in the light-hole band it is usually significantly *greater*, resulting in the incorrect relationship $|m_{lh}^*| < |m_e^*|$.

The situation is improved for the spin-orbit models since in the light-hole mass expression the couplings to both the conduction-band and the s^*-states now enter at only 2/3 their former strength, whereas in the electron mass expression (in the limit of small spin-orbit coupling) the electron-light-hole and electron-split-off-hole couplings together are about equal to the no-spin-orbit electron-light-hole coupling. Another difference in the light-hole expression is the appearance of the light-hole-"heavy" conduction-band coupling, the last term of (33). While this term is negative, it is generally does not make up for the reduction (due to the 2/3 prefactors) in the light-hole-s^* couplings because of both its usually sizable denominator gap and the prefactor of 1/3. Thus the light-hole curvature often remains somewhat above 2/3 of that due to the electron-light-hole coupling, limiting the light-hole mass to $|m_{lh}^*| < (3/2)\,|m_e^*|$. This is not a problem for many materials, but it can make it difficult to simultaneously fit the electron- and light-hole-masses of others, most notably InP[34].

Though improved, the spin-orbit models can still reproduce $|m_{lh}^*| < |m_e^*|$, although this tends to be more of a problem with the nearest-neighbor model than the second-near-neighbor model, and is generally worse for materials with larger gaps at Γ. This situation can arise because both positive terms of (32) scale vary approximately inversely with the gap at Γ while only that of the electron-light-hole term of (33) does. Because the positive terms of (33) are usually smaller in size than the negative terms of (32), we can again have $|m_{lh}^*| < |m_e^*|$ when the gap at Γ gets sufficiently large that the negative terms of (33) become greater in magnitude than the electron-split-off-hole term of (32) (i.e., about half the size of the electron-light-hole coupling). In practice this is mainly a problem with materials like AlAs and AlSb, and less so with the second-near neighbor model due to the extra freedom its additional parameters provide.

The foregoing observations on the relative capabilities of the various tight-binding models may be profitably applied towards choosing the best model for a heterostructure calculation. If hole states are important, as in a valence-band quantum well or a resonant interband tunneling diode, one of the spin-orbit models is required. (Below we shall see that the second-near neighbor is probably the better choice as it affords more freedom in fitting the hole masses along different directions.) For a conduction-band device, such as a GaAs/AlAs quantum well or RTD the no-spin-orbit sp^3s^* model might at first seem the best choice since it is less computationally intensive than either spin-orbit model. Rather surprisingly, however, it turns out that a spin-orbit model would be a far better choice; this stems from the above-noted tendency of the no-spin-orbit model to reproduce $|m_{lh}^*| < |m_e^*|$. Correctly reproducing the dispersion of the imaginary band linking the conduction and light-hole valence bands requires that m_e^* and m_{lh}^* be properly fit, and it is this dispersion which, to a large extent, determines the attenuation in the barrier. Simply put, when $|m_{lh}^*|$ is too small, this imaginary band does not protrude sufficiently far into imaginary k_z-space, resulting in an overly transparent barrier.

Fig. 1 displays the different dispersions for this complex band as produced by the nearest-neighbor sp^3s^* model with and without the spin-orbit interaction. The parameters, appropriate for $T = 300K$, are taken from ref. 25 and the complex bands are calculated using the generalized eigenproblem method[35]; the in-plane wavevector $\mathbf{k}_{||} = k_x\mathbf{e}_x + k_y\mathbf{e}_y = \mathbf{0}$. The spin-orbit model has $m_e^* = 0.182$, $m_{lh}^* = -0.135m_0$, and $E_g = 3.025$ eV, while the no-spin-orbit model has $m_e^* = 0.141$, $m_{lh}^* = -0.0786m_0$, and $E_g = 3.026$ eV. The very different light-hole masses result in rather different imaginary bands. As indicated by the vertical arrows, the attenuation constant for an electron tunneling from GaAs will be smaller in the no-spin-orbit case, resulting in an overly-transparent barrier and transmission resonance energies (or for a quantum well, bound states) which are too low. Hence including the spin-orbit interaction can be of significant benefit even for conduction-band devices.

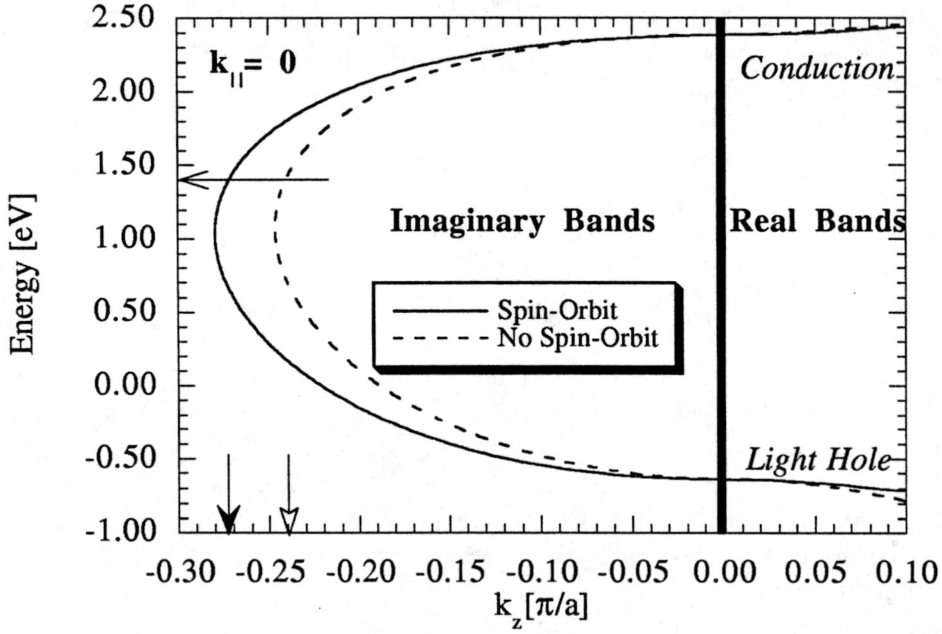

Fig. 1. Conduction and light-hole bands along with the imaginary band linking them for AlAs (at $T = 300K$, $\mathbf{k}_{||} = \mathbf{0}$), as reproduced by the nearest-neighbor sp^3s^* model with and without the spin-orbit coupling; parameters are from ref. 25. The real bands are plotted in the positive k_z part of the graph, the imaginary bands in the negative. The horizontal arrow at 1.4 eV indicates a typical energy for an electron tunneling from GaAs through an AlAs barrier; the vertical arrows indicate the imaginary wavevectors for the spin-orbit (solid) and no-spin-orbit (open) models.

So far we have only considered effective masses along [001]; for hole devices it is important to correctly model the warped valence bands along [110] and [111]. As with the [001] masses analytic formulas can be of great value and fortunately they are obtainable under the same assumptions as before for the heavy- and light-hole bands along [110] and [111]. Here, though the algebra is a bit more tedious: $\nabla_{\mathbf{k}} H(\mathbf{k})$ does not lift the degeneracy and we must therefore diagonalize the curvature matrix in the four-fold degenerate space $\{|LH;1\rangle, |HH;1\rangle, |LH;2\rangle, |HH;2\rangle\}$. The heavy-hole mass along [111] is rather simple, and the corresponding light-hole mass is related to it by a sum rule[27]:

$$\frac{m_0}{m^{(hh)}_{[111]}} = \left(\frac{2m_0}{\hbar^2}\right)\left(\frac{a}{4}\right)^2\left(\frac{1}{3}\right)\Big[4\left(\rho_a^{l,+}\right)^2\left(V_{xc,yc} - 2V_{xc,xc} - U_{xc,xc}\right) + 4\left(\rho_c^{l,+}\right)^2\left(V_{xa,ya} - 2V_{xa,xa} - U_{xa,xa}\right)$$
$$-\rho_a^{l,+}\rho_c^{l,+}\left(V_{x,y} + 3V_{x,x}\right)\left(V_{x,y} - V_{x,x}\right)/V_{x,x}\Big] \tag{43}$$

$$\frac{m_0}{m^{(lh)}_{[111]}} = \frac{m_0}{m^{(lh)}_{[001]}} + \frac{m_0}{m^{(hh)}_{[001]}} - \frac{m_0}{m^{(hh)}_{[111]}} \tag{44}$$

where we have included interactions up to second-near neighbor and the [001] light- and heavy-hole masses are given in (37) and (38), respectively; all terms in (42) have the same meanings as in the other second-near-neighbor formulas. Interestingly, it turns out that only *three* of the six masses (light- and heavy-hole along [001], [110], and [111]) are independent, for *both* [110] masses are simply functions of the [001] masses, (37)-(38), and the [111] heavy-hole mass, (42)[27]:

$$\frac{m_0}{m^{(hh)}_{[110]}} = \frac{1}{2}\left(\frac{m_0}{m^{(lh)}_{[001]}} + \frac{m_0}{m^{(hh)}_{[001]}}\right) + \sqrt{\frac{1}{16}\left(\frac{m_0}{m^{(lh)}_{[001]}} - \frac{m_0}{m^{(hh)}_{[001]}}\right)^2 + \frac{3}{4}\left[\frac{1}{2}\left(\frac{m_0}{m^{(lh)}_{[001]}} + \frac{m_0}{m^{(hh)}_{[001]}}\right) - \frac{m_0}{m^{(hh)}_{[111]}}\right]^2} \quad (45)$$

$$\begin{aligned}\frac{m_0}{m^{(lh)}_{[110]}} &= \frac{1}{2}\left(\frac{m_0}{m^{(lh)}_{[001]}} + \frac{m_0}{m^{(hh)}_{[001]}}\right) - \sqrt{\frac{1}{16}\left(\frac{m_0}{m^{(lh)}_{[001]}} - \frac{m_0}{m^{(hh)}_{[001]}}\right)^2 + \frac{3}{4}\left[\frac{1}{2}\left(\frac{m_0}{m^{(lh)}_{[001]}} + \frac{m_0}{m^{(hh)}_{[001]}}\right) - \frac{m_0}{m^{(hh)}_{[111]}}\right]^2} = \\ &= \frac{m_0}{m^{(lh)}_{[001]}} + \frac{m_0}{m^{(hh)}_{[001]}} - \frac{m_0}{m^{(hh)}_{[110]}} \end{aligned} \quad (46)$$

Note the absence of tight-binding parameters in (44)-(45).

As before, the effective mass formulas (42)-(45) tell us much about the capabilities of the models. First, note that the *only* tight-binding parameters *not* appearing in the [001] light- and heavy-hole expressions are the *second* near-neighbor interactions $V_{xa,ya}$ and $V_{xc,yc}$. Thus the second-near neighbor model affords more freedom for selecting the [111] heavy-hole mass independent of the [001]- masses. This does not mean that the second-near neighbor model is required to achieve this fit, for it is certainly possible to find nearest-neighbor parameterizations which do as well. It is, however, possible that achieving such a fit in the nearest-neighbor model may well entail significant compromises in the gaps and masses at other points in the Brillouin zone (e.g., the X-valleys). Beyond this, note that although two new second-near neighbor parameters enter (42), they allow us to independently specify only one more mass. These limitations are rather surprising given the widely held belief that even the nearest-neighbor sp^3 model can accurately fit the *valence* bands of most semiconductors.

In addition to the zone-center properties discussed above, it is also possible to obtain analytic *information* about a model at other points in the Brillouin zone. (Note that, due to the matrix dimension, it is generally impossible or impractical to obtain analytic mass formulas even at other symmetry points in the Brillouin zone, much less arbitrary **k**.). The most important example is the nearest-neighbor sp^3s^* model without the spin-orbit interaction. It is straightforward to show that for wavevectors lying on the diamond-shaped faces of the Brillouin zone, e.g. $\mathbf{k} = k_x\mathbf{e}_x + (2\pi/a)\mathbf{e}_z$, all bands are flat: *all* $E_n(\mathbf{k})$ relations are independent of k_x. This is demonstrated without explicitly finding the band dispersions by either calculating the characteristic polynomial for this wavevector and showing that it is independent of k_x[11], or by introducing a change of basis[25]. This of course means that the model reproduces an *infinite* transverse X-valley mass *at* the zone face. In practice, even when the X-valley minima occur somewhat inward from the zone edges, the transverse mass is still roughly 10 times too large (about $2.5 - 4.0m_0$ versus about $0.25m_0$).

We have thus developed several useful equations and techniques for calculating band curvatures in bulk tight-binding models in order to better parameterize them to reproduce the effective masses of interest. We have seen that at zone center it is often possible to obtain analytic inverse effective mass formulas and that these expressions reveal much about the capabilities of each of the tight-binding models considered. Some of these revelations are quite unexpected, such as the inability of the no-spin-orbit, nearest-neighbor sp^3s^* model to simultaneously fit both the electron and light-hole masses. In addition, the formulas serve as practical guides, showing which parameters will be most effective at adjusting the various masses, thereby ensuring the maximum accuracy possible from a given tight-binding model, applied to either bulk or heterostructure calculations.

5. HETEROSTRUCTURE SYSTEMS

The lack of translational symmetry on the level of a bulk primitive unit cell means that the approach to heterostructure calculations must be fundamentally different from that of bulk systems, for the Born-von Karman boundary conditions cannot be applied in the direction(s) in which there is no symmetry. Here we consider interface systems, with one or more interfaces occurring along the z-direction. The full crystal potential now possesses symmetry only in the $x-y$ plane: $V\left(\mathbf{r}+\mathbf{R}_{||}\right)=V(\mathbf{r})$ where $\mathbf{R}_{||}$ is a direct lattice vector in the $x-y$ plane. Hence the Hamiltonian is a function of only a two-dimensional wavevector, $\mathbf{k}_{||}=k_x\mathbf{e}_x+k_y\mathbf{e}_y$. Instead of the Bloch basis, (4), we employ the planar-orbital basis, composed of Bloch sums in the $x-y$ plane:

$$\left|\alpha;\,\mu;L;\,\mathbf{k}_{||}\right\rangle=\frac{1}{\sqrt{N}_{||}}\sum_{j=1}^{N_{||}}\exp\left[i\mathbf{k}_{||}\cdot\left(\mathbf{R}_{j||}+\mathbf{u}_{L||}+\mathbf{v}_{\mu||}\right)\right]\left|\alpha;\,\mu;L;\,\mathbf{R}_{j||}+\mathbf{u}_{L||}+\mathbf{v}_{\mu||}\right\rangle \quad (47)$$

where $N_{||}$ is the number of unit cells in the $x-y$ plane, the subscript $||$ denotes a vector lying in the $x-y$ plane, L is the layer index, and $\mathbf{u}_{L||}$ is a layer-dependent offset vector (e.g., in a face-centered cubic lattice where *alternating*, not adjacent, planes of lattice sites are aligned). All other symbols have the same meanings as in (4). For zincblende, the z-coordinate of a layer (relative to the 0-th layer which is placed at $z=0$) is $z_L=La/2$, where a is the conventional unit cell cube edge. The planar-orbital basis states are orthonormal:

$$\left\langle\alpha';\,\mu';L';\,\mathbf{k}'_{||}\left|\alpha;\,\mu;\,L;\mathbf{k}_{||}\;\alpha;\,\mu;\,L;\mathbf{k}_{||}\right.\right\rangle=\delta_{\alpha',\alpha}\delta_{\mu',\mu}\delta_{L',L}\delta_{\mathbf{k}'_{||},\mathbf{k}_{||}} \quad (48)$$

The wavefunction is then a superposition of states (46):

$$\left|\Psi_{\mathbf{k}||}\right\rangle=\sum_{\alpha',\mu',L'}C_{L'}^{\alpha',\mu'}\left|\alpha';\,\mu';L';\,\mathbf{k}_{||}\right\rangle \quad (49)$$

Note that the limits on the layer index L' are in general $-\infty$ and $+\infty$. Since cyclic boundary conditions are not applied in z, the Hamiltonian in the planar-orbital basis is in principle of infinite dimension. We next consider finding those boundary conditions which terminate the Hamiltonian, showing how to obtain them for heterostructure problems where the boundary layers are bulk-like.

5.1. *Boundary Conditions and Interface States*

The typical heterostructure consists of a relatively small active region, throughout which the electrostatic potential may vary and having one or more interfaces, terminated by left- and right-semi-infinite bulk-like regions, extending to $z=-\infty$ and $+\infty$, respectively (fig. 2). In the vicinity of the interface between the active and one of the terminating bulk-like regions there are, in addition to the usual propagating (extended) states, interface (evanescent) states which decay going into the semi-infinite bulk. These decaying states have complex k_z wavevectors and hence their dispersions – more properly $k_z(E)$ relations – are referred to as the complex bands of the terminating material. (An example is plotted in

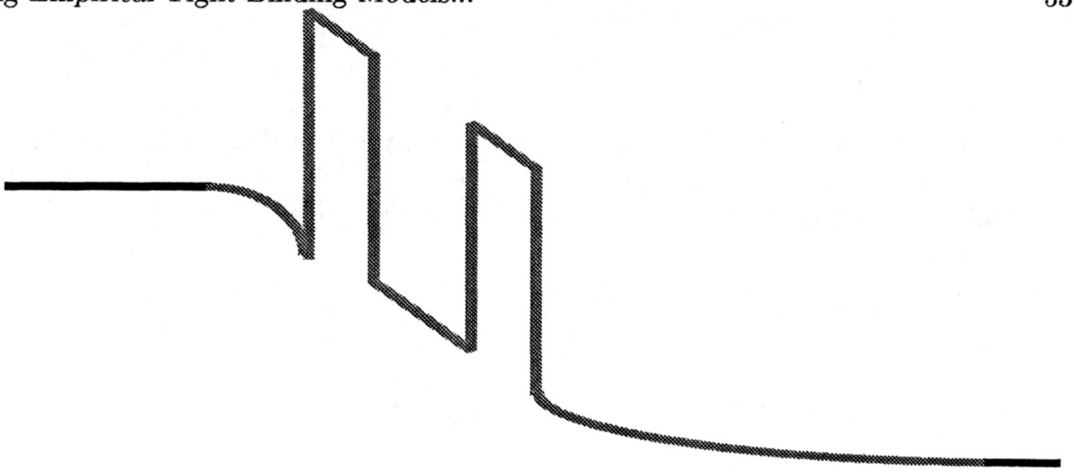

Fig. 2. Schematic conduction band-edge diagram for a typical resonant-tunneling diode under bias. The active region is shaded while the adjacent parts of the emitter (left) and collector (right) semi-infinite bulk regions are graphed in solid lines.

Fig. 1.) Because these states are characteristic of a semi-infinite bulk-like region it is natural to terminate the Hamiltonian there by expanding the wavefunction as an appropriate linear combination of these states. The problem, then, is to calculate these expansion states.

The procedure is best illustrated with a specific example. Here we consider the second-near-neighbor, spin-orbit sp^3s^* model[35]; the expressions are rather simpler for the nearest-neighbor model[11, 21] (see also subsection B below). We introduce notations for 10-element vectors,

$$\mathbf{C}_L^{\mu} = \left[C_L^{s\mu+}, C_L^{x\mu+}, C_L^{y\mu+}, C_L^{z\mu+}, C_L^{s*\mu+}, C_L^{s\mu-}, C_L^{x\mu-}, C_L^{y\mu-}, C_L^{z\mu-}, C_L^{s*\mu-}\right]^T \tag{50}$$

where the superscript T denotes the transpose, μ is the atom type and + and - denote orthonormal spin states. These could be the customary S_z-eigenstates ($\uparrow$ and $\downarrow$, however there are some algorithmic advantages to using eigenstates of spin quantized in the $x-y$ plane (e.g., S_y)[35]. We also introduce 10×10 coupling matrices, defined by:

$$\left[\underline{H}_{(L',L)}^{(\mu',\mu)}\right]_{(\alpha',\alpha)} \doteq \left\langle \alpha'; \mu'; L'; \mathbf{k}_{||} \left| H \right| \alpha; \mu; L; \mathbf{k}_{||} \right\rangle \tag{51}$$

Requiring the wavefunction (48) to be a solution of the Schrödinger equation,

$$[H - \mathbf{1}E] \left| \Psi_{\mathbf{k}_{||}} \right\rangle = 0 \tag{52}$$

yields pairs of matrix equations:

$$\left[\underline{H}_{(L-1,L)}^{(a,a)}\right]^{\dagger} \mathbf{C}_{L-1}^{a} + \left[\underline{H}_{(L-1,L)}^{(c,a)}\right]^{\dagger} \mathbf{C}_{L-1}^{c} + \left[\underline{H}_{(L,L)}^{(a,a)} - \underline{1}E\right] \mathbf{C}_{L}^{a} + \underline{H}_{(L,L)}^{(a,c)} \mathbf{C}_{L}^{c} + \underline{H}_{(L,L+1)}^{(a,a)} \mathbf{C}_{L+1}^{a} = \mathbf{0} \tag{53}$$

$$\left[\underline{H}_{(L-1,L)}^{(c,c)}\right]^{\dagger} \mathbf{C}_{L-1}^{c} + \left[\underline{H}_{(L,L)}^{(a,c)}\right]^{\dagger} \mathbf{C}_{L}^{a} + \left[\underline{H}_{(L,L)}^{(c,c)} - \underline{1}E\right] \mathbf{C}_{L}^{c} + \underline{H}_{(L,L+1)}^{(c,a)} \mathbf{C}_{L+1}^{a} + \underline{H}_{(L,L+1)}^{(c,c)} \mathbf{C}_{L+1}^{c} = \mathbf{0} \tag{54}$$

In bulk $\underline{H}^{(\mu',\mu)}_{(L-1,L)} = \underline{H}^{(\mu',\mu)}_{(L,L+1)}$, and the eigenstates of the interface system are found by setting $\mathbf{C}^{\mu}_{L+1} = \lambda_+ \mathbf{C}^{\mu}_{L}$ (forward eigenproblem) or $\mathbf{C}^{\mu}_{L} = \lambda_- \mathbf{C}^{\mu}_{L+1}$ (reverse eigenproblem); so long as $\lambda_\pm \neq 0$ it is obvious that $\lambda_- = 1/\lambda_+$. Introducing these definitions into (52)-(53) initially gives a 20 × 20 second-order eigenproblem, which, using the definitions once more, is readily converted into a 40 × 40 first-order eigenproblem for either λ_- or λ_+:

$$\lambda_+ \begin{bmatrix} \underline{H}_{(L,L+1)} & \underline{0} \\ \underline{0} & \underline{1} \end{bmatrix} [\mathbf{C}_L \, \mathbf{C}_{L-1}] = \begin{bmatrix} -\underline{H}_{(L,L)} & -\underline{H}^{\dagger}_{(L,L+1)} \\ \underline{1} & \underline{0} \end{bmatrix} [\mathbf{C}_L \, \mathbf{C}_{L-1}] \tag{55}$$

$$\begin{bmatrix} \underline{H}_{(L,L+1)} & \underline{0} \\ \underline{0} & \underline{1} \end{bmatrix} [\mathbf{C}_{L+1} \, \mathbf{C}_{L}] = \lambda_- \begin{bmatrix} -\underline{H}_{(L,L)} & -\underline{H}^{\dagger}_{(L,L+1)} \\ \underline{1} & \underline{0} \end{bmatrix} [\mathbf{C}_{L+1} \, \mathbf{C}_{L}] \tag{56}$$

where

$$\underline{H}_{(L,L+1)} \doteq \begin{bmatrix} \underline{H}^{(c,c)}_{(L,L+1)} & \underline{H}^{(c,a)}_{(L,L+1)} \\ \underline{0} & \underline{H}^{(a,a)}_{(L,L+1)} \end{bmatrix} \tag{57}$$

$$\underline{H}_{(L,L)} \doteq \begin{bmatrix} \left[\underline{H}^{(c,c)}_{(L,L)} - \underline{1}E\right] & \left[\underline{H}^{(a,c)}_{(L,L)}\right]^{\dagger} \\ \underline{H}^{(a,c)}_{(L,L)} & \left[\underline{H}^{(a,a)}_{(L,L)} - \underline{1}E\right] \end{bmatrix} \tag{58}$$

$$\mathbf{C}_L \doteq [\mathbf{C}^{c}_{L} \, \mathbf{C}^{a}_{L}] \tag{59}$$

In each semi-infinite region the wavefunction is a superposition of those interface states which decay going into and those extended states which propagate going into the region, the expansion coefficients to be determined by solving the Schrödinger equation. On the right these are interface states which decay as $z \to +\infty$ ($|\lambda_+| < 1$) and positive velocity extended states; on the left, the reverse, states which decay as $z \to -\infty$ ($|\lambda_-| < 1$) and negative velocity extended states. For tunneling problems there is also an injected state, a positive (negative) velocity propagating state coming from the left (right) bulk semi-infinite region.

Having stated the method for finding the expansion states, it is instructive to examine the physical content of (54) - (55) before discussing computational practicalities. This is most easily done within the context of the transfer matrix method[4], although for numerical and algorithmic reasons it is better to formulate the problem as a generalized eigensystem. Consider (54) and call the matrix on the left $\underline{M}_1$ and that on the right $\underline{M}_2$. Assume now that $\underline{M}_1$ is nonsingular (invertible), so that we may write,

$$\begin{bmatrix} \mathbf{C}_{L+1} \\ \mathbf{C}_L \end{bmatrix} = \lambda_+ \begin{bmatrix} \mathbf{C}_L \\ \mathbf{C}_{L-1} \end{bmatrix} = \underline{M}_1^{-1} \underline{M}_2 \begin{bmatrix} \mathbf{C}_L \\ \mathbf{C}_{L-1} \end{bmatrix} = \underline{T} \begin{bmatrix} \mathbf{C}_L \\ \mathbf{C}_{L-1} \end{bmatrix} \tag{60}$$

where $\underline{T}$ is called the (bulk) transfer matrix since it relates the coefficients on one pair of layers to those on another. The relation $\mathbf{C}^{\mu}_{L} = \lambda_+ \mathbf{C}^{\mu}_{L-1}$ of course expresses the fact that beyond layer $(L-1)$ the material is bulk-like, and the eigenstates of $\underline{T}$ in (59) are simply the propagating and evanescent states characteristic of the semi-infinite bulk material. These eigenstates serve as the expansion states since in the semi-infinite regions the Hamiltonian away from the interface is once more that of a bulk material. (Recall that in an n-th nearest-neighbor model Hamiltonian matrix elements more than n-neighbors following the interface are simply those of the bulk.)

As attractive as the transfer-matrix description of the problem is, it cannot find the bulk eigenstates for singular matrices $\underline{M}_1$ and $\underline{M}_2$, as often happens at certain values of

$\mathbf{k}_{||}$ or for certain parameter sets. For example, when some coupling parameters are zero a transfer matrix may not exist even though the *bulk* (three-dimensional) Hamiltonian is perfectly well-behaved. Similarly, propagating states exist at *all* points in the Brillouin zone, yet the transfer-matrix method cannot find them at those $\mathbf{k}_{||}$ for which a transfer-matrix does not exist. The *physics* is not the source of the difficulty; instead it is the transfer-matrix approach itself[35]. To see this, consider that case of singular $\underline{M}_1$ and $\underline{M}_2$, so that the kernels of these matrices contain non-zero vectors (i.e., $\lambda_- = 0$ and $\lambda_+ = 0$). These vectors represent states which decay immediately (or, in the reverse direction, blow up immediately), so that an implementation based on (59) obviously will not work. (The solutions still possess the necessary properties, however; see ref. 35 for details on the second-near-neighbor, spin-orbit sp^3s^* model.) Here we concentrate on the practical difficulty, namely that since states which decay immediately are part of the wavefunction expansion in each semi-infinite bulk region we require a numerical method which can compute them.

Fortunately, all of the eigenvectors (even those which immediately decay or blow up) are obtainable when the problem is solved as a generalized eigensystem and routines from EISPACK[36] and other sources are available. Using the EISPACK routines first requires that we transform the the system to a real generalized eigensystem; the transformation, which *doesnot* alter the dimension of the problem, is given in ref. 35 for the second-near-neighbor, spin-orbit sp^3s^* model. That such a transformation exists should be no surprise, for we must find for each positive-velocity extended state one of negative velocity. Rather than solving $\underline{A}\mathbf{x} = \lambda\underline{B}\mathbf{x}$, the algorithm[36] finds the solutions to

$$\beta\underline{A}\mathbf{x} = \alpha\underline{B}\mathbf{x} \tag{61}$$

where, for $\beta \neq 0$ $\lambda = \alpha/\beta$. Eigenvectors corresponding to "infinite" eigenvalues (those in ker $\underline{B}$ thus have $\beta \approx 0$ *while those of zero eigenvalue (inker* $\{\underline{A}\}$*) have* $\alpha \approx 0$*). Therefore a single calculation suffices to find solutions of both (54) and (55). The generalized eigensystem approach to the problem of finding the expansion states has proven useful in both very simple, deliberately singular, tunneling calculations[37] as well as in more typical cases[27].*

5.2. *Tunneling Problems*

Now that we have the Hamiltonian in the planar-orbital basis and a numerically stable method for finding the boundary conditions which terminate it, we can study tunneling problems. The numerous methods available for tunneling problems belong to one of two families: direct transmission coefficient calculations[11, 12] or recursive Green function calculations[20]. Direct methods are simpler to implement for fully-coherent, $\mathbf{k}_{||}$-conserving models, whereas only the recursive Green function approach can readily handle non-$\mathbf{k}_{||}$-conserving processes which might be included in the model (e.g., phonons, interface roughness, ionized impurities, and disorder)[20]. The recursive Green function approach also permits a simplified treatment of phonon scattering by way of an optical potential. However, since our chief purpose is to develop methods and procedures by which tight-binding models may be optimized for heterostructure calculations, it is sufficient to focus on fully coherent tunneling models, and we therefore restrict our attention to direct transmission coefficient calculations. As with the generalized eigenproblem method for determining the boundary conditions (Sec. III.A above) the transmission coefficient calculation is best illustrated with a specific example. Because the resulting equations can become quite unwieldy as interactions when more distant neighbors are included the discussion is easier to follow for simpler models. We therefore employ the no-spin-orbit sp^3s^* model in this subsection.

For this model we have 5- and 10-element (as opposed to 10- and 20-element) vectors; note the difference in the order of sub-vectors in $\mathbf{C}_L$ versus (58):

$$\mathbf{C}_L^{\mu} \doteq [C_L^{s\mu}, C_L^{x\mu}, C_L^{y\mu}, C_L^{z\mu}, C_L^{s*\mu}]^T \tag{62}$$

$$\mathbf{C}_L \doteq \begin{bmatrix} \mathbf{C}_L^a \\ \mathbf{C}_L^c \end{bmatrix} \tag{63}$$

Here the Schrödinger equation in the planar-orbital basis, (50)-(51) yields simpler pairs of equations due to the nearest-neighbor only coupling:

$$\left[\underline{H}_{(L-1,L-1)}^{(a,c)}\right]^{\dagger} \mathbf{C}_{L-1}^a + \left[\underline{H}_{(L-1,L-1)}^{(c,c)} - \underline{1}E\right] \mathbf{C}_{L-1}^c + \underline{H}_{(L-1,L)}^{(c,a)} \mathbf{C}_L^a = \mathbf{0} \tag{64}$$

$$\left[\underline{H}_{(L-1,L)}^{(c,a)}\right]^{\dagger} \mathbf{C}_{L-1}^c + \left[\underline{H}_{(L,L)}^{(a,a)} - \underline{1}E\right] \mathbf{C}_L^a + \underline{H}_{(L,L)}^{(a,c)} \mathbf{C}_L^c = \mathbf{0} \tag{65}$$

The eigenproblem leading to the boundary conditions is likewise simpler:

$$\begin{bmatrix} -\left[\underline{H}_{(L-1,L-1)}^{(a,c)}\right]^{\dagger} & -\left[\underline{H}_{(L-1,L-1)}^{(c,c)} - \underline{1}E\right] \\ \underline{0} & -\left[\underline{H}_{(L-1,L)}^{(c,a)}\right]^{\dagger} \end{bmatrix} \begin{bmatrix} \mathbf{C}_{L-1}^a \\ \mathbf{C}_{L-1}^c \end{bmatrix} = \begin{bmatrix} \underline{H}_{(L-1,L)}^{(c,a)} & \underline{0} \\ \left[\underline{H}_{(L,L)}^{(a,a)} - \underline{1}E\right] & \underline{H}_{(L,L)}^{(a,c)} \end{bmatrix} \begin{bmatrix} \mathbf{C}_L^a \\ \mathbf{C}_L^c \end{bmatrix} \tag{66}$$

The expansion states are found by requiring $\mathbf{C}_L^{\mu} = \lambda_+ \mathbf{C}_{L-1}^{\mu}$ and solving the generalized eigenproblem (65) for bulk as in sub-section A above.

The procedures for solving the Schrödinger equation (63)-(64) are of two types: methods based upon transfer matrices or those which result in large, sparse linear systems. Historically, the transfer matrix method[4] was the first employed and it is conceptually the easiest to understand. Unfortunately, it also tends to be numerically unstable in the extreme, and so fails for all but the shortest of systems; it was this instability problem which prompted the development of other methods[11, 12] in 1991-92. (It also should be noted that the aforementioned difficulties associated with obtaining the boundary conditions via transfer matrices were not fully appreciated until much later, in 1996[35, 37].) We shall study the transfer matrix method[4] first, showing why it becomes numerically unstable, then progress to the others which result in large, sparse linear systems[11, 12], discussing the method of ref. 11 before that of ref. 12 since the derivation of the former clearly indicates just how the numerical instabilities are removed. Before turning to the methods, however, we must terminate the Hamiltonian; this step is of course the same for all approaches.

We calculate the transmission probability for an electron incident on the active region from the left. Label the last emitter side bulk-like layer 0 and the first collector bulk-like layer N and assume that the expansion states have been found in each region (see III.A above). (Note that the emitter and collector bulk-like regions need not be the same: there is often a potential drop and the materials may even be different. At the energy chosen there must of course be propagating states on both sides for tunneling to occur.) On the emitter side call the expansion states $\mathbf{u}_{0,6}, \ldots \mathbf{u}_{0,10}$, and on the collector side call them $\mathbf{u}_{N,1}, \ldots \mathbf{u}_{N,5}$. We restrict our attention to the case where the incident wave is a single bulk-emitter eigenstate; call this incident state (positive-velocity extended state) $\mathbf{u}_{0,1}$. The wavefunction in layer 0 is thus:

$$\mathbf{C}_0 = \mathbf{u}_{0,1} + \sum_{j=6}^{10} b_j \mathbf{u}_{0,j}, \quad \mathbf{u}_{n,j} \doteq \begin{bmatrix} \mathbf{u}_{n,j}^a \\ \mathbf{u}_{n,j}^c \end{bmatrix} \tag{67}$$

Similarly, in the layer N, the wavefunction is:

$$\mathbf{C}_N = \sum_{j=1}^{5} b_j \mathbf{u}_{N,j} \tag{68}$$

In making these expansions we assume that the bulklike eigenproblems yield eigenbases: here that there are 10 linearly independent eigenvecotrs in each of the expansion layers 0 and N. While there is no guarantee that this will always be the case – the eigenproblem is not Hermitian – such difficulties in practice tend to be confined to regions very near band extrema, where for example, the forward- and reverse-propagating eigenstates become almost identical. In addition to this assumption we must differentiate between the states in (67) which decay and those which propagate (in the positive direction), so in the layer N label the propagating states 1, 2,..., $f \leq 5$. Having specified the boundary conditions, we may now use these expressions in the Schrödinger equation (63)-(64) to terminate the Hamiltonian.

In the transfer matrix method[4] we observe that if the matrices in (65) are nonsingular, we may invert the right-hand matrix and multiply it by the matrix on the left to obtain the transfer matrix equation:

$$\mathbf{C}_L = \underline{T}_{L,L-1} \mathbf{C}_{L-1} \tag{69}$$

The wavefunction is then "transferred" from one bulk-like region, 0, to the other, N, by successive (left-) multiplication of transfer matrices:

$$\mathbf{C}_N = \prod_{L=1}^{N} \underline{T}_{L,L-1} \mathbf{C}_0 \tag{70}$$

Inserting boundary conditions (66) and (67) results in a set of 10 inhomogeneous equations in 10 unknowns (the b_j). The transmission probability is simply a sum of the ratios of the transmitted fluxes to the incident flux:

$$\mathcal{T} = \sum_{j=1}^{f} |b_j|^2 \frac{\left|v_j^{(N)}\right|}{\left|v_1^{(0)}\right|}, v_j^{(L)} \doteq \frac{\partial E_n^{(L)}}{\partial k_z^{(L)}} \tag{71}$$

where the velocities can be calculated from the appropriate bulk Hamiltonians using (17) above. Although this method is straightforward and intuitively appealing, it is highly unstable due to the multiplication of transfer matrices in (69). A simple example amply illustrates the problem: Consider tunneling through an unbiased barrier of material X surrounded by semi-infinite bulk-like regions of material Y. Upon transfer into the barrier via $\underline{T}_{1,0}$ the wavefunction (66) becomes a superposition of all eigenstates of bulk-X. Now there are generally growing and decaying transfer matrix eigenstates in addition to any propagating ones, and those which grow most quickly will come to dominate the wavefunction after several multiplications of the bulk-X transfer matrix, as in (69). (Had computers infinite precision there would be no difficulty, for the solution of the linear system would return very small coefficients for these states.) For a thick barrier these growing states push the propagating and decaying states into the numerical noise and hence one obtains incorrect values for the b_j and erroneous transmission coefficients (of, say, 1000!). A different method is therefore mandatory for studying realistic heterostructures.

The first published method for alleviating this problem is due to Boykin, van der Wagt, and Harris[11] and is based upon transfer matrices. It arises from the observation that for a given calculation a transfer across S layers generally can be done accurately, where the number of layers, S, depends upon the precision of the computation (e.g., larger for double- than for single-precision). Therefore, we successively transfer across no more than the "safe" number of layers:

$$\mathbf{C}_S = \underline{T}_{S,0}\mathbf{C}_0 \tag{72}$$

$$\mathbf{C}_{2S} = \underline{T}_{2S,S}\mathbf{C}_S \tag{73}$$

finally ending up at layer N with

$$\mathbf{C}_N = \underline{T}_{N,mS}\mathbf{C}_{mS}, \quad N - mS \leq S \tag{74}$$

where m is an integer and the (left-) product of transfer matrices is defined as:

$$\underline{T}_{L+M,L} \doteq \prod_{j=L}^{L+M-1} \underline{T}_{j+1,j} \tag{75}$$

Replacing $\mathbf{C}_0$ and $\mathbf{C}_N$ in (71) and (73) by their respective expansions (66) and (67) and solving these equations *simultaneously* avoids the compounding of transfer matrix products in (69) above, thus conferring numerical stablity; the result is a large, sparse, linear system:

$$\begin{bmatrix} \underline{A}_1 & \underline{1} & \underline{0} & \cdots & \cdots & \underline{0} \\ \underline{0} & -\underline{T}_{2S,S} & \underline{1} & \underline{0} & \cdots & \underline{0} \\ \underline{0} & \ddots & -\underline{T}_{3S,2S} & \underline{1} & \ddots & \vdots \\ \vdots & \ddots & \ddots & \ddots & \ddots & \underline{0} \\ \underline{0} & \cdots & \underline{0} & \ddots & \ddots & \underline{1} \\ \underline{A}_N & \cdots & \cdots & \cdots & \underline{0} & -\underline{T}_{N,mS} \end{bmatrix} \begin{bmatrix} \mathbf{b} \\ \mathbf{C}_S \\ \mathbf{C}_{2S} \\ \vdots \\ \vdots \\ \mathbf{C}_{mS} \end{bmatrix} = \begin{bmatrix} \underline{T}_{S,0}\mathbf{u}_{0,1} \\ \mathbf{0} \\ \vdots \\ \vdots \\ \vdots \\ \mathbf{0} \end{bmatrix} \tag{76}$$

$$\mathbf{b} \doteq \begin{bmatrix} b_1 \\ \vdots \\ b_{10} \end{bmatrix} \tag{77}$$

$$\underline{A}_1 \doteq [\mathbf{0}| \quad \mathbf{0}| \quad \mathbf{0}| \quad \mathbf{0}| \quad \mathbf{0}| \quad -\underline{T}_{S,0}\mathbf{u}_{0,6}| \quad \cdots \quad | -\underline{T}_{S,0}\mathbf{u}_{0,10}] \tag{78}$$

$$\underline{A}_N \doteq [\mathbf{u}_{N,1}| \quad \cdots \quad |\mathbf{u}_{N,5} \quad |\mathbf{0} \quad |\mathbf{0} \quad |\mathbf{0} \quad |\mathbf{0} \quad |\mathbf{0}] \tag{79}$$

The solution of (75) yields the b_j, which with (70) give the transmission probability, just as before. This method has proven to be highly stable in practive and permitted the first calculations for realistic heterostructures (including space-charge regions) in the no-spin-orbit sp^3s^* basis[3].

The method of Ting, Yu, and McGill[12] continues the process of ref. 11, only without forming transfer matrices. It is thus numerically stable for the same reasons as the method of Boykin, van der Wagt, and Harris[11], and furthermore is usable even in those cases where a transfer matrix does not exist[37]. Because this method is a direct solution of the Schrödinger equation, (63)-(64), and thus advances but a single layer per pair of block-rows the resulting sparse linear system is considerably larger than that of ref. 11 for the same heterostructure. In compensation, however, there are no matrix multiplications, (74), as with the former. Here we present a slight variant of the original method of ref. 12; as before, we terminate the Hamiltonian at layers 0 and N. The termination at $L = 0$ follows from substituting (66) into (63)-(64) with $L = 1$:

$$\underline{B}_0 \mathbf{b}_0 + \underline{H}^{(c,a)}_{(0,1)} \mathbf{C}^a_1 = -\left[\underline{H}^{(a,c)}_{(0,0)}\right]^\dagger \mathbf{u}^a_{0,1} - \left[\underline{H}^{(c,c)}_{(0,0)} - \underline{1}E\right] \mathbf{u}^c_{0,1} = \mathbf{I}^{(a)} \tag{80}$$

$$\underline{B}_1 \mathbf{b}_0 + \left[\underline{H}^{(a,a)}_{(1,1)} - \underline{1}E\right] \mathbf{C}^a_1 + \underline{H}^{(a,c)}_{(1,1)} \mathbf{C}^c_1 = -\left[\underline{H}^{(c,a)}_{(0,1)}\right]^\dagger \mathbf{u}^c_{0,1} = \mathbf{I}^{(c)} \tag{81}$$

where $\underline{B}_0$ is a 5×5 matrix with columns $\mathbf{B}_{0,j}$, $j = 6, 7, ..., 10$:

$$\mathbf{b}_0 \doteq \begin{bmatrix} b_6 \\ \vdots \\ b_{10} \end{bmatrix} \tag{82}$$

$$\underline{B}_0 \doteq [\,\mathbf{B}_{0,6}| \;\; \mathbf{B}_{0,7}| \;\; \cdots \;\; |\mathbf{B}_{0,10}\,], \quad \mathbf{B}_{0,j} \doteq \left[\underline{H}^{(a,c)}_{(0,0)}\right]^\dagger \mathbf{u}^a_{0,j} + \left[\underline{H}^{(c,c)}_{(0,0)} - \underline{1}E\right] \mathbf{u}^c_{0,j} \tag{83}$$

$$\underline{B}_1 \doteq [\,\mathbf{B}_{1,6}| \;\; \mathbf{B}_{1,7}| \;\; \cdots \;\; |\mathbf{B}_{1,10}\,], \quad \mathbf{B}_{1,j} \doteq \left[\underline{H}^{(c,a)}_{(0,1)}\right]^\dagger \mathbf{u}^c_{0,j} \tag{84}$$

Similarly, the termination at layer N results from substituting (67) into (63)-(64) with $L = N$:

$$\left[\underline{H}^{(a,c)}_{(N-1,N-1)}\right]^\dagger \mathbf{C}^a_{N-1} + \left[\underline{H}^{(c,c)}_{(N-1,N-1)} - \underline{1}E\right] \mathbf{C}^c_{N-1} + \underline{B}_{N-1} \mathbf{b}_N = \mathbf{0} \tag{85}$$

$$\left[\underline{H}^{(c,a)}_{(N-1,N)}\right]^\dagger \mathbf{C}^c_{N-1} + \underline{B}_N \mathbf{b}_N = \mathbf{0} \tag{86}$$

$$\mathbf{b}_N \doteq \begin{bmatrix} b_1 \\ \vdots \\ b_5 \end{bmatrix} \tag{87}$$

$$\underline{B}_{N-1} \doteq [\,\mathbf{B}_{N-1,1}| \;\; \mathbf{B}_{N-1,2}| \;\; \cdots \;\; |\mathbf{B}_{N-1,5}\,], \quad \mathbf{B}_{N-1,j} \doteq \underline{H}^{(c,a)}_{(N-1,N)} \mathbf{u}^a_{N,j} \tag{88}$$

$$\underline{B}_N \doteq [\,\mathbf{B}_{N,1}| \;\; \mathbf{B}_{N,2}| \;\; \cdots \;\; |\mathbf{B}_{N,5}\,], \quad \mathbf{B}_{N,j} \doteq \left[\underline{H}^{(a,a)}_{(N,N)} - \underline{1}E\right] \mathbf{u}^a_{N,j} + \underline{H}^{(a,c)}_{(N,N)} \mathbf{u}^c_{N,j} \tag{89}$$

Equations (79)-(80) and (84)-(85) give the first and last pairs of block rows, respectively, of a linear system; (63)-(64) give the remaining pairs. The result is a large, banded, linear system:

$$\begin{bmatrix} \underline{B}_0 & \underline{H}^{(c,a)}_{(0,1)} & \underline{0} & \cdots & \cdots & \underline{0} \\ \underline{B}_1 & \left[\underline{H}^{(a,a)}_{(1,1)} - \underline{1}E\right] & \underline{H}^{(a,c)}_{(1,1)} & \underline{0} & \cdots & \underline{0} \\ \underline{0} & \ddots & \ddots & \ddots & \ddots & \vdots \\ \vdots & \ddots & \ddots & \ddots & \ddots & \underline{0} \\ \vdots & \cdots & \underline{0} & \left[\underline{H}^{(a,c)}_{(N-1,N-1)}\right]^\dagger & \left[\underline{H}^{(c,c)}_{(N-1,N-1)} - \underline{1}E\right] & \underline{B}_{N-1} \\ \underline{0} & \cdots & \cdots & \underline{0} & \left[\underline{H}^{(c,a)}_{(N-1,N)}\right]^\dagger & \underline{B}_N \end{bmatrix} \begin{bmatrix} \mathbf{b}_0 \\ \mathbf{C}^a_1 \\ \mathbf{C}^c_1 \\ \vdots \\ \mathbf{C}^c_{N-1} \\ \mathbf{b}_N \end{bmatrix} = \begin{bmatrix} \mathbf{I}^{(a)} \\ \mathbf{I}^{(c)} \\ \mathbf{0} \\ \vdots \\ \vdots \\ \mathbf{0} \end{bmatrix} \tag{90}$$

After solving this equation we use the b_j in (70) as before to find the transmission probability. Like the method of ref. 11 this approach has proven to be highly stable and both can easily handle active regions of 2000Åor more.

There is one more numerically stable method suited to a tight-binding basis, that of Schulman and Ting[38]. This method is closely related to that of refs. 11 and 12, and is indeed essentially a reordering of the basis in the method of ref. 11 so as to produce a banded linear system[38]. The size of the resulting system is thus the same as that of ref. 11 (i.e., smaller than that of ref. 12), but as with the method of ref. 11, matrix multiplications are necessary. Although it might at first seem that the smaller linear systems in the methods of refs. 11 and 38 implies less computational effort than with the approach of ref. 12, it turns out that when the matrix multiplications required in refs. 11 and 38 are taken into account the operation count is comparable for all[12, 38].

5.3. *Other Heterostructure Calculations*

The methods discussed above can be adapted to other heterostructure problems, however there are some subtleties involved in the implementation. For example, to find the bound states of a quantum well, we would omit the incident state $\mathbf{u}_{0,1}$ from the layer 0 expansion in (66), resulting in homogeneous linear systems for (75) and (89). The task then becomes one of searching for those energies, E, for which the large, sparse matrices in these systems become singular. (For bound states E must of course be restricted to lie in the gaps of the semi-infinite barrier materials so that only evanescent states exist in the barriers. Transmission over the quantum well occurs when there are both propagating and evenescent states in the barriers, and here the eigenvalues will be complex with the real part of E giving the resonance energy and the imaginary part the resonance width or lifteime.) Numerically this is not simple, for the expansion states $\mathbf{u}_{0,j}$ and $\mathbf{u}_{N,j}$ are themselves functions of E and we are therefore faced with a very large, nonlinear, eigenproblem which must be solved iteratively. Fortunately the method of Bowen, et al.[21] is well-suited to such an open system problem and can efficiently find the eigenvalues of interest for large systems such as (75) and (89).

5.4. *Recent Successes*

Through its automatic and natural inclusion of bandstructure effects throughout the Brillouin zone the empirical tight-binding method, when properly employed and parametrized, brings out much more physics than is possible with simpler approaches. One of the first areas of focus was the effect of the *longitudinal* X-valleys on tunneling in GaAs-AlAs heterostructures[10, 11, 13, 21]. Next, the method was applied to resonant interband tunneling diodes (RITDs), in which both conduction- and valence-band states are important[39]. A significant finding of the RITD work work is that the transmission probability is a strong function of the in-plane wavevector $\mathbf{k}_{||}$, due to the rich valence-band mixing present in quantum-confined heterostructures[39]. More surprising was the result that the $\mathbf{k}_{||}$-dependence of the transmission cannot be ignored for highly nonparabolic conduction-band tunneling heterostructures (e.g., InAs-AlSb or InGaAs-InAlAs)[14-18]. The conclusion of all these works is that in calculating the current-voltage characteristics one must take into account the $\mathbf{k}_{||}$-dependence of the transmission, so that it is at least necessary to integrate over both energy and $\left|\mathbf{k}_{||}\right|$, unlike in effective-mass models[40]. (Recently, Schulman[41] has shown how to include nonparabolicity effects in the effective-mass framework.) Finally, when properly parametrized tight-binding calculations include Hartree self-consistency and tunneling from the accumulation region notch they can achieve truly remarkable agreement with experiment[18, 20]. Especially noteworthy and illuminating is the work of Bowen, et al.[18] which systematically shows how including or omitting various effects (Hartree self-consistency, explicit integration over $\mathbf{k}_{||}$, etc.) alters the current-voltage characteristic. Thus, the tight-binding method has proven to be a very valuable tool for heterostructure modeling.

6. Conclusions

We have seen that realizing the full potential of the tight-binding method requires that the underlying model be both carefully chosen and parametrized, and that attention must be paid to numerical issues in heterostructure calculations in particular. First, we have shown that in calculating quantities as simple as the effective mass the incompleteness of the tight-binding basis must be explicitly included. Second, through our derivation of analytic expressions for the zone-center effective masses for III-V materials, we have seen that not all tight-binding models are capable of reproducing the bulk bands of a given material with equal accuracy. One surprising result is that it is usually better to employ a spin-orbit model even for calculations involving only conduction-band states, since the spin-orbit interaction is essential for properly reproducing the light-hole mass and thus the correct dispersion of the imaginary band linking the light-hole valence- band to the conduction-band. We have also studied in some detail the application of the method to heterostructrure problems, finding that care is required in calculating both the expansion states (i.e., the complex bandstructure) as well as transmission coefficients in tunneling problems. Finally, we have discussed some recent, successful applications of the tight-binding method to heterostructure calculations which show the great power and accuracy of the method when properly employed.

References

[1] J. C. Slater and G. F. Koster, Phys. Rev. **94**, 1498 (1954).

[2] Walter A. Harrison, *Electronic Structure* (Dover, New York, 1989); *Elementary Electronic Structure* (World Scientific, New Jersey, 1999).

[3] P. Vogl, Harold P. Hjalmarson, and John D. Dow, J. Phys. Chem. Solids **44**, 365 (1983).

[4] D. H. Lee and J. D. Joannopoulos, Phys. Rev. B **23**, 4988 (1981); J. Vac. Sci. Technol. **19**, 355 (1981).

[5] G. C. Osbourn and D. L. Smith, Phys. Rev. B **15**, 2124 (1979).

[6] J. N. Schulman and Yia-Chung Chang, Phys. Rev. B **31**, 2056 (1985).

[7] Yia-Chung Chang and J. N. Schulman, Phys. Rev. B **31**, 2069 (1985).

[8] Yia-Chung Chang, Phys. Rev. B **25**, 605 (1982).

[9] Yia-Chung Chang and J. N. Schulman, Phys. Rev. B **25**, 3975 (1982).

[10] Kenneth V. Rousseau, K. L. Wang, and J. N. Schulman, Appl. Phys. Lett. **54**, 1341 (1989).

[11] Timothy B. Boykin, Jan P. A. van der Wagt, and James S. Harris, Jr., Phys. Rev. B **43**, 4777 (1991).

[12] D. Z.-Y. Ting, E. T. Yu, and T. C. McGill, Phys. Rev. B **45**, 3583 (1992).

[13] Timothy B. Boykin and James S. Harris, Jr., J. Appl. Phys. **72**, 988 (1992).

[14] Timothy B. Boykin, R. E. Carnahan, and R. J. Higgins, Phys. Rev. B **48**, 14 232 (1993).

[15] Timothy B. Boykin, R. E. Carnahan, and K. P. Martin, Phys. Rev. B **51**, 2273 (1995).

[16] Timothy B. Boykin, Phys. Rev. B **51**, 4289 (1995).

[17] Timothy B. Boykin, J. Appl. Phys. **78**, 6818 (1995).

[18] R. Chris Bowen, Gerhard Klimeck, Roger K Lake, William R. Frensley, and Ted Moise, J. Appl. Phys. **81**, 3207 (1997).

[19] G. Klimeck, R. Lake, R. Chris Bowen, W. R. Frensley, and T. Moise, Appl. Phys. Lett. **67**, 2539 (1995).

[20] R. Lake, G. Klimeck, R. Chris Bowen, and D. Jovanovic, J. Appl. Phys. **81**, 7845 (1997).

[21] R. Chris Bowen, William R. Frensley, Gerhard Klimeck, and Roger K Lake, Phys. Rev. B **52**, 2754 (1995).

[22] Aldo Di Carlo and Paolo Lugli, Semicond. Sci. Technol. **10**, 1673 (1995).

[23] Timothy B. Boykin, Phys. Rev. B **52**, 16 317 (1995).

[24] J. P. Loehr and D. N. Talwar, Phys. Rev. B **55**, 4353 (1997).

[25] Timothy B. Boykin, Gerhard Klimeck, R. Chris Bowen, and Roger Lake, Phys. Rev. B **56**, 4102 (1997).

[26] Timothy B. Boykin, Phys. Rev. B **56**, 9613 (1997).

[27] Timothy B. Boykin, Lisa J. Gamble, Gerhard Klimeck, and R. Chris Bowen, Phys. Rev. B **59**, 7301 (1999).

[28] E. I. Blount, in *Solid State Physics*, edited by Frederick Seitz and David Turnbull (Academic, New York, 1962), vol. 13, p. 305.

[29] L. C. Lew Van Voon and L. R. Ram-Mohan, Phys. Rev. B **47**, 15 500 (1993).

[30] M. Graf and P. Vogl, Phys. Rev. B **51**, 4940 (1995).

[31] Neil W. Ashcroft and N. David Mermin, *Solid State Physics* (Saunders, Philadelphia, 1976), App. E.

[32] Walter A. Harrison, *Solid State Theory* (Dover, New York, 1979), p. 140.

[33] D. J. Chadi, Phys. Rev. B **16**, 790 (1977).

[34] For data on InP see: Sadao Adachi, *Physical Properties of III-V Semiconductor Compounds: InP, InAs, GaAs, GaP, InGaAs, and InGaAsP* (Wiley, new York, 1992), pp. 90-93.

[35] Timothy B. Boykin, Phys. Rev. B **54**, 8107 (1996).

[36] B. S. Grabow, J. M. Boyle, J. J. Dongarra, and C. B. Moler, *Matrix Eigensystem Routines: EISPACK Guide Extension* (Springer-Verlag, New York, 1977), Secs. 2.1.9, 2.1.10, 2.3.1, 2.3.2, 7.1.

[37] Timothy B. Boykin, Phys. Rev. B **54**, 7670 (1996).

[38] J. N. Schulman and D. Z.-Y. Ting, Phys. Rev. B **45**, 6282 (1992).

[39] M. S. Kiledjian, J. N. Schulman, K. L. Wang, and K. V. Rousseau, Phys. Rev. B **46**, 16012 (1992).

[40] R. Tsu and L. Esaki, Appl. Phys. Lett. **22**, 562 (1973).

[41] J. N. Schulman, Appl. Phys. Lett. **72**, 2829 (1998).

SEE? I'VE TOLD YOU A THOUSAND TIMES THAT WORKING IN WATER-MELON AVALANCHES WOULDN'T BE GOOD FOR GETTING ANY FUNDS!!
COMPLEX SYSTEMS LAB.
NSF
99

Vortex Avalanches in Type II Superconductors: the Sandpile Perspective

E. Altshuler *

I am never content
until I have constructed a mechanical model
of the subject I am studying.
If I succeed in making one, I understand;
otherwise I do not.

Lord Kelvin

Abstract

A good mechanical model has the unique power to connect our intuition with the most abstract physical phenomena. Although purely mechanical systems may look old fashioned in contemporary Physics, the so called *granular state* has focused much attention in the last decade. *Avalanche dynamics* is one of the most fascinating features of some granular systems, and is present in other physical contexts, like *vortices in the slowly-driven critical state of type II superconductors.* After an introduction to the phenomenology of type II superconductivity and to sandpile physics, *key* experiments on vortex avalanches performed in the last five years are reviewed. Some new experiments on avalanche dynamics in superconducting, magnetic and other systems are suggested, in the philosophy of exploring mechanical analogies as far as possible –earthquakes included. Finally, some theoretical findings reported in the last three years on vortex dynamics achieved with the help of a simple cellular automaton (CA) model are reviewed, and further extensions of the model are proposed. This CA contains the essential ingredients that unify, at least to some extent, avalanche dynamics displayed by the superconducting and mechanical systems discussed in the paper.

1. Introduction

Maxwell's "molecular vortices" in aether [1] constitute, perhaps, the most fruitful mechanical analog of electromagnetic phenomena ever conceived. It not only allowed him to understand the production of a magnetic field by an electric current and vice-versa as early as 1892, but led him to formulate the hypothesis of the existence of a displacement current and the epoch-making idea of electromagnetic waves.

Even when Electromagnetism seemed complete at the turn of the century with the experimental work of Faraday and Maxwell's equations, electromagnetic fields in the Solid State would prove to be a fertile ground for devising ingenious mechanical analogies. The

*Superconductivity Laboratory, IMRE-Physics Faculty, University of Havana 10400 Havana, Cuba

opening of the new era of Superconductivity in 1911, when Heike Kammerling Onnes reported the disapearance of electrical resistance below 4.2 K in mercury [2], is one example. While mechanical models have helped the understanding of superconductive phenomena, mechanical (or fluid-mechanical) systems displaying a close resemblance with superconductivity have seen their physics clarified with the help of superconducting theories, or are just developing in parallel. This is the case of superfluids and, particularly, of sandpiles, which are entering the XXI Century with a whole agenda of unanswered questions.

2. A Superconductivity primer

The first mechanical analog of Superconductivity is Superfluidity, a phenomenon that has only been observed for liquid helium [3]. The basic characteristic of both is their ability to sustain particle currents (being either superelectrons or superfluid particles) at a constant velocity for extended periods of time without any driving force. These currents are the only known examples of the motion of macroscopic systems without being quickly destroyed by dissipative processes. To describe the condensed particles in either system, we use a complex macroscopic wavefunction of the form:

$$\Psi(\vec{r},t) = \sqrt{n^*(\vec{r},t)}e^{i\theta(\vec{r},t)} \tag{1}$$

where n^* is either the number density of super-electrons or the density of the superfluid, and θ is the phase. It is straightforward from 1 that the amplitude squared of the wavefunction equals n_s. Although the analogy can be pushed further, we will present the next results only for the case of superconductors. Let us postulate that our macroscopic wavefunction obeys the Schrödingerlike equation

$$i\hbar\frac{\partial\Psi(\vec{r},t)}{\partial t} = \frac{1}{2m^*}\left(\frac{\hbar}{i}\nabla - q^*\vec{A}(\vec{r},t)\right)^2\Psi(\vec{r},t) + q^*\phi(\vec{r},t)\Psi(\vec{r},t) \tag{2}$$

where m^* and q^* are the superlectrons' mass and charge, respectively*, $\vec{A}$ is the vector potential, and ϕ is the scalar potential.

Simple manipulations involving equations 1 and 2, under the assumption of position and time independent n^*, give the so-called London equations [5]:

$$\frac{\partial}{\partial t}\left(\Lambda\vec{J}_s(\vec{r},t)\right) \approx \vec{E}(\vec{r},t) \tag{3}$$

$$\nabla\times\left(\Lambda\vec{J}_s(\vec{r},t)\right) = -\vec{B}(\vec{r},t) \tag{4}$$

where $\Lambda = \frac{m^*}{N^*q^{*2}}$, $\vec{E}$ and $\vec{B}$ are the electric field and magnetic field induction in the superconductor, and $\vec{J}_s$ is the superelectron current density. The first of these equations indicate that, if a constant current density is established along the superconductor, no voltage drop appears, as expected for a normal metal. In fact, this was the phenomenon observed by Kammerlingh-Onnes in 1911 for mercury when cooled below what he called the *critical temperature*, T_c. The second equation accounts for the so called Meissner-Oschenfeld

*The standard microscopic theory of Superconductivity developed in 1957 by Bardeen, Schrieffer and Cooper [4] postulates that the superelectrons arrange themselves in pairs, so m^* and q^* are twice the mass and the charge of "normal" electrons, respectively. This picture seems to hold not only for the low temperature, but also for the new high temperature superconductors effect [6]. These authors observed that, when a superconducting bulk sample was submitted to a constant magnetic field, and its temperature lowered below T_c, the average magnetic induction into the sample practically switched to zero. When solving equation 4 one can see that, in fact, the magnetic induction decays exponentially from the applied induction at the sample boundaries to virtually zero at its interior within a characteristic length, λ, known as the *London penetration depth.* This is due to the appearance of a shielding current which also decays exponentially from the surface within λ.

The combination of 1 and 2 also gives a third equation known as *fluxoid quantization law,* which reveals the quantum nature of the macroscopic wavefunction:

$$\oint_C \vec{J}_s(\vec{r},t) \cdot d\vec{l} + \int_S \vec{B}(\vec{r},t) \cdot d\vec{s} = n\Phi_0, \qquad n = 0, 1, 2... \tag{5}$$

where C is a closed path into the superconductor limiting a surface S which includes a non-superconducting inclusion piercing the superconductor in the direction of the applied magnetic field. The left hand of 5 is called *fluxoid,* and the equation says that it can only take integer values of the *flux quantum,* $\Phi_0 = 2,07 \cdot 10^{-15} Tm^2$. If C is wide enough, we can neglect the first term of the fluxoid, and then we can safely talk about *flux* quatization. The phenomenon was first observed by Doll and Näbauer [7] and by Deaver and Fairbank [8] in 1961. Equations 3, 4 and 5 are good enough to describe phenomenologically a *type I* superconductor.

Another consequence of postulates 1 and 2 when applied to a junction consisting in a thin insulating layer sandwiched between two superconducting electrodes is the so-called *Josephson effects* [9,10].A real junction can be modeled as an "ideal" junction in parallel with a capacitor (associated to the effect of the finite-area superconducting electrodes) and a resistor (associated to the "normal" electrons flowing through the junction due to thermal excitation and/or to excess applied current).The solution of this equation can be beautifully visualized by means of a mechanical analogue suggested by Anderson in 1964 [11] and worked out by Sullivan and Zimmerman in 1974 [12].

3. Vortices and the Critical State

3.1. *Vortices in ideal superconductors*

A more general approach to Superconductivity is the Ginzburg-Landau (G-L) theory [13]. It can be regarded as the application of Landau's theory of second-order phase transitions to the case of superconductors, resulting in the so-called Ginzburg-Landau equations. In fact, *equation 2 is just a linearized form of one of them.* G-L theory can be conveniently modified to account for the case of superfluid He, giving the so-called Ginzburg-Pitaevsky equations [14]. It is perhaps the prediction of the existence of type II superconductors the most important result of G-L theory. Let us define the parameter

$$\kappa = \frac{\lambda}{\xi} \tag{6}$$

where ξ is the so-called *coherence length* (a lengthscale characterizing the maximum spatial variation of the superconducting wavefunction). Then, type I and type II superconductors correspond to the conditions $\kappa \leq 1/\sqrt{2}$ and $\kappa \geq 1/\sqrt{2}$, respectively. But the striking differences between type I and type II superconductors are best represented in their $H - T$ diagrams.

Figure 1a (left)displays the two phases corresponding to type I superconductors, separated by a line known as the *thermodynamic critical field*, $H_c(T)$. Under that line we find the "Meissner" phase ($\vec{B} = 0$ into the superconductor), while above there is the "Normal" phase, where the material is completely penetrated by the applied field.

Figure 1a (right) displays the diagram for a "low T_c" type II superconductor, in which three phases are observed, separated by two lines, $H_{c1}(T)$ and $H_{c2}(T)$, known as the *lower* and *upper* critical fields, respectively. While below $H_{c1}(T)$ and above $H_{c2}(T)$ type II superconductors behave qualitatively identical to type I, a new phase known as *mixed state* appears between them. In this region, the sample is *partially* penetrated by the applied field. Based on the G-L theory, Abrikosov proved, in 1957 that this penetration occurs as an hexagonal lattice of *vortices*, which are "mesoscopic" objects whose structure and basic properties will be discussed below [15]. Their existence was directly proved by Essman and Träubein in 1967 through Bitter decoration experiments [16].

Figure 2a shows a rough diagram of a vortex. It consists of a normal cylinder (or *core*) of radius ξ paralell to the direction of the applied field which is surrounded by "vorticial" superconducting currents decaying within a radius λ in such a way that they can be regarded as tiny concentric coils producing a magnetic induction in the same direction of the applied field (An alternative picture is that the vorticial currents prevent the Meissner phase outside the vortex to be "invaded" by the field inside it!).

Apart from G-L theory, a simple argument allows to calculate the flux associated to a single vortex. Considering the experimental fact that vortices in the mixed state arrange in an hexagonal lattice, it is easy to see that the mean magnetic induction of the system is

$$\langle B \rangle = \frac{2}{\sqrt{3}\alpha^2}\Phi_v \tag{7}$$

where α is the intervortex distance and Φ_v is the magnetic flux associated to each vortex. If we now apply the flux quatization condition 5 along a suitable path around the vortex, we get $\Phi_v = \Phi_o$. Then, *the magnetic flux associated to one vortex is a single flux quantum.*

Knowing the vortex structure, it is easy to get an intuitive picture of the mixed state close to its boundaries. Near $H_{c1}(T)$, the lattice constant is very large. As one approaches $H_{c2}(T)$, vortices become increasingly crowded, and, very close to $H_{c2}(T)$, $\alpha \to \xi$: the whole system is nearly in the normal state.

Insisting in our phenomenological approach, we can get reasonable approximations for $B(r)$ and $J_s(r)$ for a single vortex by conveniently modifying London equation 3. The results are qualitatively illustrated in Figure 2b,c assuming cylindrical coordinates. We can also get the following expression for the *electromagnetic energy per unit length of the vortex* [17]:

$$E_v = \frac{\Phi_0^2}{4\pi\mu_0\lambda}K_0\left(\frac{\lambda}{\xi}\right) \tag{8}$$

where $K_0(\frac{\lambda}{\xi})$ is the zeroth-order modified Bessel function, that can be approximated by $ln(\frac{\lambda}{\xi})$ if $\lambda \gg \xi$. We can also estimate the interaction energy per unit length between two paralell vortices with fields pointing in the same direction (we will call them just *paralell* vortices)[17]:

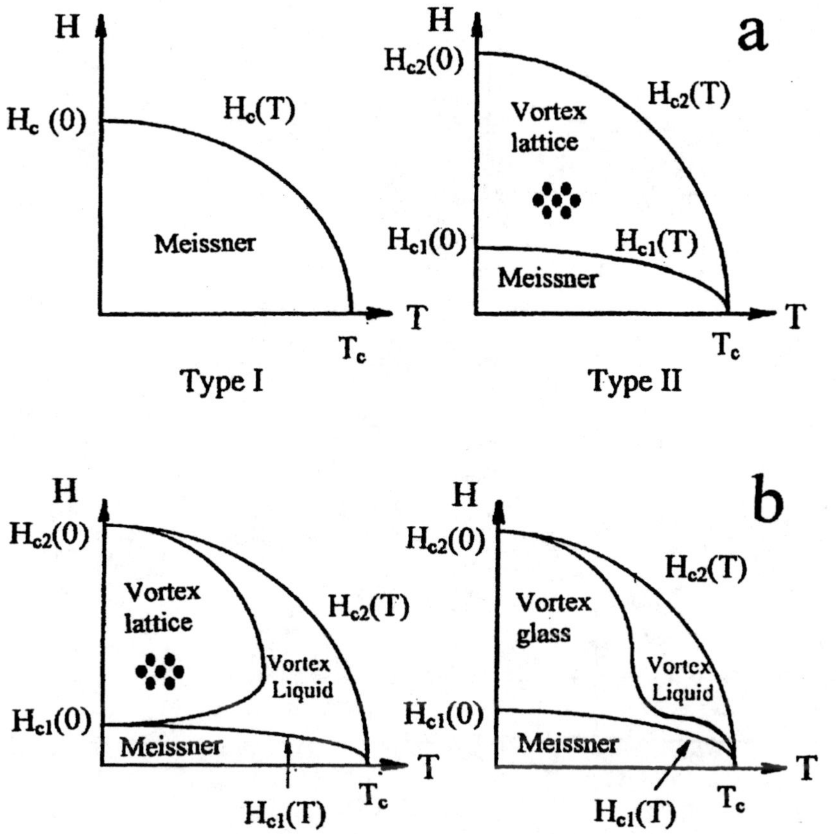

Figure 1: Magnetic phase diagrams for (a) low-T_c and (b) high-T_c superconductors. In (a), type I and type II superconductors correspond to the left and right graphs, respectively. In (b), two possible phase diagrams are represented for high-T_c materials.

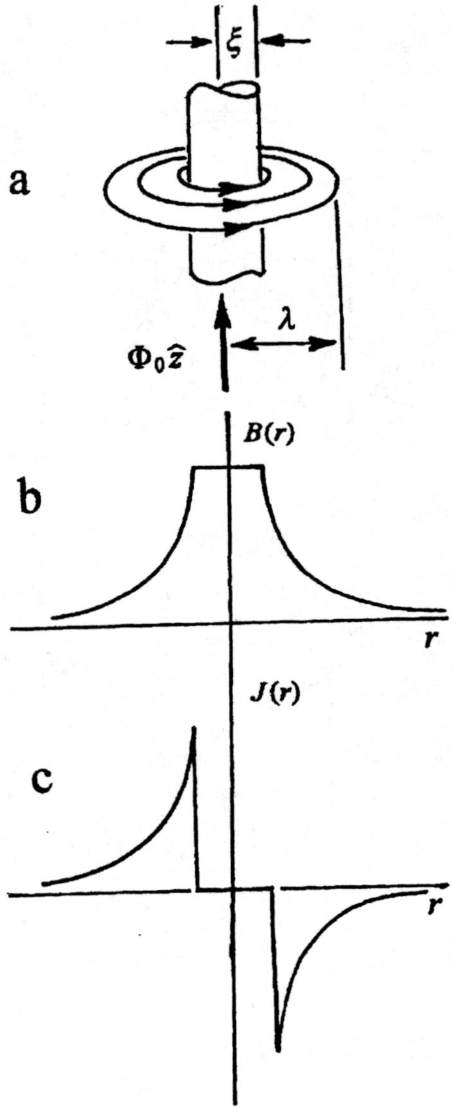

Figure 2: The isolated vortex. (a) Spatial representation of the vortex core and the vorticial supercurrents. (b) Simplified graph of the magnetic induction associated to the vortex. (c) Simplified graph of the supercurrents associated to the vortex.

$$E_{vv} = 2E_v - \frac{\Phi_0^2}{4\pi\mu_0\lambda^2} K_0(\frac{l}{\xi}) \qquad (9)$$

where l is the intervortex distance . It can be easily seen that the first term of the right corresponds to the simple superposition of the individual vortex energies per unit length, so the last term is the interaction energy. Its minus sign indicates that *two paralell vortices repel each other.* If we calculate the force of interaction per unit vortex length through the gradient of E_{vv} in the direction of the interaction, we get:

$$f_{vv} = -\frac{\Phi_0^2}{2\pi\mu_0\lambda^3} K_1(\frac{l}{\xi}) \qquad (10)$$

where $K_1(\frac{l}{\lambda})$ is the modified Bessel function of the first order. Again, the minus sign indicates repulsive interaction.

This interaction can be also visualized coming back to our superfluid analogy. The upper part of Figure 3a shows the current lines of two superconducting vortices, and the lower part displays the flow pattern resulting from their superposition (the crosses indicate that the field enters the paper). Following elementary hydrodynamic arguments, a high pressure region appears between the vortices (due to the low transversal fluid velocity in that region), while low pressure regions appear at both sides (related to the high transversal fluid velocity there). This pressure map makes the vortices to repel each other. In fact, *superfluid vortices do exist in helium* [18].

Besides the three-dimensional "straight" ones discussed above, a rich zoology of vortices have been studied. One of these are the "doughnut-shaped" or "circular" vortices firstly discussed by Tinkham [19] and quite intensively re-visited in the nineties [20,21,22,23,24,25]. In contrast with straight ones, *circular vortices are unstable at zero applied field* because of their tendency to "shrink out" looking for energy minimization. As in the case of straight vortices, this effect can be also visualized dynamically: the vorticial doughnut can be sectioned into straight vortices of infinitesimal length. Then, each infinitesimal vortex attracts its "anti-paralell" mate at the other extreme of the corresponding diameter, provoking the shrinking of the doughnut. Other peculiar types of vortices appear in anisotropic superconductors, where λ and ξ are different along two perpendicular directions of the crystal. This can produce vortices with ellipsoidal instead of circular crossection, for example. The "High Temperature Superconductors" (or "HTc's") discovered by Bednorz and Müller in 1986 [26] and rapidly enriched and developed by others [27,28,29,30] are "extreme" anisotropic superconductors. In fact, they can be modelled as arrays of paralell superconducting layers along the ab planes of the structure (called "CuO planes") alternating with non-superconducting layers. When a current flows perpendicular to these layers, the system can be treated as a stack of Josephson junctions. If a magnetic field is applied perpendicular to the CuO-planes ($H \perp ab$), *two-dimensional* "pancake" vortices appear laying in the CuO planes which eventually form stacks in the direction of the field (Figure 4a) [31,32,33,34]. If the field is applied paralell to the ab planes ($H \parallel ab$), vortices appear involving both supercurrents along the ab planes, and Josephson currents perpendicular to the ab planes (Figure 4b) [35,36,37,38,39,40,41].

Let us return to simple three-dimensional straight vortices and examine their interaction with a transport current of density $\vec{J}$. Considering that a vortex carries a flux quantum, it can be shown that it experiences a "Lorentz-like" force per unit length of the form:

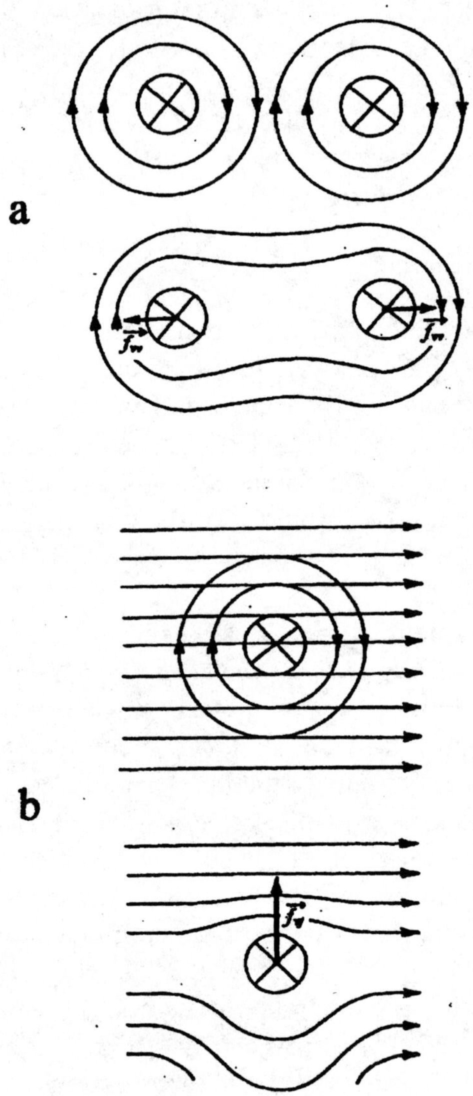

Figure 3: (a) Current superposition which illustrates the origin of the repulsive interaction between two vortices. (b) Current superposition which illustrates the origin of the force exerted on a vortex by a transport current.

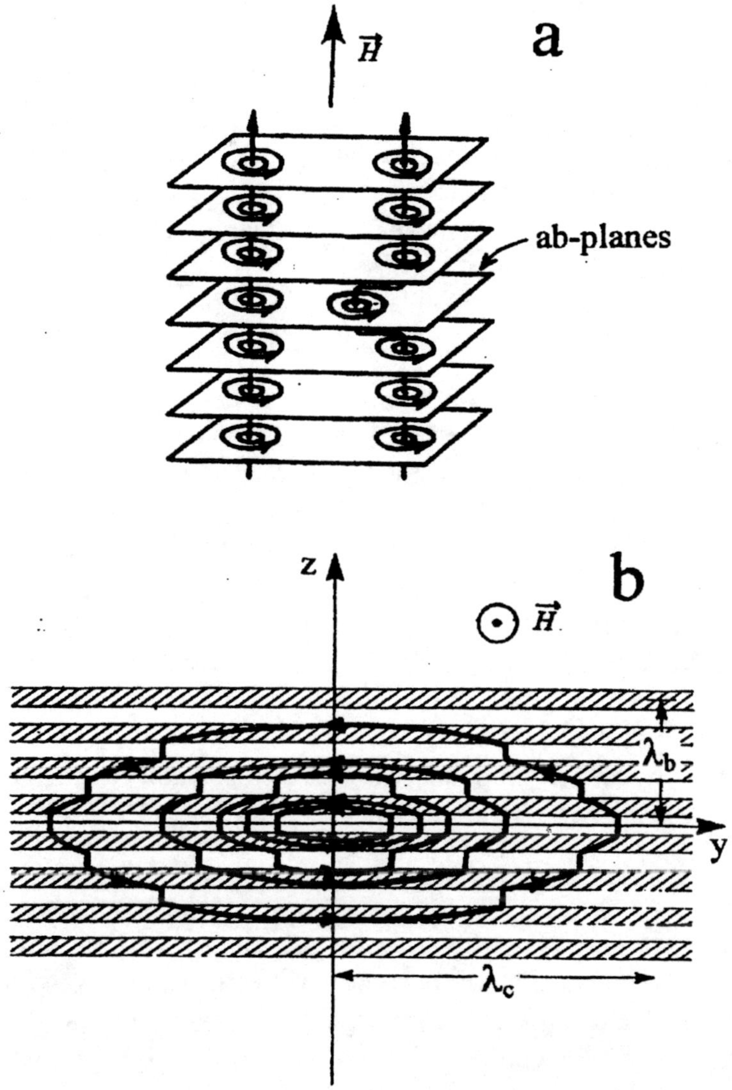

Figure 4: (a) Arrays of pancake vortices associated to a magnetic field applied perpendicularly to the "ab planes" of a high-T_c superconductor. (b) Josephson vortex associated to a magnetic field applied paralell to the "ab planes" of a high-T_c superconductor.

$$\vec{f_{vj}} = \vec{J} \times \Phi_0 \vec{i_z} \tag{11}$$

where $\vec{i_z}$ is the unit vector in the direction of the vortex magnetic field. Again, our hydrodynamic analog allows us to visualize the situation. The upper part of Figure 3b displays the straight current flow lines and the vorticial lines, while the lower part shows their superposition. Then, high and low pressure regions develop on the bottom and top sides of the vortex, respectively, causing a force pointing upwards. This, of course, follows the vector product of equation 11 if the flow is identified with $\vec{J}$ and the vortex comes out of the paper. The movement of a vortex caused by this force is known as *flux flow*. As demonstrated by Bardeen and Stephen [42], *flux flow is a dissipative process* because of the existence of a viscous drag per unit length opposed to $\vec{f_L}$, related to the scattering of superelectrons at the core boundary when it moves. Phenomenologically, the magnitude of the drag force can be roughly expressed as

$$f_D = \eta v_v \tag{12}$$

where η is a drag coefficient which depends on the core details, and v_v is the vortex velocity.

Let us integrate now all vortex interactions for the case of a sample in the mixed state submitted to an external magnetic field. First, the Abrikosov lattice can be thought as the *equilibrium* result of the vortex-vortex interactions into the sample –which try to push them off the sample– plus the interaction between these vortices and the external field (thought as a "vortex continuum" outside the sample) –which tries to push the vortices inside–. If we now introduce a transport current pointing upwards, and we assume that the field comes out of the paper, each vortex will suffer an extra force of identical magnitude and direction, so *the whole lattice will move to the right.*

All in all, each vortex in the lattice experiences a *net* force per unit length of magnitude:

$$f_T = J\Phi_0 - \eta v_v \tag{13}$$

Note that the (equilibrium) vortex-vortex forces associated to its neighbors cancel out due to the symmetry of the Abrikosov lattice so they do not appear in 13. A naive conclusion from this is that the application of a transport current in the mixed state would immediately produce dissipation, so the critical current —defined as the maximum applied current that a type II superconductors in the mixed state can stand without dissipation— would be zero. This is not the case in the experiment, as we will see in the next section.

3.2. *Vortices in real superconductors*

3.2.1. *Pinning and $J-V$ curves*

Point defects, dislocations, second-phase inclusions and other defects make real superconductors inhomogeneous, in contrast to "ideal" superconductors. Let us model an inhomogeneity as a non-superconducting infinite cylinder of radius ξ and consider its interaction with a vortex paralell to the cylinder.

It is not difficult to see that it is energetically favorable for the vortex core to set just on the defect, since, in principle, the creation of the normal core in the Meissner sea has an energy cost which is "relieved" if the core finds an "already normal" location. If this is the situation, it is said that the vortex is *pinned* by the defect, which is then called *pinning center*. The energy needed for extracting a vortex from a pinning center is known as *pinning energy*. The approximate pinning energy per unit vortex length can be expressed, in the case of the defect described above, as:

$$U_P = \frac{1}{2}\mu_0 H_c^2 \pi \xi^2 \tag{14}$$

where H_c is the thermodinamical critical field. It should be noted that other inhomogeneities such as cylindrical defects with radius bigger or smaller than ξ, spherical defects, or planar ones, are "less efficient" pinning centers (i.e., have smaller pinning energies). The minimum force needed to extract a vortex from a pinning center is called *pinning force*. The pinning force per unit vortex length for our model defect can be readily estimated from 14 as:

$$f_P = \frac{U_P}{2\xi} \tag{15}$$

Of course, the real picture must include the whole Abrikosov lattice, for which we have to solve the problem of the *summation of pinning forces*, which is highly nontrivial in the presence of random pinning [43]. To get rid of this complication, let us assume the very unrealistic situation of one pinning center per vortex, and equal pinning strengths for all centers. Then, we can define the *critical current density* in the mixed state of a type II superconductor with pinning that which produces a Lorentz-like force strong enough to depin the lattice, i.e.:

$$f_P = J_c\Phi_0 \Longrightarrow J_c = \frac{f_P}{\Phi_0} = \frac{U_P}{2\xi\Phi_0} \tag{16}$$

This result solves the pretended contradiction at the end of the last section. It also indicates that, for type II superconductors with pinning, the total force can be expressed as:

$$f_T = J\Phi_0 - \eta v_v - J_c\Phi_0 \tag{17}$$

and the vortex velocity in the steady state (i.e., $f_T = 0$), as:

$$v_v = \frac{1}{\eta}(J - J_c)\Phi_0 \tag{18}$$

The application of Faraday's law indicates that the voltage dissipation due to vortex motion is proportional to its velocity, so it can be written as:

$$V = (J - J_c)\rho_{ff} h \tag{19}$$

where h is the distance between voltage probes in the direction of the transport current, and ρ_{ff} is called *flux flow resistivity*

In conclusion, if we apply an increasing transport current through the extremes of a real type II superconductor in the mixed state with the field applied perpendicular to the current, and we measure the voltage drop through a second pair of (inner) contacts, as shown in Figure 5a, we find a $J - V$ curve like that shown in figure 5b (dotted line).

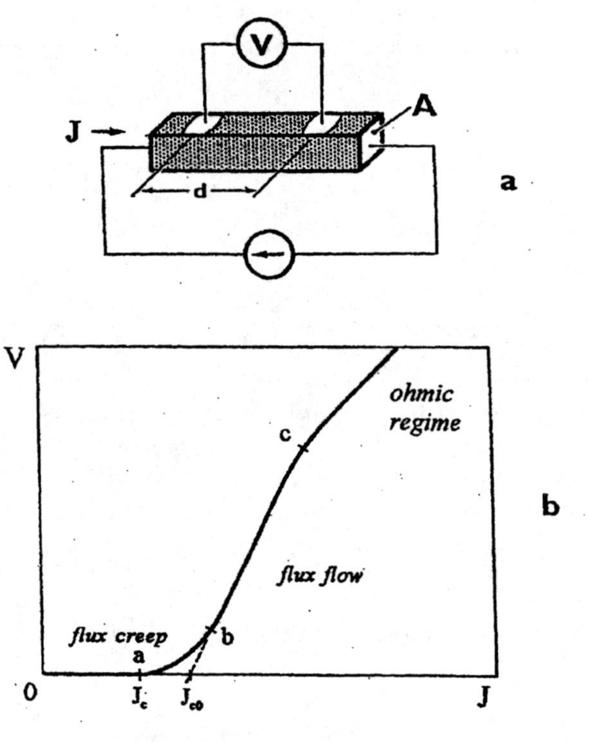

Figure 5: (a) Illustration of the "four probe arrangement". (b) Typical J-V curve measured with the arrangement depicted in (a) on a low-T_c superconductor.

3.2.2. *The critical state*

Let us examine how vortices distribute into a type II superconducting slab of infinite dimensions along y and z submitted to a magnetic field $-\mathbf{z}H$, as represented at the right of Figure 6. As the field exceeds H_{c1}, vortices enter through the surface.

Assuming that each vortex inside the slab is located at a pinning center, it suffers two opposite effects: a "magnetic pressure" pushing it into the superconductor, and a pinning force opposite to its . This is illustrated in the top view of the slab depicted at the upper part of Figure 6 (the slab has been represented as finite along x and also along y for didactical reasons). This competition provokes a nonuniform distribution of vortices, i.e., the vortex density is highest at the boundaries, and decreases when we get into the sample. A practical way to represent this situation is to substitute the effect of the individual vorticial currents by the effective "macroscopic" currents resulting from their superposition –as represented in Figure 6– that can be thought of responsible for the " Lorentzlike" force acting on the vortex cores.

Charles Bean proposed, back in 1964 [44] that the entire system would organize itself in such a way that those currents would match the critical current density of the superconductor so the magnetic induction at a given point $\vec{r}$ is given by Ampere's law:

$$\left|\nabla \times B(\vec{r})\right| = \mu_0 J_c \tag{20}$$

It is worth noting that $B(\vec{r})$ is a *mesoscopic* quantity in the sense that it is the average magnetic induction over many vortices around the position $\vec{r}$, but it is not the thermodynamic induction averaged over the whole sample. This model is often referred in the literature as the *critical state model*. In its original approach, Bean also assumed that J_c was independent from the local induction, i.e., from $\vec{r}$. Under these conditions, it is easy to see that the current and magnetic field induction of our slab along x are the ones depicted in Figure 6 (note that the gap of the magnetic induction at the sample's surface is due to the presence of the London diamagnetic currents, which are not commonly taken into account in critical state models).

The module in equation 20 implicitly expresses a third hypothesis related to the "memory" of the material regarding how the magnetic field was applied. This hypothesis can be expressed like this: *the variations in the external field affect the vortex distribution from the sample boundaries to its interior, in such a way that different sections of the flux profile have slopes proportional either to J_c or to $-J_c$.* This nontrivial fact is better illustrated through a practical example as that depicted in the upper part of Figure 6, where the sample has been submitted to a field increase from zero to a certain H_m, and then to a field removal. Observe that the shielding capacity of the sample is not given by an increase of the magnitude of the shielding current –since $J = J_c$ everywhere–, but by the *volume* available for the flow of this current. Then, the maximum shielding capacity is reached in diagram 3 (see Figure 7), in which the external field has reached the value $H = H^* + H_{c1}$, where H^* is often called *full penetration field*. When the external field starts to diminish, the critical state at the "inner core" remains unchanged, while the vortices start to abandon the sample through the boundaries. This finally gives a profile like that depicted in diagram 8. However, it should be noted that the history can be very different if a lower H_m value is applied. The thermodynamic magnetization –i.e., that measured in typical magnetometric experiments– can be calculated, for our geometry, as:

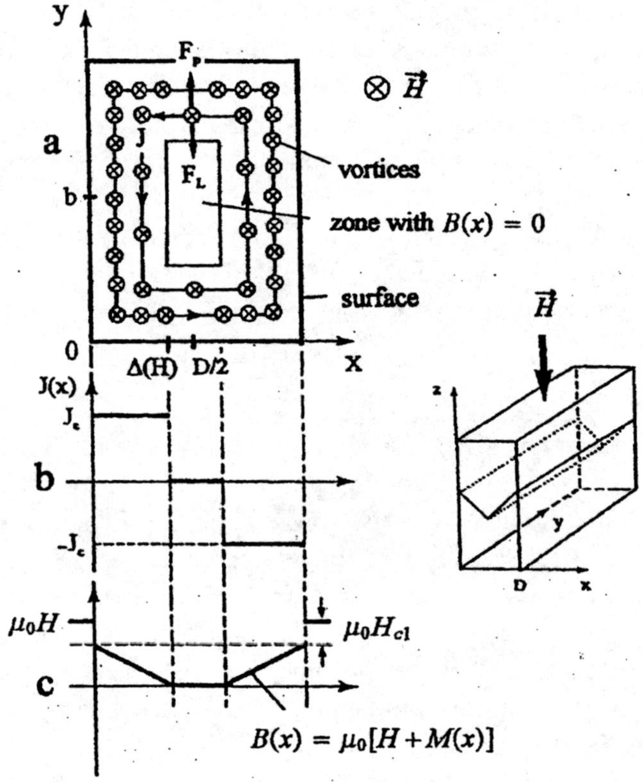

Figure 6: The critical state in type II superconductors. (a) Top view of a superconducting slab submitted to a magnetic field entering the page in which vortices are represented by crosses. (b) Current profile and (c) magnetic induction profile associated to the vortex arrangement depicted in (a). At the right, a simplified three dimensional representation of the magnetic induction profile into the sample. H_{c1} has not been neglected, as in the original model. [44]

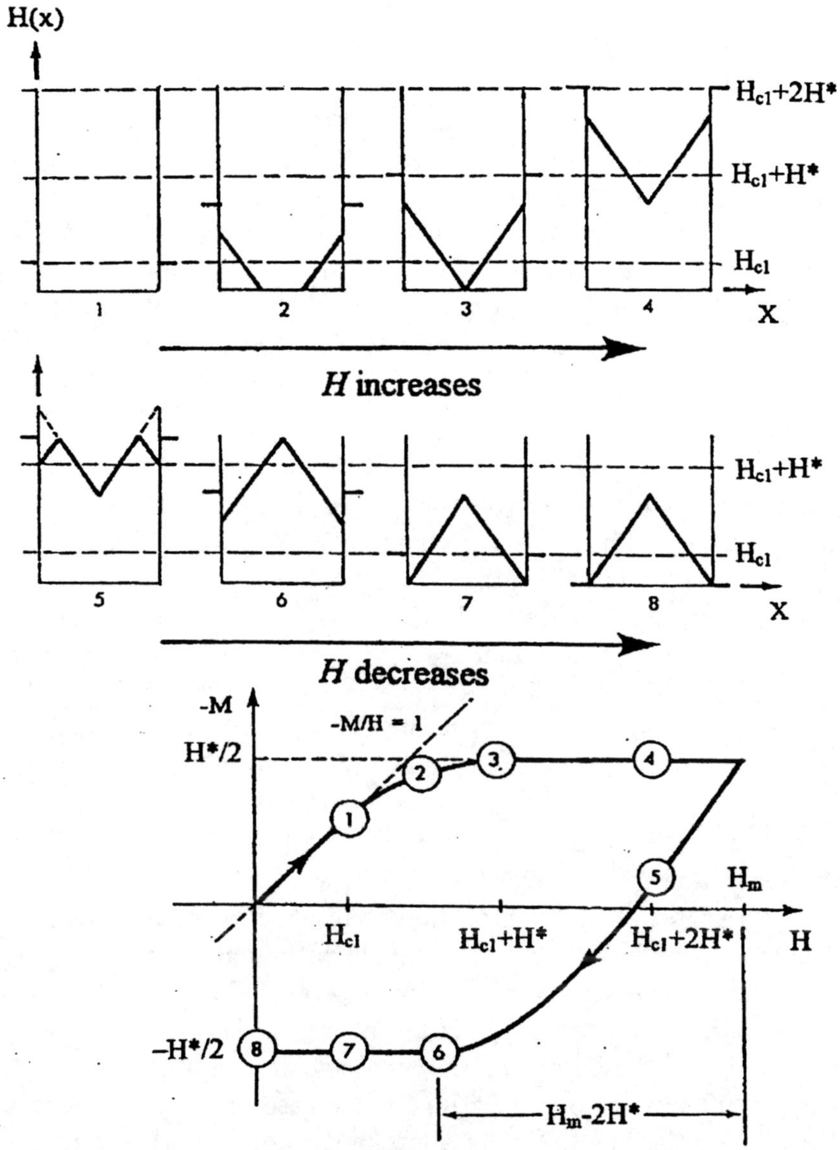

Figure 7: Magnetization derived from the critical state. Eight magnetic field critical state profiles corresponding to eigth values of the external field are illustrated at the upper part of the figure, and the corresponding points in the magnetization curve are depicted below. H_{c1} has not been neglected, as in the original model [44].

$$M(H) = \frac{\langle B(x) \rangle}{\mu_0} - H = \frac{\int_0^D B(x)dx}{\mu_0 D} - H \tag{21}$$

for the particular magnetic history of Figure 7, the resulting $MvsH$ diagram is shown on its lower part.

It is worth noting that Bean's hypothesis $J_c \neq J_c(H)$ is not essential to critical state models. In fact, different $J_c(H)$ dependencies have derived in other models such as the "Kim model" [45] and the "exponential model" [46]. In our opinion, the critical state model is a major triumph of physical intuition with a certain " engineering" flavor, which was originally conceived without the knowledge of the existnce of vortices. Even when it has been extensively enriched for decades [47], the elucidation of the precise dynamics involved in its *establishment* –specially that related with vortex avalanches– as a scientific challenge both from the experimental [48,49,50,51,52,53,54,58,55,56,57] and from the theoretical [59,60,61,62,63,64,65,66,67,68,69,70,71,72] points of view. We will discuss in detail some of these works below.

Flux Creep

The previos decription of the critical state is basically correct at zero Kelvin. At finite temperatures, thermal activation can make the vortices –or vortex bundles– jump out of their pinning centers, redistributing into the sample in such a way that the Bean's profile changes in time, and, consequently, the shielding currents and sample magnetization. This phenomenon –known as *flux creep*– was firstly observed by Kim et al on the relaxation of persistent currents in NbZr tubes [74] and explained by Anderson and Kim [76] Their model –widely known as *Anderson-Kim model*– assumes that vortices jump off the pinning centers at an average frequency given by the Arrhenius relation $f = f_0 e^{-U/kT}$ were f_0 is an "effective attempt frequency", k is Boltzmann's constant, T is the temperature, and U is the pinning energy. Since the vortices are submitted to a Lorentz force in the critical state, it is clear that they basically jump out of the pinning centers in the direction of the Lorentz-like force. This situation can be represented by " tilting" the pinning wells in such a way that vortices "see" a lower potential barrier in the direction of the Lorentz-like force and a higher barrier in the opposite. The potential landscape that thermal activation encounters is often called "washboard potential", and is represented in the upper part of Figure 8. In the original Anderson-Kim approach [76] the pinning energy takes the form $U(J) = U_0(1 - J/J_{c0})$ where U_0 is the pinning potential in the absence of Lorentz-like forces (i.e., no flux gradient) and J_{c0} is the critical current density at $T = 0K$. In the case of low thermal activation ($U_0 >> kT$), these hypotheses give the following approximate formula for the relaxation of the shielding currents in the critical state:

$$J = J_{c0}[1 - \frac{kT}{U} ln(\frac{t}{t_0})] \tag{22}$$

where $t_0 \sim 1/f_0$. Taking into account equations 20 and 21, a similar decay can be derived, in a first approximation, for the macroscopic magnetization:

$$M = M_0[1 - \frac{kT}{U} ln(\frac{t}{t_0})] \tag{23}$$

where M_0 is the initial magnetization. This formula is only valid for "Fully penetrated samples", i.e., flux profiles like those depicted in diagrams 3 or 4, Figure 7.

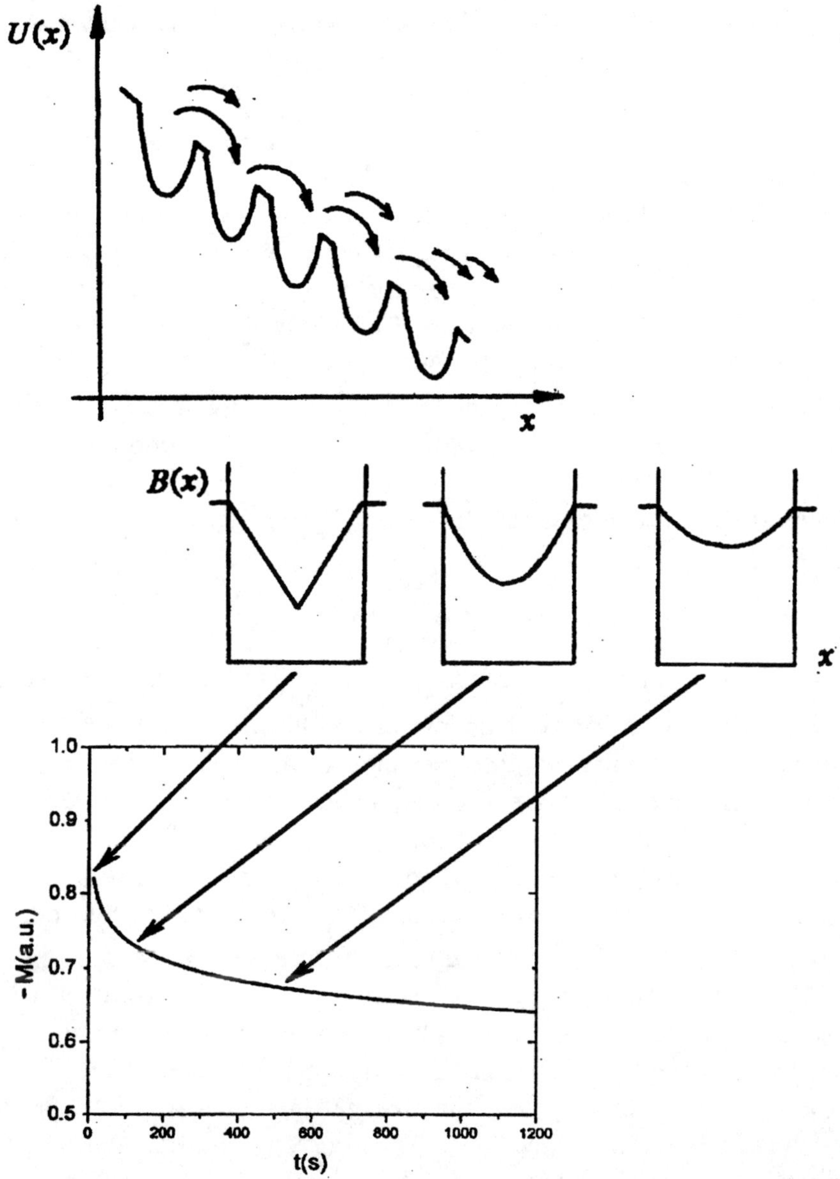

Figure 8: The upper part illustrates the "washboard" potential of pinning centers which provokes vortices to *creep* when thermally activated (vortex movement is represented by arrows). The resulting temporal evolution of the magnetic induction into the superconductor is qualitatively illustrated in the middle, which is connected to a *flux creep* measurement with illustrative purposes.

The temporal relaxation of the magnetization is observed experimentally both in low and in high T_c superconductors. Figure 8 shows, in its middle section, how Bean's profiles "relaxes" in time due to thermally activated vortex jumps from the high to the low flux density regions in the sample. In Figure 8, a correspondence has been established between this process a fit to a relaxation experiment on a YBCO crystal performed with a Vibrating Sample Magnetometer [77] with didactical purposes. Although the Anderson-Kim approximation given by formula 23 is quite accurate for $U >> kT$ –which is a common situation in low-T_c supercondcutors–, typical results for High T_c are better described by nonlinear dependencies of the pinning potential such as $U(J) \propto [1-(J/J_{c0})]^{3/2}$ [78], $U(J) \propto U_0[(J/J_{c0})^{\mu}-1]$ [79] and $U(J) \propto U_0 ln(J_{c0}/J)$ [80] (observe that the last two pinning energies diverge as J approaches 0, which can be understood on the basis of *collective pinning theory.* In fact, High Temperature superconductors gave new life to flux creep phenomena, due to the very high creep rates they exhibit, called "giant flux creep" after Yeshurum and Malozemoff [81]. For an excellent review of the subject, the reader can consult [82].

Finally, it is worth noting that flux creep effects are also present in transport measurements. For example, in the case of an homogeneous type II superconductors, they are responsible for the exponential "rounding" of the J-V curves before J_{c0}, as depicted in Figure 5. More complex phenomena can take place in the case of inhomogeneous superconductors [83].

Flux Jumps

As said before, the *attainment* of the critical state when ramping up (or down) the applied magnetic field is far from being well understood. One of the related phenomena that has attracted most interest due to its great technological relevance for the design of superconducting magnets, is the appearance of the so called *flux jumps.* They have been observed both in low T_c [84,85,86,87] and in High T_c superconductors [88,88,89,90,91,92,93,94,95,96,97]. The current interest in the phenomenon is illustrated by a very recent study of flux jumps in Nb films which reveals their correlation with the shape of the magnetization curve [57].

Flux jumps can be defined as thermodynamic catastrophes leading to the full or partial loss of magnetization in the superconductor. These events are the result of a positive feedback loop: as the external field is increased, vortices move to "reacommodate" in the sample, provoking heat dissipation through a Bardeen-Stephen mechanism [42]. If the magnetic diffusivity of the material is higher than the thermal one, this situation produces a temperature rise which, in turn, diminishes the critical current density, allowing the entrance of extra magnetic flux, which closes the feedback loop. It is intuitively clear that this process depends on the "intrinsic" thermal properties of the sample (specific heat and thermal conductivity), as well as on some "extrinsic" ones (sample size and quality of the thermal link with the sample holder). It also depends on how the critical current density depends on temperature and applied field, and how fast the external field is ramped.

The following conditions can give rise to the appearance of flux jumps: *"big" samples with poor thermal links with the sample-holder, magnetic fields applied at high ramp rates, strong temperature dependence of the critical current density, poor thermal conductivity, and low specific heat.* Taking into account that the specific heat increases with temperature for high temperature superconductors (see, for example, reference [97]), *one concludes that very low temperatures are a nice scenario for flux jumps in* $HT_c's$.

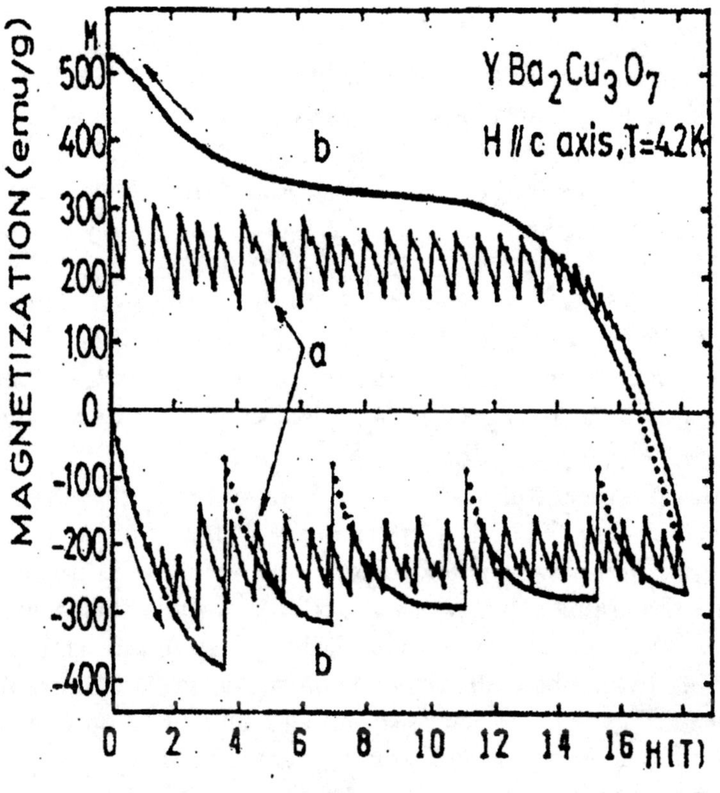

Figure 9: Magnetization loop showing *flux jumps* taken from [90]. (a) Sample in liquid He bath. (b) Sample in exchange gas.

As an example, Figure 9 shows magnetization loops with flux jumps measured by Guillot and coworkers [90] on a "large" YBCO crystal at 4.2 K with the magnetic field applied along the c-axis at a ramp rate of 300 Oe/s. Flux jumps are more dramatic for curve (b) (sample cooled with exchange gas) than for curve (a)(sample immersed in liquid helium), illustrating the effect of the thermal linkage of the sample. Some of these properties are contained in a *stability parameter*, β, usually defined to characterize thermally-triggered flux jumps [98].

It is important to note that, despite that no systematic special effort has been historically devoted to the study of their size and temporal statistics, flux jumps are regarded by most authors as a "rather periodic" event [88]. Some researchers have even defined a "frequency of magnetothermal oscillations" to describe flux jumps for a HT_c granular sample [97].

4. Sandpiles: new physics for old phenomena

In general, granular matter behaves differently from solids, liquids or gases. So it is not strange that some of the brightest minds in Physics of the XVIII and XIX centuries such as Coulomb [99], Faraday [100] and Reynolds [101] devoted some efforts to the understanding of granular systems. The strange fact is that the last decade of the XX century has concentrated most of the historical efforts to approach granular matter from a physical perspective, which has been described in several reviews [102,103,104,105,106,107,108,109,110,111,112,113,114,115]. This is connected, to some extent, with the use of sandpiles as the paradigm of *Self Organized Criticality* (SOC), a theory proposed by Bak and coworkers in 1987 [116] to explain a wide spectrum of natural phenomena, specially the so called "$1/f$" noise in many physical systems... even when the applicability of the paradigm cannot be fully demonstrated on *real* sandpiles.

Sandpiles constitute a beautiful mechanical analog of the critical state of type II superconductors, an idea first sketched by DeGennes in his 1966' classic book "Superconductivity of Metals and Alloys" [59]. As sand is added on a flat horizontal surface, the gravitational force tends to bring the grains down, while the frictional forces between grains prevent its action. The result of this competition is the building of a *pile*, characterized by an *angle of repose* or *critical slope*, which basically depends on the effective intergrain friction, and the friction between the grains and the flat surface. Of course it also depends on gravity, but it is usually a constant in "conventional" experiments (however, the rotation of the piles around a vertical axis, their immersion in some fluid, the tilting of the flat surface on which the pile is built, or an –expensive!– experiment in the Moon are examples of how to control also the "gravitational" forces). This situation remainds us of the description of the critical state of superconductors of section 2.2.2: the grains can be regarded as the vortices, the role of pinning is played by the intergrain friction, and the Lorentzlike force is analogous to the gravitational one.

Even when "everyday" conical sandpiles of "inverted V" shape are good enough for describing the analogies with the critical vortex state, more elaborated piles must be built to resemble typical increasing-field Bean's profiles. This was didactically exploited by Campbell and Evetts in 1972 [43] to illustrate the influence of the sample *shape* in the resulting critical state profiles. The top picture of Figure 10 shows our own version of the sandpile critical profile equivalent to Diagram 4, Figure 7. It was obtained by evenly adding sand, through its two longer edges, to a rectangular box with open top and bottom faces which represents the superconducting sample: the sand profile inside the box mimics the critical state into the superconductor. The middle picture in Figure 10 represents the situation when the box is lowered relative to the sand it contains, allowing some of it to exit through the edges. The sand landscape into the box now resembles diagram 5 of Figure 7. Observe how the finiteness of our box produces critical slopes in two perpendicular directions.

To finish the presentation of the "static" analogies between sandpiles and the critical state of superconductors, it should be noted that two-dimensional (2D) sandpiles (we are using this term to describe conical piles or the ones formed into our sandbox) generally have quite linear profiles, which nicely adapt to the original *Bean* critical state model [44]. In the case of one-dimensional (1D) piles (we are using this term to describe piles formed between two vertical plates separated by the size of one grain), "kink" profiles and "logarithmic tails" are observed [117], as well as other complex structures [118]. To the author's knowledge, no 1D-critical state has been studied in the case of vortices to check if the analogy still works.

Not only static, but also some *dynamic* properties can be compared between sandpiles and critical vortex states. One of them is *relaxation.* Jaeger and Nagel, for example, measured in 1989 the temporal evolution of the angle of repose of a *slowly driven* sandpile [†]submitted to vibrations produced by an audio speaker [119], trying to mimic the effect of thermal activation on the critical state of a superconductor. The angle of repose decayed *logarithmically* in time within a range of vibration intensities, as the magnetization does in typical flux creep experiments (see section 2.2.2). The authors used an analogy of Anderson-Kim's flux creep model [76] to explain their results. It is still an open question, though, which property of the loudspeaker vibration (i.e., amplitude, frequency, mechanical power injected into the system, etc.) should be identified with thermal activation in vortex creep experiments. Moreover, the analogue of *temperature* is not very clear in the case of a sandpile [120] from a formal thermodynamic perspective. In any case, this example illustrates how the superconducting knowledge can shed some light on granular phenomena, as we suggested in the second paragraph of this paper. The bottom picture of Figure 10 illustrates the effect of gentle vibrations on the "decreasing field" sand profile shown in the middle picture. Observe the decreasing of the slopes, and the "rounding" of the edges, reminescent of the intermediate diagram of Figure 8 dealing with flux creep in type II superconductors (the "emergence" of a few big impurities which were originally buried in the sand illustrates one of the most characteristic phenomena of granular media: the so-called "Brazil nut seggregation" [121]).

But perhaps the most motivating dynamical analogy between sandpiles and the superconducting critical state was also drafted by DeGennes back in 1966 [59]: *the critical slope of a sandpile to which grains are added is manitained through sand avalanches, and it seems reasonable to expect a similar dynamical behavior for the case of the critical state of a superconductor submitted to an increasing magnetic field.*

In his book "How Nature Works–the science of Self Organized Criticality" [122], Bak gives a vivid description of the formation of a sandpile and its avalanche dynamics which can be considered a refinement of our own one given above. In his description, our sandpile is an *open system*, since we will construct the pile by adding grains of sand. The flat surface on which the sand is added represents an *equilibrium state* –energy minimum of the system–,

[†]The term is used in the sense that a new grain is added to the pile only after the relaxation provoked by the previous addition has concluded.

Figure 10: Sandpile analogues of two different stages of Bean's critical state model. Top picture: increasing field. Note similarity with the diagram at the right of Figure 6. Middle picture: decreasing field. Bottom picture: middle profile after "thermal" activation.

but the intergrain friction prevents the sand from adopting this state. In spite of this, the grains start to form a pile. In the first stages, they just "stick" to the grains they have landed on. When the pile gets steeper, the grains start to topple from the landing locations to lower levels, but still provoking just *local* avalanches. The toppling events make the system *dissipative*, since heat is released due to friction. There is a moment, though, in which the average number of grains added to the pile matches that abandoning the pile off the edges, i.e., the system has reached a *stationary state.* This state is *critical* in the sense that it can be only maintained if the different parts of the pile are able to "communicate" by means of *many sized avalanches*, which eventually can span the whole pile...a fact that could not be foretold from the properties of the individual grains! *But the key feature that makes this system a paradigm of Self Organized Criticality is that a critical state "self-organizes" in the sense that it is reached with no regard of the details of the specific experiment, such as the shape of the grains or the roughness of the horizontal surface.* At least for the case of *theoretical* piles.

In fact, Bak and coworkers demonstrated in 1987 that, for a simple computational model representing the dynamics of a sandpile the following *power laws* described the avalanche size distribution:

$$P(s) \sim s^{-\tau} \tag{24}$$

and the avalanche lifetime distribution:

$$P(t) \sim t^{-\alpha} \tag{25}$$

where s and t are the avalanche sizes and lifetimes, respectively. This means that many-sized avalanches occurred in the system. Even when we are not going to describe the model in detail for the sake of brevity, it is worth noting that *all* the avalanches are considered, and not only those associated to grains falling off the edge, and that inertial effects are not taken into account. In the case of 2D piles, Bak and coworkers found $\tau \approx 0.98$ and $\alpha \approx 0.42$, respectively. A remarkable early attempt to probe the *universality* of these exponents (i.e., their robustness when the details of the simulation are changed) was sonly published by Kadanoff and coworkers [123]. From his work, one may infer that 2D sandpile models tend to show SOC behavior, but 1D models do not.

The noise spectrum associated to a lifetime distribution given by equation 25 can be calculated as $S(\omega) \sim \omega^{-2+\alpha}$. This led the authors to propose SOC as the dynamical explanation of the "$1/f$ noise" observed in many natural phenomena.

An important issue both in simulations and in real experiments with sandpiles is the problem of the *finiteness* of the system. This results in the departure from the power laws in the big avalanche region, also called "cutoff". One could test if a finite system is critical by measuring the avalanche distribution for different system sizes, and check if the different distributions collapse into a single curve when applying to the data a *critical size scaling ansatz* such as:

$$P(s, L) = L^{-\beta} f(s/L^{-\nu}) \tag{26}$$

where L is the size of the system, f is the *scaling function* and β and ν are critical exponents [124].

The explicit urging for experiments by Bak and coworkers in 1987 [116] managed to bring real sand into the laboratories. The "avalanche" started with the "rotating drum" experiment [119], which didn't show power law avalanche distributions – neither resembled too much the original sandpile model– and continued with the work of Held and coworkers on real sand 2D piles to which grains were slowly added on top [125]. Although finding power-law distributed avalanches in a certain range, they observed large avalanches causing oscillations of the slope of the pile. As pointed out in [122], this experiments still had an important difference with the original SOC model: only the off-the-edge avalanches were recorded. The existence of inertial effects in real sandpiles was a second major difference with the very simple origial model. After several other experimental studies [126,127,128,129,130,131]

an important experimental improvement was made in 1996 by the "Oslo group" [132,133], when measuring *also the internal avalanches* of slowly-driven cuasi-1D ricepiles with the help of a video camera. For elongated rice grains –i.e., when the frictional inter-grain forces suppressed the inertial effects– SOC behavior was found.

Figure 11 shows the finite size scaling of their data. The dependence on the grain shape –a detail of the system– on the SOC behavior, though, indicates that the sandpile model is not as robust as initially supposed. The off-the-edge avalanche behavior of 1D piles of balls studied by our group suggest that SOC may also depend on the *nature* of the *base* of the piles in such a way that bases which induce "ordered" profile doesn't pass the finite size scaling test, while those with bases inducing "disordered" profiles do. Figure 12 displays two extreme cases of this behavior: in the upper half, bases provoking a smooth pile profile give "uncollapsable" avalanche size distributions for three different base lengths. In the lower half, a base provoking a very disordered profile gives a nice collapse when the ansatz (26) is applied to the avalanche distributions for three different base lengths. This result is most intriguing, considering that inertial effects are not purposedly suppressed in our experiment [118].

It is not strange, then, that the next step was to look for SOC on a physical system basically analogous to sandpiles, but free –at least in a much bigger extent than sandpiles– of inertial effects: *the critical state of type II superconductors.*

5. Avalanches in the vortex state

Experiments: present and future

Field et al succeeded in 1995 in the first conscious experiment to directly establish the analogy between sand and vortex avalanches [52]. They used a setup basically similar to those previously used in references [134,135,136]: a 1800 turn pickup coil was coaxially set into a thin-walled tube of the low temperature superconductor $Nb_{0.47wt}Ti_{0.53wt}$ with 6mm of outer diameter, 0.25mm wall thickness and 3.4 mm length. Following the authors, *this geometry guaranteed a close analogy to [conical] sandpiles.* An external magnetic field was applied along z-axis with different ramp speeds, and the voltage induced at the pickup coil amplified, and recorded by a computer. The equivalent sandpile analog of this setup would be an horizontal ring with a vertical cylindrical wall along its outer perimeter, and no wall along the inner one. The sand would be continually added along the perimeter of the wall from the outside. Then, after establishing a critical profile, avalanches would start to drop across the inner perimeter of the ring. The basic idea of Field's experiment was to detect

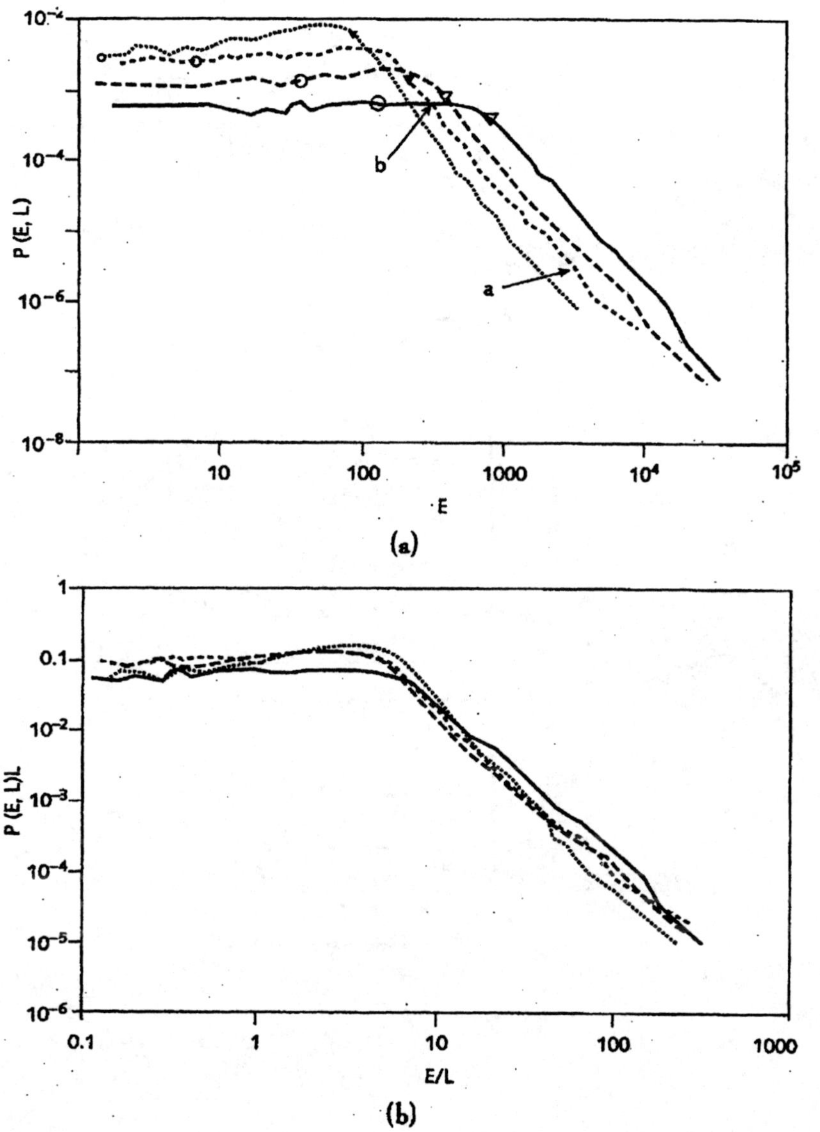

Figure 11: (a) Avalanche energy distributions for four base sizes in a ricepile, after Frette and coworkers [132]. (b) Data collapse after applying a finite size scaling ansatz [132] .

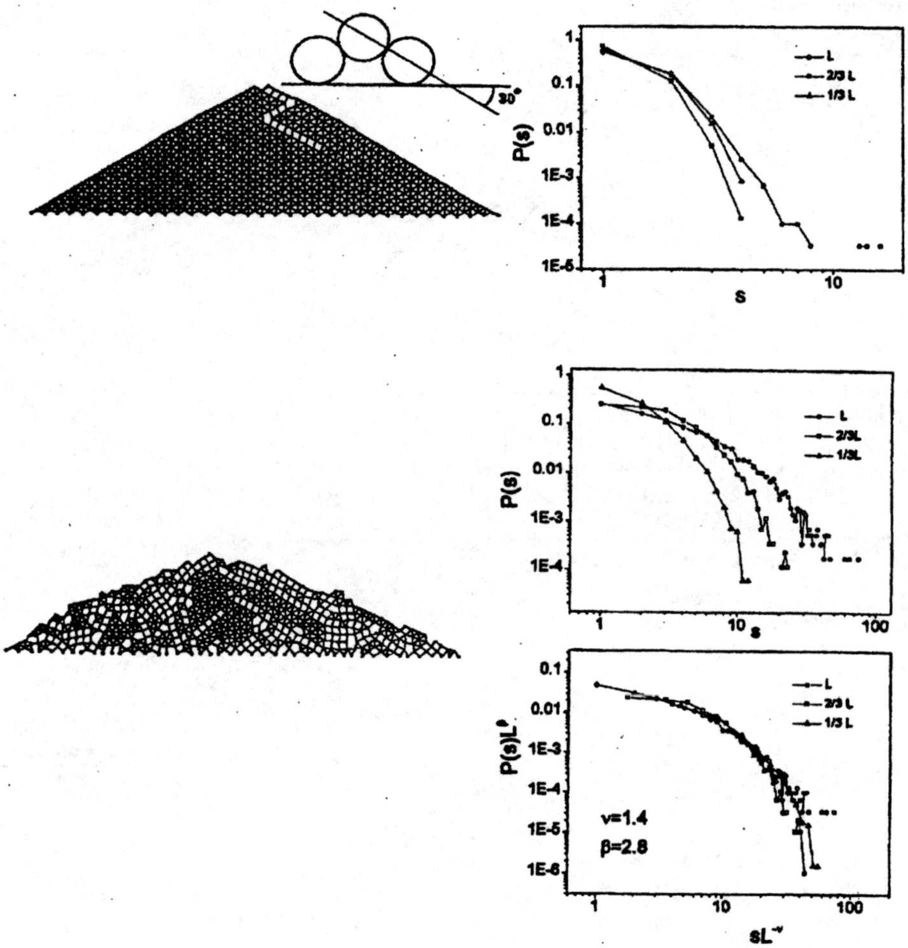

Figure 12: Base-dependent avalanche behavior in 1D piles of balls. Upper part: well ordered piles (left) produced by a conveniently calculated base show avalanche size distributions (right) which do not collapse when trying finite size scaling. Lower part: disordered piles associated to disordered bases (left) show good collapse of the avalanche size distributions measured for different base lengths (right).

the equivalent vortex avalanches by means of the pick up coil. Then, if a "flux avalanche" of length l "drops" inside the cylinder, it will induce a voltage on n turns of the pick up coil, were $n = \frac{l}{L}N$ (L being the length of the pick up coil and N the total number of turns). Then, the detected voltage will be

$$V = \frac{l}{L}N\frac{d\phi_{actual}}{dt} \tag{27}$$

Field and coworkers defined their avalanche size as an "effective bundle volume" given by

$$s \approx l\phi_{actual} = \frac{L}{N}\int V dt \tag{28}$$

which is a convenient definition in view of the fact that it was not possible to determine l from the available data. In the 1968 experiment by Heiden and Rochlin, a $1-mm$ coil was used but, even in that case, the authors expressed their preoccupation about the existence of a geometrical factor tending to increase the number of counts due to bundles linking only part of the coil [135].

Figure 13a displays the temporal response of the voltage at the pick up coil for a field window of $30Oe$ centered at $7.55kOe$ and measured at a sweep rate of $5Oe/s$ (figures b and c are horizontal enlargements of Figure 13a). Following the authors, two contributions can be identified: a flat background related to a continuous entrance of flux at the rate of the external field ramp, and *pulses related to sudden entrances of flux bundles, i.e., avalanches.* The latter are estimated to represent 3 percent of the total flux income. Although the authors calculate that the number of vortices associated to avalanches range from 50 to 5000, this is pervaded by the indetermination in the bundle length, l.

Figure 14a shows the avalanche size distributions reported in [52] For the three fields studied and $5Oe/s$ sweep rate, *they can be described by power laws* for more than one decade. This situation is not common in sandpile experiments, where a departure from the pure power-law behavior is seen [125,128,130]. Some authors interpret this situation as a consequence of the finite size of their systems, which is not a very important handicap in the vortex avalanches here described, since, following [52], the number of vortices in the wall is typically a million times greater than those involved in avalanche events). The authors didn't find the relaxation oscillator behavior observed by others in 2D sandpiles [125] which is related by them to the lack of inertial effects in vortices. As mentioned earlier, this situation makes vortex dynamics closer to sandpile models [116,123]. The observed variation of the power law exponents with the applied field in the range from -1.4 to -2.2 is explained in [52] to the different inter-vortex and inter-pin distances attained at the different field values. This may be considered reminiscent of the influence of grains friction, shape [132] or even of the type of base [118] on analogous exponents in the case of sandpile dynamics. Finally, the authors report the power spectrum of the observed avalanches, defined as the squared Fourier transform of a field sweep similar to that of Figure 14b. While a $1/f$ behavior is seen for slow ramps, a faster ones give a Lorentzian peak, and a $1/f^2$ spectrum at high frequencies. The authors conclude from this fact that the critical behavior is achieved only very near the marginally stable state, when the system is slowly driven (typically $5Oe/s$ in their experiments). Interestingly enough, Heiden and Rochlin found *exponential* distributions of "flux jumps" entering their cylinder independently from the field sweep

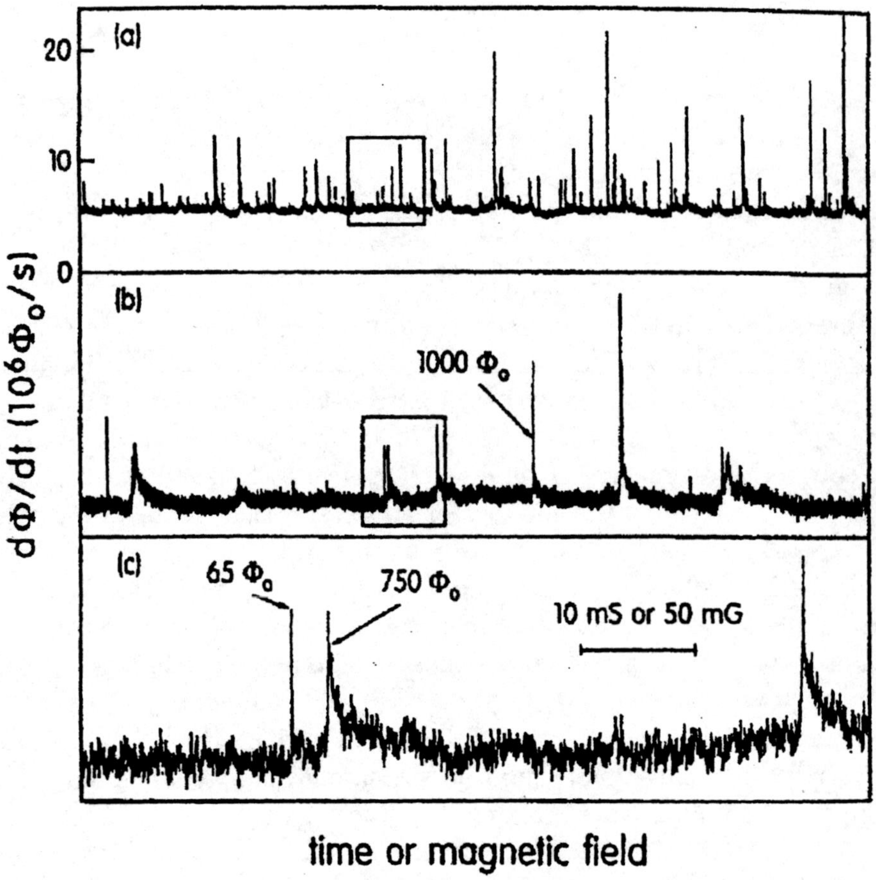

Figure 13: Response of the pick up coil of Field and coworker's experiment [52] which demon strates the avalanche behavior of vortices in low-T_c tubes.

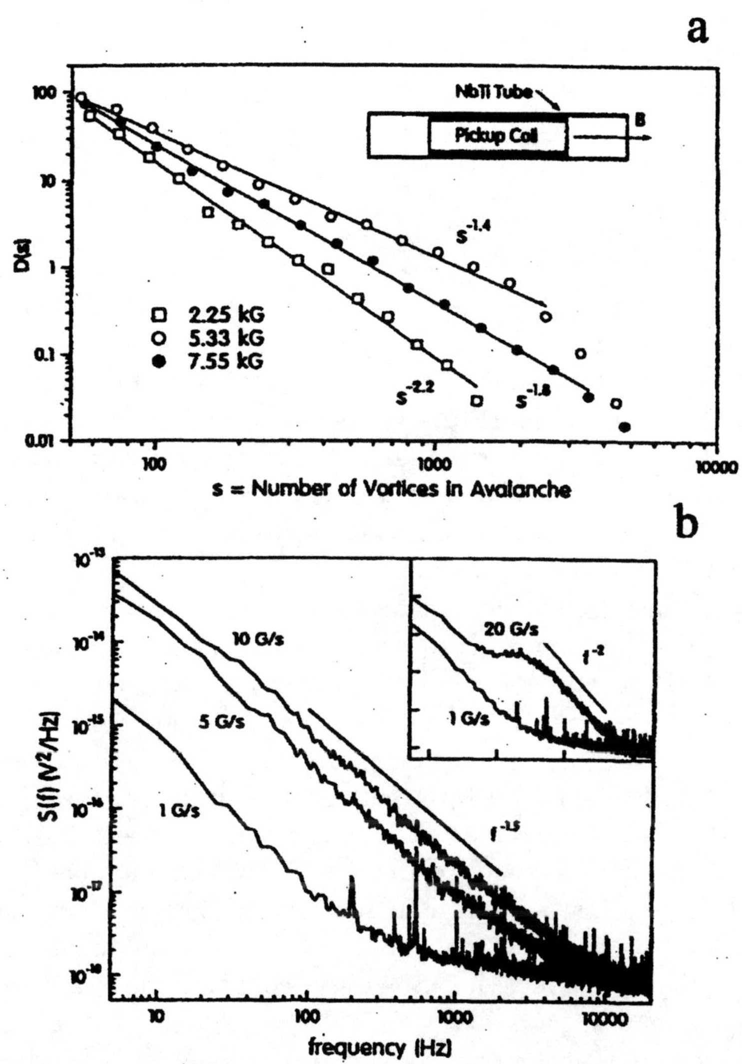

Figure 14: (a) Avalanche size distributions in the experiment of Field and coworkers [52] for different field windows. (b) Avalanche power spectrum reported by the same authors.

rate, at least in the range from 10 to $100 Oe/s$ in their 1968' pioneering experiment [135]. Even when their "flux jumps" distributions included a higher number of events than those presented in figure 14a, the exponential behaviour was found for less than a decade, and the authors believed that their statistics was still poor for the assessment of the exponential behavior.

Let us turn back to the problem of the "avalanching" flux bundle length. One could estimate this length using the theory of Collective Pinning [137,138]. By treating the vortex lattice plus random pinning as an elastic system , this theory assumes that vortices elastically deform to "accommodate" to a distribution of pinning centers *within a characteristic pinning length along z* roughly given by [139,140]:

$$L_c \approx \xi \sqrt{\frac{J_0}{J_c}} \tag{29}$$

where J_c and J_0 are the measured and the depairing critical current densities, respectively, and ξ is the coherence length. These vortex segments –or vortex bundle segments, at higher fields– are the "objects" which may "jump" to a neighbor metastable state within Collective Pinning Theory [139]. If we substitute typical values of these parameters for a low-T_c superconducting alloy at temperatures under $5K$ [43] ,*we get a correlation length of the order of a few microns.* When considering very poorly pinned materials –i.e., very small J_c– the correlation length increases to tenths of a millimeter. Is that the typical *avalanche* z-axis bundle length in the experiment of references [52,135]?

In our opinion, revisiting the 1965' experimental set up of Wischmeyer [136] would be a fruitful exersise to get, at least, an approximate response to that question, and, generally speaking, more insight into Field and coworkers' experiment. It also consisted in a low-T_c cylinder submitted to an axial magnetic field , but *three* detection coils were coaxially mounted, as depicted in the lower left section of Figure 15. Two of them were located at the inner wall of the cylinder in such a way that their adjacent ends were a distance $d = 2.5mm$ apart, and the third at its outer wall. The lower right section of Figure 15 shows measurements obtained by Wischmeyer with his set up which are roughly equivalent to those depicted in the upper right section of Figure 15, except that *three* signals are detected. While the temporal evolution of the voltage of the coils inside and outside the tube are highly correlated, the signals corresponding to the coils into the cylinder are poorly correlated.

Winschmeyer results indicate that not only the flux exit through the cylinder wall, but even the flux entry, may be characterized by peak events!. This can be understood for system-spanning avalanches –a big exit of vortices off the edge implies a decrease of vortex density at the entry– or even for avalanches occurring close to the entry boundary. But if it happens to *all* the avalanches, it means that some kind of surface barrier prevents the applied field to *enter* the system smoothly. The data from [136] is not good enough to discern among these possibilities. They also give a clue about the length of Field's avalanching objects: a few millimiters seems to be its upper limit (definitively 2.5 mm in [136]). Let us imagine an experiment with Winschmeyer's setup in which the distance d can be diminished until correlation between the pulses detected by the two inner coils is observed. The corresponding value of d can be interpreted as the *maximum length* of the bundles crossing the wall. A statistical analysis of these correlations for even smaller

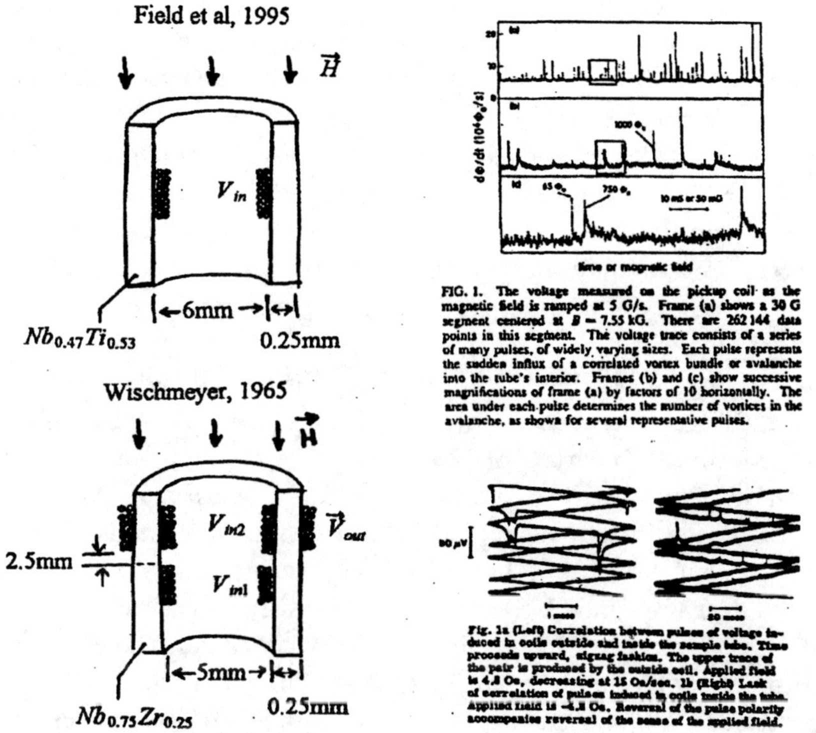

Figure 15: Upper section: set-up by Field and coworkers (left), and resulting pickup voltage (right) [52]. Lower section: set-up by Wischmeyer (left), and resulting voltages at the three pickups involved in the experiment [136] .

distances could even give an idea of the distribution of flux bundles lengths in the z-direction and then, would allow us to learn more about the *flux* sizes of the avalanches. We even can learn how to control the size of the avalanching objects (related to vortex "rigidity" [43] by tuning the measuring temperature. Of course, if expression 30 gives a true estimation of the avalanching bundles' length, this experiment would never detect correlation between the outputs of the inner coils, at least for the d values reasonably affordable in the experiment.

As can be induced from the last paragraph, besides inertia, *temperature* is another important difference between the avalanche behavior of superconductors and of sandpiles. In general, avalanche experiments on sandpiles are not performed in the presence of mechanical vibrations. So, a perfectly equivalent experiment on superconductors should be performed *at zero Kelvin.* Although it is not possible, common sense indicates that the avalanche behavior of superconductors would approach that of sandpiles only at very low temperatures. But this happens not to be the case, as demonstrated by the experiments of Zieve et al [53], Nowak et al [54], and Behnia et al [58].

The 1996's report of Zieve and coworkers [53] used two experimental novelties in the study of vortex avalanches: state-of-the-art magnetic field sensors, and High T_c materials ‡ The former consisted in the use of photolithographically prepared miniature Hall probes to detect the magnetic induction very close to the sample. This technique had been gaining popularity among the superconducting experimental community in the nineties [142,143,144,145,146,147,148,149].

Zieve and coworkers studied untwinned $YBa_2Cu_3O_{7-\delta}$ single crystals with a critical temperature of 92 K. They ramped an external field applied parallel to the smaller dimension of the crystals, and measured the *internal* avalanches on the samples by mounting them on a miniature Hall probe with a typical active area of $3\,\mu m \times 5\,\mu m$. This resulted in a sensitivity of a single flux quantum, a dramatic improvement as compared with Field and coworkers' experiment [52]. Another interesting feature of the micro Hall probes is that they allowed the measurement of "internal" avalanches –which can be compared with the rice piles experiments reported in [132]– in contrast with the off-the-edge vortex avalanches reported before [52]. Following the authors, the results were basically similar when the sensor was positioned at different spots on the crystal surface, of typical dimensions of $1.05 \times 0.5 \times 0.005\,mm^3$. The avalanche dynamics was explored at *very low* temperatures ($0.1\,K - 0.4\,K$, approximately), and hysteretic effects were studied by ramping the external magnetic field from zero to around $80\,kOe$, and then back to zero, at a sweep rate of $7\,Oe/s$.

Even when Zieve and coworkers' experiment eventually introduced demagnetization artifacts, their geometry allowed –in principle– a better control of the z-axis length of the avalanching objects than in cylinder experiments. Let us estimate L_c in their experiment with the help of formula 30. We first calculate the depairing critical current for a typical YBCO crystal with the magnetic field applied along the c-axis at $0K$ by substituting $\xi_{ab} = 30$ and $\lambda_{ab} = 3nm$ in the following Ginzburg-Landau result [17]:

$$J_0 = \frac{\Phi_0}{3\sqrt{3}\pi\mu_0\lambda^2\xi} \tag{30}$$

‡The first observation of vortex avalanches using a similar setup was reported in 1993 by Seidler and coworkers [50]. We instead discuss in detail reference [53], since it can be regarded as a continuation of [50] with a heavy accent on the dynamical origin of the avalanches and their sandpile analogy, as we will see.

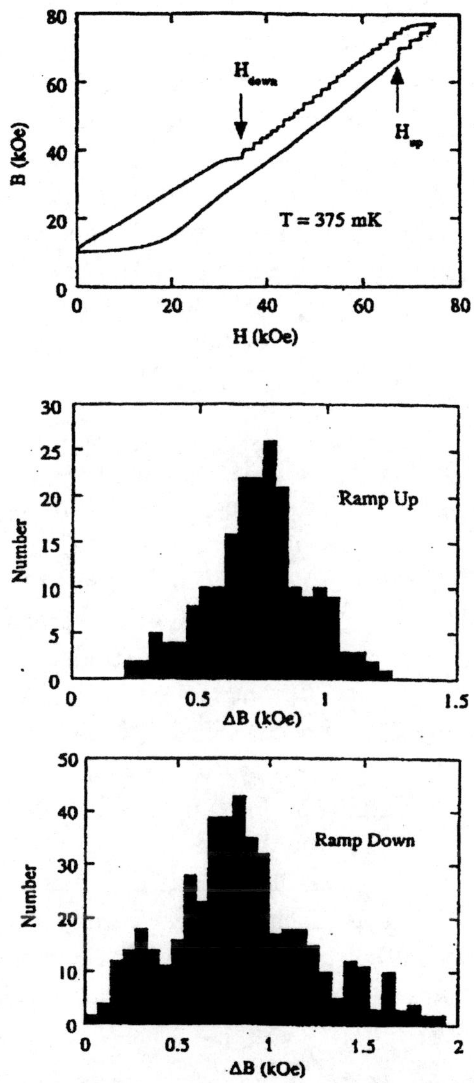

Figure 16: Top graph: Typical hysteresis loop measured by Zieve and coworkers [53] on a YBCO crystal in which steps corresponding to vortex avalanches are visible. Intermediate and lower graphs: avalanche sizes distributions derived from the up and down field ramps in the same experiment [53].

which gives $J_0 \approx 3.6 \times 10^{11} A/cm^2$. A rough estimate of J_c based on the hysteresis loop reported in [53] (upper section of Figure 16) and the crystal dimensions also reported in [53] gives a lower limit for J_c of $10^6 A/cm^2$, a value typically reported for YBCO crystals form magnetic measurements (see, for example, reference [90]). If we substitute this value and J_0 in formula (30), we can estimate an upper limit of of $20\mu m$ for L_c, *which is bigger than the reported crystal thickness of* $5\mu m$. This is a good point in favor of Zieve and coworkers experiment [53]. Besides the crystal thickness, a second relevant length related to vortex dimensionality appears in the case of any layered superconductor: the interlayer spacing. As pointed out in section 2.1, this situation provokes the appearance of "pancake" vortices on the superconducting layers. While for low enough temperatures and fields these vortices are strongly correlated as 3D strings, (which can be treated as "conventional" vortices), above a certain line in the H-T plane the correlation can be broken, and the 2D pancakes move independently [140]. *This complicates the dynamics for any experiment to detect vortex avalanches on High* T_c *superconductors, unless we make sure to work in the "vortex lattice" or "vortex glass" regions of the phase diagrams depicted at the lower section of Figure 1.* This condition seems to be also fulfilled in [53] due to their very low temperature window.

A conceptually simple experiment would help to get rid of any doubt about the dimensionality of the avalanching objects for any geometry involving a magnetic field perpendicular to a superconducting crystal or foil: it would consist in attaching not one, but *a couple of paralell micro Hall probes situated at the two wide surfaces of the YBCO crystal* and analyze the *synchronism* of the vortex avalanches measured by either probe. This experiment can be regarded as the "magnetic relative" of the six-terminal flux transformer set-ups that has been used for studying different vortex phase transformations through I-V curves in HT_c superconductors [150,151,152,153,154,155,156,157]. From the instrumental point of view, it is definitively closer to the experiment by Lee and coworkers [158]. They measured magnetic flux noise at opposite sides of YBCO and BSCCO crystals and films in the thickness range between 0.075 and 30 microns *with two SQUID's.* The authors concluded that the correlation of vortex motion at both sides of the sample depends on its thermomagnetic history and pinning details. Their experiment has two important differences with the one with micro-Hall probes. Firstly, while the latter is "local" ($10 \times 10\mu m^2$), the former is "global" ($180 \times 180\mu m^2$ SQUID hole area). Secondly, the latter measures the vortex movement correlation associated to a field ramp, while the former probes the thermally activated flux motion it *in time* at zero field.

The upper graph of figure 16 contains some of the most interesting findings reported in [53]. First, vortex avalanches are observed as steps when the magnetic field is ramped up and down. Second, they appear at a certain field value, H_{up}, and disappear at a *different* value, H_{down}, which are independent from the field ramp rate. H_{down} showed to be particularly stable against variations of the temperature and field ramp histories. The results of several well conceived experiments made the authors reach the conclusion that *the difference between* H_{up} *and* H_{down} *,i.e., between the ramp directions, is an intrinsic property of the system than cannot be "suppressed" by changing the experimental conditions.* The second major finding of Zieve and coworkers is that the avalanche size distributions in both ramp directions are not power-laws, but sharply peaked, and centered around 750 flux quanta (see the two lower graphs in Figure 16). This doesn't fit into the SOC scheme.

These distributions could be expected from thermally triggered flux jumps, as described in section 2.2.2. However, the authors dissipate such doubts with several arguments. For instance, their avalanches are observed in the "high" field region, while thermally triggered flux jumps are typical of the "low" field region, and their steps never reach the line $B = H$ –i.e., the magnetization never vanishes– which constitutes a common "catastrophe" in thermally triggered flux jumps. They also do not depend too much on field rates, as expected for thermal instabilities.

Then, the authors conclude that their avalanches have a true *dynamic* origin, and try to give an intuitive picture of the observed hysteretical behavior through the sandpile analogy. As described in section 3, the mechanical process to fill a box with sand to get the "V-shaped" Bean's profile is different form that required to resemble the decreasing-field Bean's profile: while in the former, sand should be *added* through the boundaries at the *higher* positions of the pile, in the latter it must be *removed* from the *lowest* positions. These two processes are quite different, since the lower positions support much of the overall pile weight, so sand removal may trigger avalanches easier (corresponding to the lower values of H_{down} as compared to H_{up}) § Zieve and coworkers extend the analogy to account also for the peaked avalanche distributions observed: in their opinion, vortex mass renormalization [140] takes place at the very low temperatures of their experiments, making vortex inertial effects significant. This brings closer their vortex system to sandpiles, which, as stated in section 3, have been found to show peaked avalanche size distributions due to inertial effects [125,128].

Nowak et al [54] performed a variation of Field and coworkers' experiment, but substituting the superconducting cylinder by Nb rings with inner and outer diameters of $15\mu m$ and $98\mu m$, respectively, and $0.5\mu m$-thickness. They used the same system as [53] for avalanche detection: one $3\mu m \times 5\mu m$-area Hall probe was positioned at the center of the hole, and a second one, $22\mu m$ off-center and directly above the ring. The sensitivity of these probes were $10\Phi_0$ and Φ_0, respectively. They allowed the measurement of the *total* flux involved in avalanches crossing the perimeter of the hole, and also those occurring *a certain position* above the superconducting ring (internal avalanches). This geometry might allow to approach the vorticial analogy to a *typical* sandpile, i.e., that excited by adding particles at the top of the pile. To do it, we just need to increase the magnetic field *only at the hole.* This can be achieved by using a miniature split ring-solenoid fitting closely over the specimen at the hole region. Internal and off-the-edge avalanches could be measured by placing Hall probes on the ring and at its outer perimeter, respectively.

Nowak and coworker's geometry allows –in principle– to control of the z-axis length of the "avalanching" objects even better than in the case of [53], since no layered structure is present in Nb. Let us estimate L_c in this experiment. We first calculate the depairing critical current for pure Nb at $0K$ by substituting $\xi = 40nm$ and $\lambda = 85nm$ [141] in 31, which gives $J_0 \approx 3.4 \times 10^7 A/cm^2$. Substituting in formula (30) this and the value $J_c = 5 \times 10^{10} A/m^2$ reported in [54] for their Nb films, gives $L_c \approx 0.1\mu m$. Although this is twice the film thickness use for fabricating the rings, this proportion might be even lower, since J_0 may be smaller than that calculated from 31, due to impurities in Nb [160]. Then, a "worse" Nb material from the point of view of pinning strength would be convenient in this experiment to *guarantee*

§To the author's knowledge, no experiment or has been conducted to check out such as important hysteretical behavior in sandpiles. Although not aimed at exploring hysteresis, recent *simulations* of Manna and Khakdhar [159] indicate power-law avalanche distributions for internal avalanches in granular media where the grains are *removed* from the bottom of a 1D pile. No inertial effects were considered.

"avalanching objects" of the same length of the film thickness. As a matter of fact, this can be easily achieved for thin-film amorphous superconductors [161].

Figure 17 shows some of the basic findings of Nowak and coworkers [54]. As in the case of [53], steps appear in the $B-H$ plane, indicating the occurrence of avalanches at different temperatures –always higher than those studied in [53]–. Interestingly enough, no trace of H_{up} and H_{down} fields similar to those reported in [53] is observed. This might be related to the much lower field values in the present experiment. The most interesting fact observed by Nowak and coworkers is the dramatic dependence of their avalanche size distributions with temperature, as summarized in Figure 18. Below $T/T_c > 0.35$, the avalanche size distribution follows a decreasing exponential, as reported in [135], and the avalanches are not "complete" (they don't reach the $H-B$ line). For $0.2 < T/T_c < 0.3$, however, it is sharply peaked around $2000\Phi_0$. The corresponding avalanches are system-spanning and "complete", which suggests thermally-triggered flux jumps. For $0.2 > T/T_c$, a broad avalanche size distribution is recovered, but it is not exponential. These observations remained the same within a ramp rate window from 2 mOe/s to 20 Oe/s, indicating that the experiments took place in the slowly driven regime.

The authors explain their data on the basis of a thermally triggered mechanism typical of flux jumps. They even define a thermal stability parameter (see section 2.2.2), for their ring geometry, and argue that the broad distribution of avalanches they observe in the range $0.3 < T/T_c < 0.4$ –and reported by others [52,135]– is related to a regime in which that parameter is marginally greater than 1. This contrasts with a SOC dynamic mechanism for vortex avalanches.

To the author's knowledge, the last experimental study of non-catastrophic vortex avalanches in slowly ramped external fields is the 1999' one by Behnia and coworkers [58]. Pulling even further the Hall micro-probe technique, these authors attempted to study in detail the spatial and temporal correlations of *internal* vortex avalanches in the x-y plane of a a $20\mu m - thick$ polycrystalline Nb film of lateral $0.8 \times 0.8mm^2$. The measurements were performed using a row of 8 equally spaced Hall micro-probes of $20 \times 5\mu m^2$ area each, able to detect $0.16\Phi_0$. Each row was 0.35 mm long, lying on the film perpendicularly to one of its edges, in such a way that the sensor closest to the film edge was 0.2 mm away from it (see the upper section of Figure 19 for an approximate sketch of the sensors arrangement). As in previous cases, let us estimate the collective creep bundle size in this experiment, based on the available data. Considering formula (31) and the values $\xi = 60nm$ and $\lambda = 150nm$ as extracted from the values of the critical fields [162], we obtain $J_0 \approx 7 \times 10^6 A/cm^2$. A rough estimation of the critical current density at $H = 1500Oe$ based on the application of Bean's model to the $B-H$ loop reported by the authors gives a lower limit of $J_c \approx 5 \times 10^4 A/cm2$. When these values are substituted in formula (30), we get an approximate upper limit for L_c of $0.7\mu m$, which is quite smaller than the film thickness. So, the "rigidity" of vortices measured in the experiment is not guaranteed. After checking the establishment of the Bean's critical state profile at $T = 4.8K$ –something difficult to assess in previous experiments due to the small amount of Hall sensors–, Behnia and coworkers ramped up the external field at $1.1Oe/s$ starting at $B = 0.15T$. They obtained, at each sensor, a $B-H$ pattern with *steps* qualitatively similar to those reported in [53] and [54], which indicated the presence of flux avalanches. *The authors estimated that 40 to 70 % of the flux increase into the sample was in the form of avalanches and not fluidlike, which contrast with the 3 % reported by*

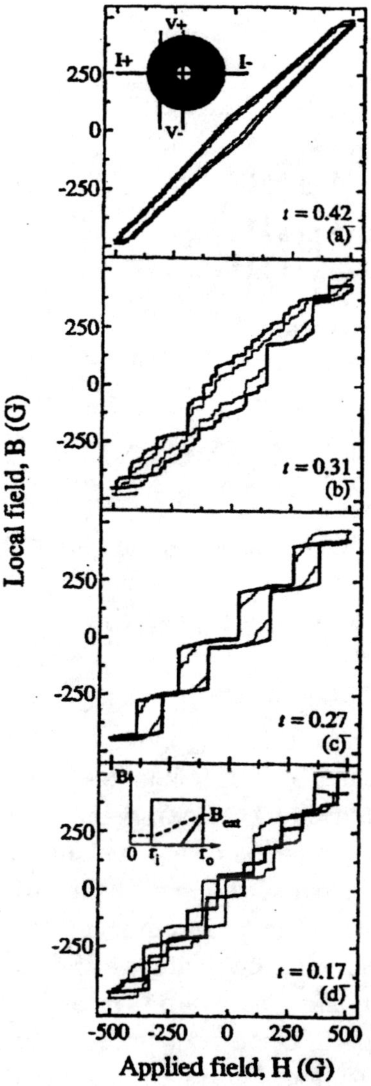

Figure 17: Hysteresis loops measured by Nowak and coworkers [54] on a Nb ring (depicted as an inset in the top graph) in which T-dependent steps representing vortex avalanches are visible ($t = T/T_c$).

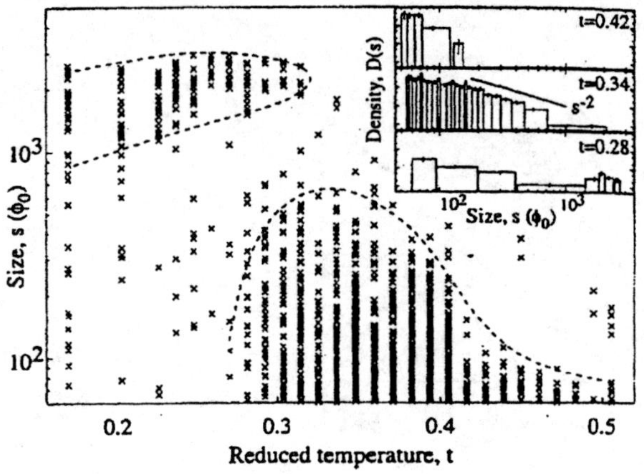

Figure 18: Representation of the avalanche statistics as measured by Nowak and coworkers, extracted from [54] .

Field et al in 1995 [52].

At the lower left of Figure 19, the avalanche size statistics is displayed for one sensor (black dots) and for the output of a whole row (columns). Both distributions follow a power law for less than a decade with a critical exponent of -2.1, within the range of those reported in [52], with the biggest avalanches comprising around 5 vortices under one probe area. However, the differences in the zone of big avalanches suggest that not all the avalanches extend through the whole length of the row (if they did, the two outputs would be exactly the same). Note that measuring the output of an entire row of sensors introduces an "artifact" qualitatively similar, within the x-y plane, to that present in reference [52] due to the extension of their pick up coil along z-axis. To further explore this finding, Behnia and coworkers analyze the spatio-temporal correlations of the avalanches measured on two micro-probes located $50\mu m$ apart, resulting in the $B-H$ diagram at the lower right of Figure 19. The avalanches at the sensor closer to the sample boundary generally occur first, and then propagate "down-the-hill" to the second sensor. They also observed that avalanches generally increased when moving in the same direction. These two facts are reflected in the right-left and up-down assymetry in the inset of the $B-H$ graph, respectively. All this findings suggest a SOC picture of the observed avalanches.

Under $T = 3.4K$, however, a completely different picture arises: catastrophic, hysteretic, avalanches with peaked size distributions are observed, in agreement with the work of Zieve and coworkers [53]. As them, Behnia and coworkers claim that such avalanches are not thermally triggered, since they demonstrated to be insensitive to the field ramp speed. They also present the $H-T$ phase diagram of their sample –a very valuable data not often reported by other authors–, and remark that the catastrophic avalanches occur in a low

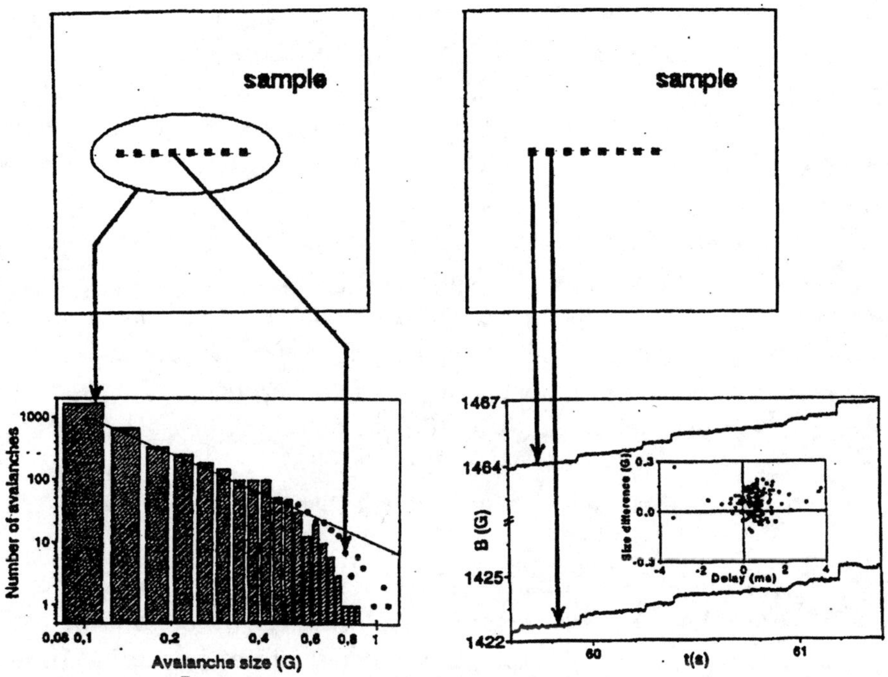

Figure 19: Avalanche size distributions (left) and spatiotemporal correlations of avalanches (right) on a Nb foil submitted to an increasing magnetic field [58].

temperature region *close to H_{c1}, and far from H_{c2}.*

The authors suggest that, close to H_{c1}, the critical state is not well established, so, as reported for sandpiles, the occurrence of "quasi-periodical" avalanches can be related to the "absence of a single critical slope". This contrasts with our own results in 1D piles (see Figure 12) and the "Oslo model" in which SOC behavior is observed only when "dynamically disordered" pile profiles are attained [118,132,133].

A quite different approach to the measurement of vortex avalanches was used by Aegerter in 1998 [56]: *instead of ramping the external field –as in the cases previously discussed– the author measures flux avalanches during the thermally activated time relaxation of the magnetization* at different temperatures. At each temperature, the author measures the *global* magnetization of a $Bi_2Sr_2CaCu_2O_8$ crystal during a 24-hour period by means of a SQUID magnetometer, after applying a certain magnetic field to the sample in zero-field-cooling conditions. However, neither the value of the applied field nor the relative orientation of the crystal are reported in reference [56]. The avalanches appeared as *spikes* in the magnetization versus time characteristics. Even when the avalanche dynamics studied in this "thermally driven" experiment cannot be compared with those reported in "field-driven" conditions [135,52,53,54,58] in a straightforward fashion, it is worth pointing out the basic observations accounted by the author: *the avalanche size distributions are described by power laws at "low" temperatures ($0.06T_c$) and at long relaxation times, while they are fitted by exponential laws at "high" temperatures ($0.8T_c$), and at short relaxation times.* When power laws arise, the author obtains an exponent of 2 in their avalanche distributions, very similar to that reported in [52] from global, field-driven experiments at $0.3T_c$. An interesting additon to Aegerter's experiments would be to detect "thermally activated" avalanches in the temperature region $T < 0.01T_c$ and check if the peaked avalanche distributions reported by Zieve and coworkers in their field-driven experiments on other high-T_c, YBCO, appear here.

Some of the experiemntal details and results reported in the avalanche-detection experiments discussed above, have been summarized in Table 1. Although their contents might be partially unnacurate –due to lack of space and difficulties in finding data in some of the articles cited– it brings a necessary global idea of the current "state of the art" in the subject. It makes clear, for example, that many voids have to be filled up to get a complete picture of the avalanche dynamics in superconductors. It may be suggested, for example, that the micro-Hall probe technique should be used to explore more thoroughly the H-T diagram of the different materials in the search of avalanche dynamics. Even when most of the cited papers are fine pieces of experimental work, they do not spot the detected dynamic regimes in the context of the H-T diagram for their specific sample. On the other hand, some experimental details not reflected directly in Table 1 also present voids. For example, the time windows explored by the different authors for estabishing their avalanche size distribution data roughly vary form a hundred seconds in [52] and [58] to several hours in [53] and [54].

Doughnut-vortex avalanches

Let us examine the experiment depicted in Figure 20: a straight current-carrying wire is located along the symmetry axis of a type II superconducting hollow cylinder with pinning. Due to the azimuthal magnetic field created by the wire, an asymmetric critical

Ref	*Geom.*	*Material*	*Sensor*	*Avalanche type*	*Trange*	*H range [kOe]*	*Avalanche distribution*
[135] (1968)	hollow cylinder	Pb-In	pickup coil	off-edge	0.6 T_c	0.550 - 0.850	exponential
[52] (1995)	hollow cylinder	Nb-Ti	pickup coil	off-edge	0.3 T_c	2.25 - 7.55	power law (slow ramps)
[53] (1996)	planar	YBCO crystal	Hall probe	internal	$T \leq 0.01\ T_c$	0-80	peaked
[54] (1997)	ring	Nb film	Hall probes	off-edge & internal	0.15 1.12 T_c-	-500 - 500	peaked/pow. law (dep. on T)
[56] (1998)	planar	BSCCO crystal	SQUID	off-edge	0.06- 0.8T_c	?	exp/pow. law (dep. on T & t)
[58] (1999)	planar	Nb film	Hall probes	internal	0.52 T_c	1.5	peaked/pow. law (dep. on H & T)

Table 1: Basic features and results of experiments on vortex avalanches discussed in the text, as extracted directly or indirectly from the respective references. Most of the experiments were performed by ramping the external field in the range from 0.002 to 100 Oe/s, except for [56], in which the field was not ramped.

state is established on the crossection of the cylindrical shell associated to a distribution of doughnut-shaped vortices like those described in section 2.1 [25]. This asymetry is connected to the fact that the wire-associated "external" field in which the cylinder wall is embedded is higher at the inner wall, and lower at the outer wall. Let us now suppose that we *slowly increase* the current on the wire.

By analogy with experiments and simulations on slab geometry in which the external field is ramped, vortex avalanches are likely to occur. In principle, their fingerprint would be *spikes* in the longitudinal voltage of the cylinder, which would eventually resemble those depicted in Figure 13. It doesn't seem to exist a trivial "sandpile" analogy of this experiment in which the avalanching objects change their length –and, following formula 8, their electromagnetic energy– as they move. Is this system expected to show SOC?.

A final divertimento: "Magnetic" and "Superconducting" earthquakes

Earthquakes are clearly one of the natural catastrophes calling for a sound physical model able to yield predictions as accurate as possible. Even though earthquakes are enormously complex phenomena involving the dynamics of disparate elements ranging from the Earth's crust to human constructions, there is one fact that makes it reasonable to tackle the problem: earthquake intensity distributions measured along many years in different regions follow a simple power law known as *Gutenberg-Richter law* [163].

Burridge and Knopoff [164] introduced a spring-block model for earthquakes basically similar to that represented in Figure 21. In the model, a tectonic fault is represented by a two-dimensional array of blocks resting on a surface with friction. The blocks are interconnected by coil springs along the y-axis, and each one is linked to a moving plate through leaf springs, which account for the forces on the material near the fault due to tectonic plate motion. When the plate is moved along x-axis, no block moves until, at least,

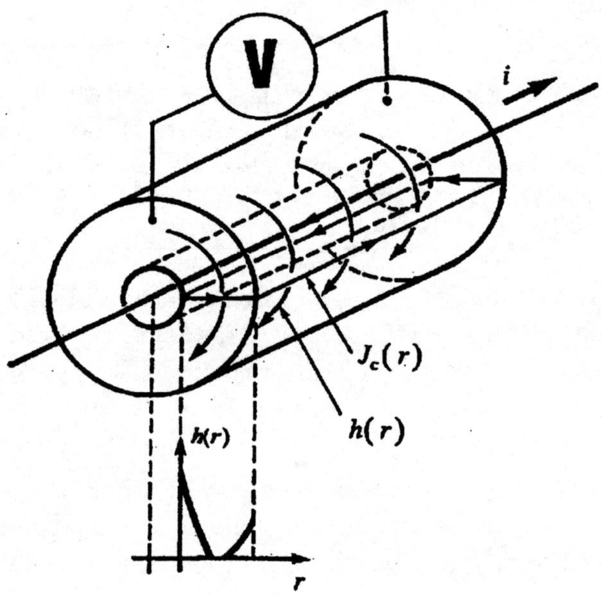

Figure 20: Type II superconducting hollow cylinder with a coaxial current-carrying wire: critical state of "doughnut-shaped" vortices [25] .

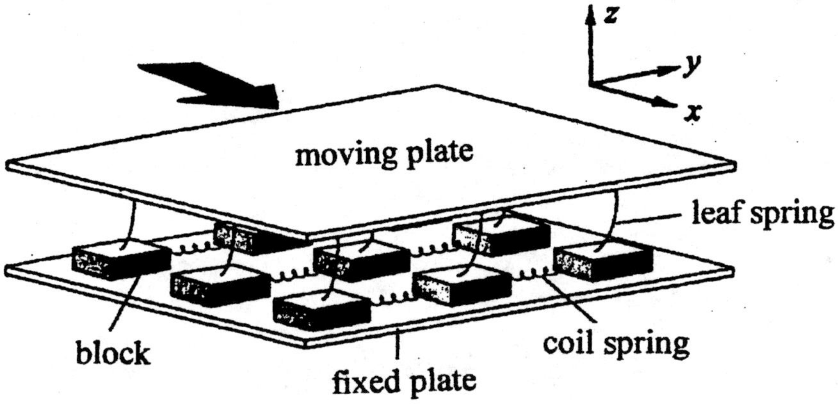

Figure 21: Block-spring model for earthquakes [122].

one of them receives a force from its leaf spring higher than a certain threshold. When this happens, the block slips in the x-direction, which increases the forces on the neighboring blocks via the coil springs, which eventually induces a "chain reaction" resulting in the displacement of several blocks. In the model, this *avalanche-like* event is identified as an earthquake, whose intensity can be defined as the total displacement of the blocks. Simulations show that the distribution of the avalanches in such model follows a power law, even when dissipation is considered (i.e., part of the elastic force is not transmitted to the neighboring blocks). This has been interpreted by many authors as a demonstration that earthquakes display SOC behavior [165,166,167,168,122]. Even more, the finite size scaling analysis of this model indicates that it belongs to the same universality class exhibited by one-dimensional granular piles [133] and interface deppining [169]. Similar results have been found when the model was translated into a mechanical experiment [170]. Interestingly enough, when inertial effects are considered (i.e., the slips events are not instantaneous), simulations indicated that power laws are only obtained for "small" earthquakes, but more or less periodic "big" earthquakes take place [171], which resembles better the rotating drum experiments with sand of reference [119] than real earthquakes.

A curious variation of the earthquake model would be to substitute the springs by magnetic forces. This can be achieved by means of the experimental setup depicted in Figure 22: a regular set "primary" permanent magnets are stuck to the upper face of a nonmagnetic slab which rests on a second plate. A second set of magnets is situated on the lower face of the latter, in such a way that they are kept in position through the attractive interaction with the "primary" set of magnets along the x-axis (the magnetic moments of both layers of magnets are in the same direction). Besides this interaction, the "secondary" magnets experience a repulsive magnetic interaction among them in the x-y plane. These interactions along z-axis and along the x-y plane play the role of the forces exerted by the leaf and coil springs of the earthquake model depicted in Figure 21 (although a perfect analogue would require coil springs also along x). The difference would basically rely in the details of the forces involved, particularly their *long range* nature, which contrasts with the short range commonly assumed. Now, we start to move the upper plate. Since the

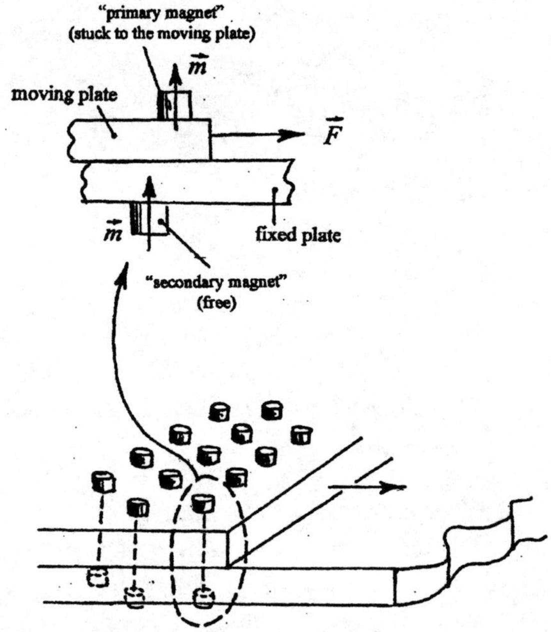

Figure 22: "Magnetic earthquakes": a variation of the mechanical earthquake model depicted in Figure 20 proposed by the author.

"secondary" magnets are submitted to a frictional force against the lower acrylic plate, avalanches can appear when the magnetic force exerted on one of these magnets surpasses the frictional threshold. The present author has performed some preliminary measurements with a very limited amount of magnets in this schedule, and slip events spanning two orders of magnitude in are easily achieved. The displacements of the secondary magnets can be measured by analyzing images recorded by a CCD camera or, even better, using highly-linear, contactless capacitive sensors. Would this system show SOC, or it will be suppressed by the nature of the magnetic forces and the strong inertia of the magnets?

We have purposely represented in Figure 22 not a squared, but an hexagonal arrangement of "primary" magnets. In fact, this set-up has many points of contact with a system of superconducting vortices piercing a pair of paralell superconducting films separated by an insulating layer, a system studied by Ivar Giaever in his elegant 1966' "flux transformer" experiments [172], in which a transport current is applied to the "primary" superconducting film, and a the voltage drop is measured along the same direction on the "secondary" film. In this system, the field- and temperature-dependent vortex-vortex interactions in the x-y plane described by formula (10) would match the coil springs in the earthquake model, while the elastic moduli for the triangular vortex lattice c_{44} (tilt) and c_{66} (shear) – which are also functions of B and T [173,174,43,140]– would account for the leaf spring action. This means that a careful tuning of the applied field and temperature would allow to scan a continuum of "spring" characteristics. In our experiment, the transport current on the "primary" film would be ramped up at a certain rate, and the time integral of the voltage measured on the "secondary" film would give the eventual flux-avalanche behavior, in a way somewhat similar to that of the 1995' experiment by Field et al [52]. A computational model for this experiment will be drafted in the next section.

Our magnet experiment can be further enriched by including several layers of nonmagnetic plates with "secondary" magnets, and measure the resulting avalanches, for example, at the bottom layer. This would mimic the situation of a high T_c crystal with a magnetic field applied perpendicular to the ab planes of the structure: the magnets correspond to the "pancake" vortices introduced in section 2.1, provided we were working in a region of the H-V phase diagram were the pancake vortices have lost their Josephson coupling, and only the magnetic interaction remains. It should be underlined, though, that a more complex elastic theory of the flux line lattice is needed to describe such system [140].

Modern techniques allowing the fabrication of normal metal, semiconductor and superconductor micro- and nano-structures invite to expand further our analogies. For example, two-dimensional arrangements of islands of Al in the normal state interconnected by tunnelling junctions in the x-y plane and capacitively coupled between them and to a ground plane [175,176,177] give another reasonably good analogy with the earthquake model. Typical experiments involving these arrangements is the measure of I-V curves when the arrangement is excited by a voltage, provoking transport of charge from island to island, through tunnelling effects [175]. In this system, vortices may correspond to charges, quenched disorder (pinning) to random offset charges in the islands, and vortex interactions, to the rules governing the tunnelling process. It must be said, however, that there is an upper limit for the islands charges, differently from the vortex scenario, where the pinning strength is virtually unbounded [178]. A rough approximation of our vorticial earthquake experiment would be to apply a voltage *to the ground plane* (i.e., a pinningless medium in the vorticial system), and measure the "charge avalanches" *at the islands arrangement.*

6. Critical State and Cellular Automata

Computer simulations have become more and more popular in contemporary science. Hence, it is not surprising that the critical state has been extensively studied by means of *MonteCarlo simulations* (particularly in connection to flux creep phenomena, but also with avalanche dynamics) [179] and *molecular dynamics simulations, MD* [60,63]. The latter is potentially the ultimate weapon in vortex-state simulations, since it consists in following the dynamics of any set of interacting particles –vortices in our case– by integrating their equations of motion [180]. However, MD simulations have the important handicap that they usually require strong computational tools if big systems are to be tackled. *Cellular automata* models (CA), on the other hand, are simpler computational models that can grasp the physical essence of many physical situations with less computer power demand.¶ A CA is a dynamical system where space, time and the states of the system are discrete. In the CA, each *cell* of a regular spatial lattice can have only one of a finite number of states, which is updated according to a *local rule*, i.e., depending only of the state of the cell and

¶Bak and coworkers' model for SOC mentioned in section 3 was a CA!

their neighbors in the previous time step.

We are going in this section to concentrate in a particular cellular automaton model introduced by Bassler and Paczuski in 1998 [68] which simplicity allows the exploration of several features of the critical state, specially its avalanche behavior. The model represents the vortex dynamics on a two-dimensional simple hexagonal lattice. A lattice site situated at x has a pinning energy randomly chosen to be $V_{pin}(x) = p$ with probability q, and $V_{pin}(x) = 0$ with probability $1 - q$. Vortices are only allowed to exist at the lattice sites, each of which can contain $m(x)$ vortices. The upper section of Figure 23 shows an example accounting for a few lattice sites, which are represented by open circles, while the dots represent vortices. Vortices at site x of the figure can move to site y only if the "force" defined as

$$
\begin{aligned}
F_{x \to y} \quad &= \quad V_{pin}(y) - V_{pin}(x) + [m(x) - m(y) - 1] \\
&\quad + r[m(x1) + m(x2) - m(y1) - m(y2)]
\end{aligned}
\tag{31}
$$

is greater than zero. In the above expression, r is a measure of the effect of the x and y sites nearest neighbors, namely $x1$, $x2$, $y1$ and $y2$ (In the specific example of Figure 21, assuming $V_{pin}(x) = V_{pin}(y)$ and $r = 0.2$, we obtain $F_{xy} = 1$). All sites are updated in paralell, and only one vortex from a particular site is allowed to move at each update. Then, if $F_{x \to y} > 0$, one vortex jumps from x to y. If *two* directions are unstable when calculating the force for a given x, it jumps to one of them selected at random.

The increase of the external magnetic field is represented by the random addition of vortices to the system through its left boundary. Periodic boundary conditions are applied to the top and bottom of the lattice. The vortices reaching the right boundary are removed (they "fall out" of the system), and no vortices are allowed to abandon it through the left boundary.

Note that the different forces acting on vortices are not calculated through equations (14)-(17) –as expected from a MD approach–. Instead, they are represented by simple CA rules chosen with physical intuition. For example, the first two terms at the right hand of the force given by expression 31 indicate that the vortices tend to move from weak to strong pinning centers, while the rest of the terms indicate that they tend to move in the direction of the (local) flux gradient due to intervortex repulsion. A more subtle matter is to discuss if the model is able to deal with the viscous force associated to vortex movement, introduced in section 2. In fact, the intrinsic discreteness of the CA implies eventual vortex "stops" at a few "pinnningless" sites in its way from one pinning center to the next one: in my opinion, this somehow mimics the finite velocity of vortices during *flux flow*.

The lower part of Figure 23 displays a typical sequence of vortex density profiles for increasing values of the external field obtained through the model described. A relatively small density of sites has been selected in the example to illustrate the vortex density fluctuations along the profile, which reproduces accurately the predictions of Bean's critical state model, illustrated in Figures 6 and 7.

But the most exciting use of the CA model is the possibility to examine *how* the critical state is reached. For that purpose, Bassler and Paczuski excited their CA by adding a vortex to a random site at the left boundary of the system, resembling an increasing magnetic field in the outside. The lattice sites were then continually updated until no more unstable sites

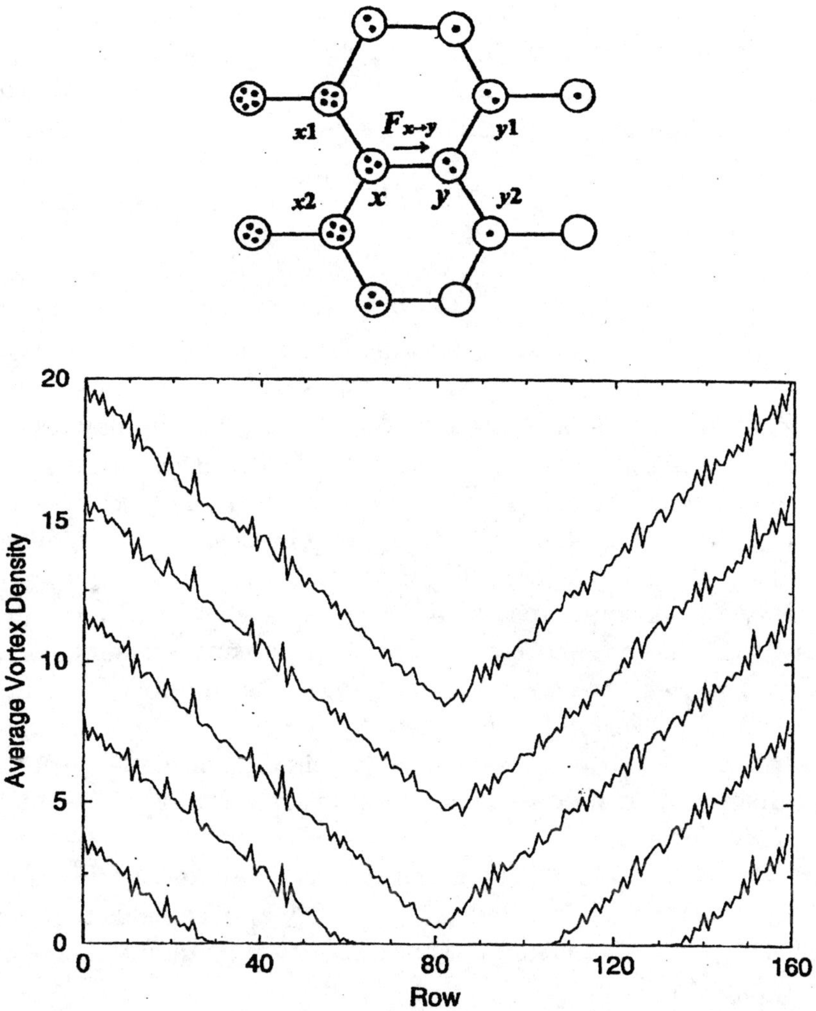

Figure 23: Illustration of the cellular automaton network proposed by Bassler and Paczuski [68] to simulate the critical state of type II superconductors. The open circles represent the network sites, and the black dots, vortices. Lower section: An example of critical state profiles for increasing magnetic field obtained through the cellular automaton.

persisted. This process was called a *vortex avalanche. The avalanche size was defined as the number of topplings following a vortex addition, while the avalanche duration was defined as the number of updatings necessary to complete one avalanche.* Once the lattice was again stable, a new vortex was added.

Figure 24a shows the times series of vortices falling off the right edge of an approximately squared system of $L = 200$ size, and parameters $r = 0.1$, $p = 5$ and $q = 0.1$, which qualitatively resembles the result of the 1995' experiment by Field et al [52] represented in Figure 13, although one must be aware of the fact that *flux* avalanches are not *directly* shown there. The following finite size scaling ansazt was use to analyze the avalanche sizes and durations distributions, respectively for systems of sizes $L = 60$, $L = 200$ and $L = 600$.

$$P(s, L) = s^{-\tau} f(s/L^D) \tag{32}$$

$$P(t, L) = t^{-\tau_t} f(t/L^z) \tag{33}$$

The data collapse –illustrated in Figure 24b,c– resulted in critical exponents $\tau = 1.63 \pm 0.02$, $D = 2.7 \pm 0.1$, $\tau_t = 2.13 \pm 0.08$ and $z = 1.5 \pm 0.1$. (note that no finite size scaling was necessary in [52] due to the huge size of the real vortex system). Similar results were obtained for a wide range of simulational parameters, indicating the robustness of the system, and SOC behavior. It must be pointed out that, even when many features between the experiment reported in [52] and these CA simulations are coherent, the model differs from the experiment in treating the vortices as pointlike particles, while real vortices are objects of unknown dimensionality, as previously discussed.

Bassler and Paczuski's CA has enjoyed a rapid popularity among the superconducting simulational community. Cruz and coworkers, for example [71] have applied the model to systems with a *periodic* distribution of pinning centers (which can be achieved experimentally with state-of-the-art ion bombardment techniques [181]), finding power law distributions of avalanche sizes and durations, as well as a new universality class for strong and dense periodic distributions of pinning centers.

Mohler and Stroud, on the other hand, used the CA to calculate the flux noise of a system analogous to that reported in [52], concluding that slowly applied fields produce "1/f" noise, while fast field ramps provoke a peak in the noise spectrum, in qualitative agreement with some experimental findings.

The experimentally inescapable problem of how thermal activation of vortices affect the avalanche dynamics has been tackled by Mulet and coworkers [73] by modifying the original CA with a MonteCarlo-type algorithm, which differs from the previous analytical [48,62,65] and MD [66] approaches to the subject. The central idea of Mulet and coworkers' model is that, for vortices in sites where $F_{x \to y} < 0$, there is still the possibility of motion with a "thermal" probability given by $P_{x \to y} \propto e^{-U(j)/kT}$, where $U(j)$ can follow different pinning models, and j is the local critical current density, which is proportional to the local gradient of the vortex density, as described in section 2.2.2. In the simulation, vortices are added as in [68] until the critical state is established, and then the system is allowed to relax via thermal activation. The simulations indicate power-distributed avalanches with critical exponents related to those found by Bassler and Paczuski in their original paper [68]. The results of these simulations are partially coherent with the relaxation experiments reported in reference 56. Both works agree in that power laws are lost as the temperature increase but, while in [73]

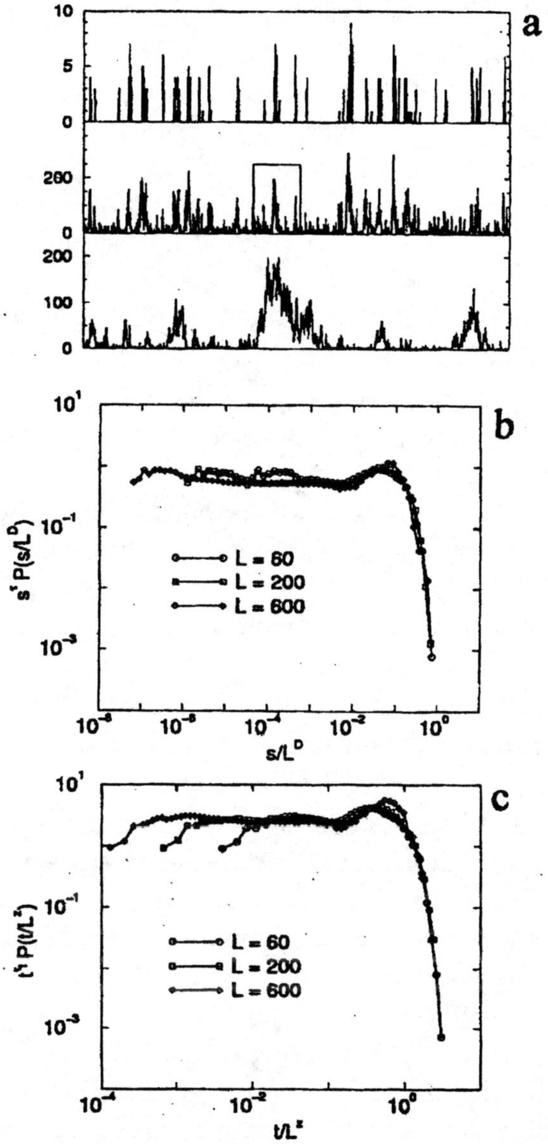

Figure 24: Top graph: temporal series of vortex avalanches reported by Bassler and Paczuski using the cellular automaton illustrated in Figure 23 [68]. Intermediate and lower graphs: data collapse after finite size scaling of the avalanche sizes and lifetimes distributions by the same authors.

many-sized avalanches are predicted from the first moments of the relaxation process, they are observed in reference 56 only at large relaxation times. The theoretical prediction in [73] comes form the fact that, in the model, a "perfect" critical state is assumed at $t = 0$. However, it must be stressed that Aegerter's experiments are performed with a SQUID [56], so only "off-the-edge" avalanches can be detected. On the other hand, its resolution is relatively small –as the author himself states.

A more straightforward application of Bassler-Paczuski's CA is the study of the effect of a transport current in the avalanche dynamics of the critical state. A basic approach to the subject is to assume that the effect of a transport current applied perpendicularly to the magnetic field on a thin slab (i.e., entering into the page in diagram 3 of Figure 7) is to "tilt" the field boundary conditions relative to the ones corresponding to the fully penetrated critical state at zero transport current [98]. Then , the effect of the transport current increase would be to add or release vortices asymetrically at the boundaries, and let the CA rules to "take charge" of the dynamics [182].

Curiously enough, most of the molecular dynamics simulations of vortices in the presence of a transport current assume a uniform Lorentzlike force acting on each vortex. The sandpile model for this would be the "rotating drum" experiment, in which a box of sand is slowly tilted [119]. This is not the approach followed in the CA model described in the last paragraph. In fact, its sandpile analogue would be to lower, for example, the right vertical wall of the box containing the "V" sandpile depicted in Figure 10a, as new sand is added at the left wall (which height would be increased to support the increasing left shoulder of the pile).

Finally, I will suggest some further qualitative ideas that may be implemented with the Bassler-Paczuski's CA, and may help the understanding of some properties of the critical state of type II superconductors.

A simple modification of the algorithm would be to establish a rule for the diminishing of the pinning potential as the vortex number at a given site increases, in order to resemble the field dependence of the critical current density in real superconductors. This should produce curved flux profiles, as those obtained, for example, from Kim's critical state model [45]. It would also eventually destroy SOC dynamics.

A less trivial modification of the algorithm would allow simulation of flux jumps. As we have seen above, they constitute an important experimental issue to rule out when the objective is to measure only the "dynamically driven" flux avalanches [53,54,58]. Including in the CA algorithm both kinds of flux avalanches may be useful to have an idea of their interaction. One qualitative approach of flux jumps could be to establish a maximum local rate of vortex topplings in the original CA algorithm (i.e., a maximum dissipation due to vortex movements that the sample can stand without increasing their temperature). If this maximum is reached, the pinning strengths of that region would be lowered by a certain factor (i.e., as the temperature increases due to excess flux motion, the local critical current density decreases). This would imply a "catastrophic" enter of flux into the sample (i.e., a flux jump). The values of the different parameters of the new automaton would represent those relevant to flux jumps: specific heat, thermal conduction, temperature dependence of the critical current density, etc. Observe that this idea does not imply a MC-flavored simulation of thermal activation like in [73].

Although surely much more computing-time consuming, some ideas can be suggested for approaching the *dimensionality* of real vortices with the use of Bassler-Paczuski CA. It is perhaps a variation of the "superconducting-earthquake" experiment proposed in the last section the one most easily reproducible with the CA. The key modification would be to use *two* paralell lattices instead of one, with the external field applied perpendicular to the lattices. Then, the force formula (32) should be modified to account for the interaction between vortices –or vortex sections– *located on different lattices. One simple possibility is to add a term expressing the "preference" of vortices to occupy sites on one network directly above (or below) occupied sites in the paralell plate.* A first step to implement such idea is to define the forces at the upper and lower networks, as:

$$\begin{aligned} F^u_{x\to y} &= V_{pin}(y)^u - V_{pin}(x)^u + [m(x)^u - m(y)^u - 1] \\ &\quad + r^u[m(x1)^u + m(x2)^u - m(y1)^u - m(y2)^u] - cm(x)^l \end{aligned} \tag{34}$$

$$\begin{aligned} F^l_{x\to y} &= V_{pin}(y)^l - V_{pin}(x)^l + [m(x)^l - m(y)^l - 1] \\ &\quad + r^l[m(x1)^l + m(x2)^l - m(y1)^l - m(y2)^l] - cm(x)^u \end{aligned} \tag{35}$$

where the superscripts "*u*" and "*l*" refer to the upper and lower networks, respectively, and *c is the inter-network or inter-layer coupling constant.* Firstly, we would examine the critical state dynamics at both networks by adding vortices at their boundaries. Once established the critical states –which should be quite similar at both networks– we would mimic the injection of an increasing transport current in the upper plate by "slowly tilting" *there* the boundary conditions, and examine the avalanche dynamics *at the lower plate.* The coupling constant, *c*, might be able to tune interesting *dynamic phases* in the style of [183]! . This simulation, however, best reproduces a flux transformer-like experiment *in which the superconducting films are substituted by infinite slabs*: it is well known that the vortex distribution of *films* in perpendicular applied fields can be extremely complex due to strong demagnetizing effects [145,184], and their modeling is eventually beyond the original Bassler-Paczuski CA model, at least close enough to the sample boundaries. However, the case of the pancake-vortex dynamics in layered supercondcutors might be easier to approach, since it basically consists in *many* paralell CA networks, which helps to get rid of extreme demagnetizing effects.

A further variation of the Bassler-Paczuski CA could be also applied to simulate charge transport in mesoscopic structures of the kind described at the end of section 4. The lattice sites would resemble very closely the metallic dots in which charges are confined, the vortices would now represent charges, and the pinning strengths would be substituted by the random offset voltages at the islands. Finally, the new force would depend on the potential difference between adjacent islands, which, in turn, would be a function of their charges. Then, if $V_i > V_j + constant$ [177], charge would be transferred from site i to site j. Otherwise, no transfer would take place. Even when these ideas are very preliminar, they could be refined to study the avalanche behaviour of such mesoscopic systems, which, as a matter of fact, has not been directly probed in the experiment.

After so many examples of how the road between sand and vortex piles can be followed in both directions –including *detours* to other systems–, we could state a final question: why not substitute vortices for grains, and use the Bassler-Paczuski CA for simulating avalanches in some sandpile experiments?

7. Conclusion

Experiments in vortex avalanches have demonstrated the validity of the sandpile mechanical analogue to understand some dynamic properties of the critical state in type II superconductors. How far this analogy can be exploited is still an open question. The temperature-dependent change in behavior of the superconducting critical state from power-law to peaked vortex size distributions observed in some experiments –possibly related to vortex inertia– is one of the most intriguing aspects of the problem. State-of-the-art experimental techniques for the measurement of the *local* magnetic induction have allowed, in the last years, to get a preliminary picture of the spatio-temporal correlations of vortex avalanches, but more work should be done, specially regarding the correlations along the direction paralell to the applied field. A variety of vortex-avalanche experiments exploring different geometries and the inclusion of transport currents is waiting for future efforts. Some experiments on sandpiles and magnet arrangements might also help to understand the dynamics of vortex avalanches. Cellular automata (CA) models –which have already shown their effectiveness in approaching vortex avalanches– constitute an elastic tool to tackle some of these problems, if local rules are carefully selected on a solid phenomenological background. There is no doubt that the zoology of "vorticial CA's" will grow fast in the next years.

Acknowledgments

I would like to thank the Editors for the liberty I enjoyed to choose the subject of this article and their continued encouragement, particularly from L.M. Gaggero. I am indebted to my former student and colleague R. Mulet for being my partner in the difficult task of tackling new scientific problems in a Third World environment along many years, and, specifically, for compulsive arguments on avalanche dynamics. I highly appreciate discussions with R. Cruz, A.J. Batista, O. Ramos, A. Vázquez, O. Sotolongo, K. Bassler, M. Paczuski, G. Reiter, A. Rimberg and Y. Paltiel; scientific advice from C.W. Chu, J.R. Clem, H. Herrmann and E. Zeldov, and experimental details provided by E. Nowak and K. Behnia. I also thank the critical reading of the manuscript by G. Reiter, R. Mulet and K.Bassler, as well as the latter and G. Gunaratne for cooperation in Figure 10. Material support during the last stage of this work is gratefully acknowledged to the ICSC-World Laboratory, the Physics Department at the University of Houston, and the Texas Center for Superconductivity. Thanks to my wife and family, writing this paper has not seriously damaged my mental health.

References

[1] J.C. Maxwell: "The theory of molecular vortices applied to statical electricity". *Philosophical Magazine*, January-February 1892

[2] H. Kammerlingh-Onnes : "Sur les résistances électriques". En la *Théorie du rayonnement et les quanta, rapports et discussions de la réunion tenue Bruxelles*, du 30

octobre au 3 novembre 1911 sous les auspices de M.E. Solvay, pgs 304-312

[3] For an excellent review, see D.R. Tilley and J. Tilley *Superfluidity and Superconductivity* (3rd. ed.), Adam Hilger, Bristol and New York, 1990

[4] J. Bardeen, L.N. Cooper and J.R. Schrieffer, *Phys. Rev.* **108** (1957) 1175-1204

[5] F. London and H. London, *Proc. Roy. Soc.* **A149** (1935) 71

[6] W. Meissner and R. Ochsenfeld, *Naturwiss* **21** (1933) 787

[7] R. Doll and M. Näbauer, *Phys. Rev. Lett.* **7** (1961) 51-52

[8] B.S. Deaver, Jr. and W.M. Fairbank, *Phys. Rev. Lett.*, **7** (1961) 43-46

[9] B.D. Josephson, *Phys. Lett.* **1** (1962) 251-253

[10] P.W. Anderson and J.M. Rowell, *Phys. Rev. Lett.* **10** (1963) 230-232

[11] P.W. Anderson, in *Lectures on the manybody problem*, Ravello, 1963 (E.R. Caianello, Ed.) vol. 2, Academic, 1964, p 113-135

[12] D.B. Sullivan and J.E. Zimmerman, *Am. J. Phys.* **45** (1974) 429-433

[13] V.L. Ginzburg and L.G. Landau, *Zh. Eksp. Teor.Phys.* **20** (1950) 1064

[14] V.L. Ginzburg and L.P. Pitaevskii, *Zh. Eksp. Teor. Fiz.* **34** (1958) 1240

[15] A.A. Abrikosov *Zh. Exsp. Teor. Fiz.* **32** (1957) 1442

[16] U. Essman and H. Träuber. *Phys. Lett.* **24A** (1967) 526

[17] T.P. Orlando and K.A. Delin *Foundations of Applied Superconductivity*, Addison-Wesley, 1991.

[18] L. Onsager, *Nuovo Cimento* **6** (1949) 249, H.E. Hall and W.F. Vinen, Phil. Mag. **46** (1955) 546

[19] M. Tinkham *Introduction to Superconductvity* McGraw-Hil, 1975

[20] V.A. Kozlov and A.V. Samokhvalov, *JETP Lett.* **53** (1991) 158

[21] V.A. Kozlov and A.V. Samokhvalov, *Physica C* **213** (1993) 103

[22] Y. A. Genenko, *Phys. Rev. B* **49** (1994) 6950

[23] R. Mulet and E. Altshuler, *Physica C* **252** (1995) 295-302

[24] F. Pérez-Rodríguez, A. Peérez- González, J.R. Clem, G. Gandolfini and M.A.R. LeBlanc, *Phys. rev. B* **56** (1997) 3473

[25] E. Altshuler and R. Mulet, *Physica C* **292** (1998) 39-47

[26] J. Bednorz and K.A. Müller, *Z. Phys.* **B64** (1986) 189

[27] M. Wu, J. Ashburn, C. Torng, O. Hor, R. Meng, L. Gao, Z. Huang and C.W. Chu, *Phys. Rev. Lett.* **58** (1987) 908

[28] H. Maeda, Y. Tanaka, M. Fukutomi, and T. Asano, *Jpn. J. Appl. Phys. Lett.* **27** (1988) 209

[29] Z. Sheng and A. Hermann, *Nature* **332** (1988) 55

[30] A. Schilling, M. Cantoni, J.D. Guo, H.R. Ott, *Nature* **363** (1993) 56

[31] L.N. Bulaevskii, Zh. Eksp. Teor. Fiz. **64** (1973) 2241

[32] K.B. Efetov, Zh. Eksp. Teor. Fiz. **76** (1979) 1781

[33] J.R. Clem and M.W. Coffey, Phys. Rev. **B42** (1990) 6209

[34] J.P. Carton, J. Phys. I (Paris) **1** (1991) 113

[35] S.N. Artemenko and A.N. Kruglov, *Phys. Lett.* **A 143** (1990) 485

[36] M.V. Feigel'man, V.B.Geshkenbein and A.I. Larkin, *Physica C* **167** (1990) 17

[37] A. Buzdin and D. Feinberg, *J. Phys.* (France) **51** (1990) 1971

[38] J.R. Clem, *Phys. Rev.* **B43** (1991) 7837

[39] K.H. Fischer, *Physica C* **178** (1991) 161

[40] S. Chakravarty, B.I. Ivlev and Yu.N. Ovchinnikov, *Phys. Rev. B* **42** (1990) 2143

[41] L.N. Bulaevskii, M. Ledvij and V.G. Kogan, *Phys. Rev. B* **46** (1992) 366

[42] J.Bardeen and M.J.Stephen, *Phys. Rev. A* **140** (1965) 1197

[43] A.M. Campbell and J.E. Evetts, *Adv. Phys.* **72** (1972) 199

[44] Ch. Bean, *Rev. Mod. Phys.* **30**, 31 (1964)

[45] Y.B. Kim, C.F. Hempstead and A.R. Strnad, *Phys. Rev. Lett.* **9** (1962) 306; P.W. Anderson and Y.B. Kim, *Rev. Mod. Phys.* **36** (1964) 39; Y.B. Kim, C.F. Hempstead and A.R. Strnad, *Rev. Mod. Phys.* **36** (1964) 43

[46] W.A. Fietz, M.R. Beasley, J. Silcox and W.W. Webb, *Phys. Rev. A* **136** (1964) 335

[47] See, for example: A.M. Campbell and J.E. Evetts, *Adv. Phys.* **72** (1972) 199; H. Ullmaier *Irreversible properties of type II superconductors*, Springer, Berlin, 1975; K. Tachikawa, K. Kitazawa, H. Maeda and T. Mstsushita (eds.) *Critical state in superconductors*, World Scientific, 1995.

[48] X.S. Ling, D. Shi and J.I. Budnik, *Physica C* **185-189** (1991) 2181

[49] Z. Koziol, P.F. de Chatel, J.M. Franse, Z. Tarnawski and A.A. Menovsky, *Physica C* **212** (1993) 133

[50] G.T. Seidler, C.S. Carrillo, T.F. Rosenbaum, U. Welp, G.W. Crabtree and V.M. Vinokur, *Phys. Rev. Lett.* **70** (1993) 2814

[51] P. Liderer, J. Boneberg, P. Brüll, V. Bujok and S. Herminghaus, *Phys. Rev. Lett.* **71** 2646; C.A. Durán, P.L. Gammel, R.E. Miller and D.J. Bishop, in *Critical state in superconductors*, K. Tachikawa, K. Kitazawa, H. Maeda, T. Matsushita (eds.), World Scientific, 1995 (p. 16); B-U. Runge, U. Bolz, J. Eisenmenger and P. Leiderer, paper presented at M2S-HTSC-VI, Houston, Texas, feb. 20-25, 2000.

[52] S. Field, J. Witt and F. Nori, *Phys. Rev. Lett.* **74**(1995) 1206

[53] R.J. Zieve, T.F. Rosenbaum, H.M. Jaeger, G.T. Seidler, G.W. Crabtree and U. Welp, *Phys. Rev. B* **53** (1996) 11849

[54] E.R. Nowak, O.W. Taylor, L. Liu, H.M. Jaeger and T.J. Selinder, *Phys. Rev. B* **55** (1997) 11702

[55] K. Kawashima, *Phys. Rev. B* **58** (1998) 490

[56] C.M. Aegerter, *Phys. Rev. E* **58** (1998) 1438

[57] P. Esquinazi, A. Setzer, D. Fuchs, Y. Kopelevich, E. Zeldov and C. Assmann, *Phys. Rev. B* **60** (1999) 12454

[58] K. Behnia, C. Capan, D. Mailly and B. Etienne. cond-mat/9902334 (25 feb 1999)

[59] P.G. DeGennes *Superconductivity of metals and alloys.* Benjamin, New York, 1996.

[60] O. Pla and F. Nori, *Phys. Rev. Lett.* **67** (1991) 919

[61] X.S. Ling and J.I. Budnik, in *Magnetic susceptibility of superconductors and other spin systems* ed. R.A. Hein, T.L. Francavilla and D.H. Liebenberg, Plenum, New York, 1991 (p. 377)

[62] C. Tang, *Physica A* **194** (1993) 315

[63] R.A. Richardson, O. Pla and F. Nori, *Phys. Rev. Lett.* **72** (1994) 1286

[64] C. Reichhardt, C.J. Olson, J. Groth, S. Field and F. Nori, *Phys. Rev. B* **52** (1995) 10441

[65] E. Bonabeau and P. Lederer. *Phys. Rev. B* **52**(1995) 494

[66] S. Spencer, H.J. Jensen, *Physica C* **264** (1996) 95

[67] R. Mulet and E. Altshuler, *Physica C* **281** (1997) 317

[68] K.E. Bassler and M. Paczuski, *Phys. Rev. Lett.* **81** (1998) 3761

[69] R. Prozorov and D. Giller, cond-mat/9901344v2

[70] G. Mohler and D. Stroud, *Phys. Rev. B* **60** (1999) 9738

[71] R. Cruz, R. Mulet and E. Altshuler, *Physica A* **275** (1999) 15

[72] K.E. Bassler, M. Paczuski and G. Reiter, *Phys. Rev. Lett.* (15 November, 1999)

[73] R. Mulet, R. Cruz and E. Altshuler, preprint (2000)

[74] Y.B. Kim, C.F. Hempstead and A.R. Strnad, *Phys. Rev. Lett.* **9** (1962) 306

[75] P.W. Anderson, *Phys. Rev. Lett.* **9** (1962) 309

[76] P.W. Anderson and Y.B. Kim, *Rev. Mod. Phys.* **36** (1964) 39

[77] E. Altshuler et al, unpublished (1990)

[78] M.R. Beasley, R. Labush and W.W. Webb, *Phys. Rev.* **181** (1969) 682

[79] M.V. Feigel'man, V.B. Geshkenbein, A.I. Larkin and V.M. Vinokur, *Phys. Rev. Lett.* **63** (1989) 2303

[80] E. Zeldov, N.M. Amer, G. Koren, A. Gupta, R.J. Gambino and M.W. McElfresh, *Phys. Rev. Lett.* **62** (1989) 3093

[81] Y. Yeshurum and A.P. Malozemoff, *Phys. Rev. Lett.* **60** (1988) 2202

[82] Y. Yeshurum, A.P. Malozemoff and A. Shaulov, *Rev. Mod. Phys.* **68** (1996) 911

[83] E. Altshuler, R. Cobas, A.J. Batista-Leyva, C. Noda, L.E. Flores, C. Martínez and M.T.D. Orlando, *Phys. Rev. B* **60** (1999) 3673

[84] B.B. Goodman and M. Wertheimer, *Phys. Lett.* **18** (1965) 236

[85] S.H. Goedmoed, C. van Kolmeschate, J.W. Metselaar and D. DeKlerk, *Physica* **31** (1965) 537

[86] S.H. Goedmoed, C. van Kolmeschate, P.H. Kes and D. DeKlerk, *Physica* **32** (1966) 1183

[87] P.S. Swartz and C.P. Bean, *J. Appl. Phys.* **39** (1968) 4991

[88] J.C. Levet, M. Potel, P. Gougeon and H. Noel, *Nature* **331**(1988)307

[89] J.L. Tholence, H. Noel, J.C. Levet, M. Potel and P. Gougeon, *Solid State Commun.* **65** (1988) 1131

[90] M. Guillot, M. Potel, P. Gougeon, H. Noel, J.C. Levet, G. Chouteau and J.L. Tholence, *Phys. Lett.* **A 127** (1988) 363

[91] M. Guillot, J.L. Tholence, O. Laborde, M. Patel, P. Gougeon, H. Noel and J.C. Levet, *Physica C* **162-164** (1989) 361

[92] K. Chen, S.W. Hsu, T.L. Chen, S.D. Lan, W.H. Lee and P.T. Wu, *Appl. Phys. Lett.* **56** (1990) 2675

[93] K. Chen, Y.C. Chen, S.W. Hsu, W.H. Lee, and P.T. Wu, *Physica C* **173** (1991) 227

[94] A. Gerber, J.N. Li, Z.Z. Tarnawsky, J.J.M. Franse and A.A. Menovsky, *Phys. Rev. B* **49** (1993) 6047

[95] L. Gao, Y.Y. Xue, R.L. Meng and C.W. Chu preprint No. 93:076 (1993), Texas Center for Superconductivity.

[96] K.H. Müller and C. Andrikidis, *Phys. Rev. B* **49** (1994) 1294

[97] L. Legrand, I. Rosenman, R.G. Mints, G. Collin and E. Janod, *Europhys. Lett.* **34** (1996) 28

[98] M.N. Wilson *Superconducting magnets*, Clarendon Press, Oxford, 1983

[99] C. Coulomb, in *Memoir de Mathematique et de Physique*, Vol. 7, Academie de Sciences, L'Imprimerie Royale, Paris (1773), p.343

[100] M. Faraday, *Phil. Trans. R. Soc. London* **52** (1831) 299

[101] O. Reynolds, *Philos. Mag.* **20** (1885) 469

[102] D. Bideau and J.A. Dodds (eds.) *Physics of Granular Media*, Les Housches Series, Nova Science Publishers, 1991

[103] H.M. Jaeger and S.R. Nagel, *Science* **255** (1992) 1523

[104] R.P. Behringer, *Nonlinear Science Today* **3** (1993) 1

[105] D. Bideau and A. Hansen (eds.) *Disorder and Granular Media*, Random Materials and Processes Series, North Holland, New York, 1993

[106] H.M. Jaeger, J.B. Knight, C.H. Liu and S.R. Nagel, *Mater. Res. Bull.* **19** (1994) 25

[107] A. Mehta (ed.) *Granular Matter an Interdisciplinary Approach*, Springer-Verlag, New York, 1994

[108] A. Mehta and G.C. Barker, *Rep. Prog. Phys.* **57** (1994) 383

[109] R.P. Behringer, *Proc. MRS* **367** (1995) 461

[110] H.H. Hayakawa, H. Nishimori, S. Sasa and Y.H Taguchi, *Jpn. J. Appl. Phys.* **34** (1995) 397

[111] H.M. Jaeger, S.R. Nagel and R.P. Behringer, *Rev. Mod. Phys.* **68** (1996) 1259

[112] J. Duran *Physics of Granular Materials*, Springer Verlag, New York, 1999

[113] L.P. Kadanoff, *Rev. Mod. Phys.* **71** (1999) 435

[114] P.G. deGennes, *Rev. Mod. Phys.* **71** (1999) S374

[115] H.J. Herrmann, *Physica A*, **263** (1999) 51

[116] P. Bak, C. Tang and K. Wiesenfeld, *Phys. Rev. Lett.* **59**, 381 (1987); C. Tang and P. Bak, *Phys. Rev. Lett.* **60**, 2347 (1988); M. Paczuski and S. Boettcher, *Phys. Rev. Lett.* **77**, 111 (1996)

[117] J.J. Alonso and H.J. Herrmann, *Phys. Rev. Lett.* **76** (1996) 4911

[118] E. Altshuler, O. Ramos, C. Martínez, L.E. Flores and C. Noda, Preprint (1999)

[119] H.M. Jaeger, C.H. Liu and S.R. Nagel, *Phys. Rev. Lett.* **62** (1989) 40

[120] C.S. Campbell and C.E. Brennen, J. Fluid. Mech. **151** (1985) 167; S.F. Edwards and R.B.S. Oakeshott, *Physica A* **157** (1989) 1080; A. Mehta and S.F. Edwards, *Physica A* **157** (1989) 1091; H.M. Jaeger and S.R. Nagel, Science **255** (1992) 1523; H.J. Herrmann, in *Disorder and Granular Media* (Bideau and Hansen, eds.), North-Holland, Amsterdam, 1993 (pag. 305).

[121] J.C. Williams, Powder Techn. **15** (1976) 245; A. Rosato, K.J. Strandburg, F. Prinz and R.H. Swendsen, *Phys. Rev. Lett.* **58** (1987) 1038; *Powder Techn.* **48** (1986) 239; P.K. Haff and B.T. Werner, *Powder Techn.* **49** (1986) 59; P. Devillard, *J. Physique* **51** (1990) 369; R. Jullien, P. Meakin and A. Pavlovitch, *Phys. Rev. Lett.* **69** (1992) 640;

[122] P. Bak "How *Nature* works", Springer Verlag, New York, 1996

[123] L.P. Kadanoff, S.R. Nagel, L. Wu and S. Zhou, *Phys. Rev. A*, **39**, 6524 (1989)

[124] See, for example, Wilde and Singh *Statistical Mechanics*, Wiley, New York, 1998.

[125] G.A. Held, D.H. Solina, D.T.Keane, W.J. Haag, P.M. Horn and G. Grinstein, *Phys. Rev. Lett.* **65**, 1120 (1990)

[126] M. Bretz, J.B. Cunningham, P.L. Kurezynski and F. Nori, *Phys. Rev. Lett.* **69** (1992) 2431

[127] P. Evesque, *Phys. Rev. A* **43** (1991) 2720

[128] J.Rosendahl, M. Vekic and J. Kelley, *Phys. Rev. Lett.* **47**, (1993) 1401

[129] P. Evesque, D. Fargaix, P. Habib, M.P. Luong and P. Porion, *Phys. Rev. E* (1993) 2326

[130] S.K. Grumbacher, K.M. McEwen, D.A. Halverson, D.T. Jacobs and J. Linder, *Am. J. Phys.* **61**, (1993) 329

[131] J. Rodendahl, M. Vekic and J.E. Rutledge *Phys. Rev. Lett.) (***73** (1994) 537

[132] V. Frette, K.Christensen, A. Malthe-Søslashrensen, J. Feder, T. Jossang and P. Meakin, *Nature*, **379**, 49 (1996)

[133] K. Christensen, A. Corral, V. Frette, J. Feder and T. Jøssang, *Phys. Rev. Lett.* **77** (1996) 107

[134] Y.B. Kim, C.F. Hempstead and A.T. Strnad. *Phys. Rev.* **131**(1963)2468

[135] C. Heiden and G.I. Rochlin, *Phys. Rev. Lett.* **21** (1968) 691

[136] C.R. Wischmeyer, *Phys. Lett.* **19** (1965) 543-545

[137] A.I. Larkin and Yu.N. Ovchinnikov, *Zh. Eksp.Teor.Fiz.* **65** (1973) 1704

[138] A.I. Larkin and Yu.N. Ovchinnikov, *J. Low Temp. Phys.* **34**(1979)409-427

[139] G. Blatter, V.B. Geshkenbein and V.M. Vinokur, *Phys. Rev. Lett.* **66** (1991) 3297

[140] G. Blatter, M.V. Feigel'man, V.B. Geshkenbein, A.I. Larkin and V.M. Vinokur, *Rev. Mod. Phys.* **66** (1993)1125

[141] R.J. Donnelly, *Cryogenics* in Physics Vade Mecum, edited by H.L. Anderson. AIP, 1981.

[142] M. Konczykowsky, L.Burchalov, Y. Yeshurun and F. Holtzberg, *Supercond. Sci. Tech.* **4** (1991) S33

[143] N. Chikumoto, M. Konczykowsky, N. Motohira, and A.P. Malozemoff, *Phys. Rev. Lett.* **69** (1992) 1260

[144] M. Darwin, J. Deak, L. Hou, M. McElfresh, E. Zeldov, J.R. Clem and M. Idenbom, *Phys. Rev. B* **48** (1993) 13192

[145] E. Zeldov, A.I. Larkin, V.B. Geshkenbein, M. Konczykowsky, D. Majer, B. Khaykovich, V.M. Vinokur and H. Shtrikman, *Phys. Rev. Lett.* **73** (1994) 1428

[146] Y. Abufalia, A. Shaulov, Y. Wolfus, R. Prozorov, L. Burlachkov, Y. Yeshurun, D. Majer, E. Zeldov and V.M. Vinokur, *Phys. Rev. Lett.* **75** (1995) 2404

[147] Y. Abufalia, D. Giller, Y. Wolfus, A. Shaulov, Y. Yeshurum, D. Majer, E. Zeldov, J.L. Peng and R.L. Greene, *J. Appl. Phys.* **81** (1997) 4944

[148] Y. Abufalia, M. McElfresh, A. Shaulov, Y. Yeshurum, Y. Paltiel, D. Majer, H. Strikman and E. Zeldov, *Appl. Phys. Lett.* **72** (1998) 2891

[149] Y. Abufalia, Y. Wolfus, M. McElfresh, A. Shaulov, Y. Yeshurum, Y. Paltiel, H. Shtrikman and E. Zeldov, *J. Appl. Phys.* **85** (1999) 5471

[150] R. Busch, G. Ries, H. Werthner, G. Kreiselmeier and G. Saemann-Ischenko, *Phys. Rev. Lett.* **69** (1992) 522

[151] H. Safar, P.L. Gammel, D.J. Bishop, D.J. Mitzi and A. Kapitulnik, *Phys. Rev. Lett.* **68** (1992) 2672

[152] H. Safar, P.L. Gammel, D.A. Huse, S.N. Majumdar, L.F. Schneemeyer, D.J. Bishop, D. López, G. Nieva and F. de la Cruz, *Phys. Rev. Lett.* **72** (1994) 1272

[153] Y.M. Wan, S.E. Hebboul and J.C. Garland, *Phys. Rev. Lett.* **72** (1994) 3867

[154] D. López, G. Nieva, F. de la Cruz, H.J. Jensen and D. O'Kane, *Phys. Rev. B* **50** (1994) 9684

[155] F. de la Cruz, D. López and G. Nieva, *Phil. Mag. B* **70** (1994) 773

[156] Yu. Eltsev, W. Holm and Ö. , *Phys. Rev. B* **49** (1994) 12333; *Physica C* **235-240** (1994) 2605

[157] Yu. Eltsev and Ö. Rapp, *Phys. Rev. B* **51** (1995) 9419

[158] T.S. Lee, N. Missert, L.T. Sagdahl, J.R. Clem, K. Char, J.N. Eckstein, D.K. Fork, L. Lombardo, A. Kapitulnik, L.F. Scheemeyer, J.V. Waszczak and R.B. Van Dover, *Phys. Rev. Lett.* **74** (1995) 2796

[159] S.S. Manna and D.V. Khakdhar, *Phys. Rev. E* **58** (1998) 39-74

[160] E. Nowak, private communication

[161] P.H. Kes, in J. Evetts (ed.) *Concise Encyclopedia of Magnetic and Superconducting Materials*, Pergamon Press, 1992.

[162] K. Behnia, private communication.

[163] B. Gutenberg and C. Richter *Seismicity of the Earth.* Princeton Univerity Press, Princeton, 1949

[164] R. Burridge and L. Knopoff *Bull. Seis. Soc. Am.* **57** (1967) 341

[165] P. Bak and C. Tang, *J. Geophys. Res. B* **94** (1989) 15635

[166] A. Sornette and D. Sornette, *Europhys. Lett.* **9** (1989) 197

[167] K. Ito and M. Matsuzaki, *J. Geophys. Res. B* **95** (1990) 6853

[168] Z. Olami, H.J. Feder and K. Christensen, *Phys. Rev. Lett.* **68** (1992) 1244

[169] H. Leschhorn, *Physica A* **195** (1993) 324

[170] H.J. Feder and J. Feder, *Phys. Rev. Lett.* **66** (1991) 2669

[171] J.M Carlson and J.S. Langer, *Phys. Rev. Lett.* **62**(1989) 2632

[172] I. Giaever, *Phys. Rev. Lett.* **15** (1965) 825

[173] R. Labush, *Phys. Stat. Solidi. (b)* **32** (1969) 439

[174] E.H. Brandt, *Phys. Stat. Solidi (b)* **35** (1969) 1027; **36** (1969) 371

[175] C. Kurdak, A.I. Rimberg, T.R. Ho and J. Clarke, *Phys. Rev. B* bf 57 (1998) 6842

[176] A.J. Rimberg, T.R. Ho and J. Clarke, *Phys. Rev. Lett.* **74** (1995) 4714

[177] A.A. Middleton and N.S. Wingreen, *Phys. Rev. Lett.* **71** (1993) 3198

[178] A.J. Rimberg, personal communication (1999)

[179] See, for example, G. Carneiro, *Phys. Rev. B* **50** (1994) 6982; R. Mulet and E. Altshuler, Phys. Stat. Solidi (b) **182** (1994) K31; R. Mulet and E. Altshuler, *Physica C* **281** (1997) 317

[180] J.M. Haile *Molecular dynamics simulations*, Wiley, New York, 1992.

[181] K. Harada, O. Kamimura, H. Kasai, T. Matsuda, A. Tonomura. V.V. Moshchalkov, *Science* **74** (1996) 1167

[182] K. Bassler et al, unpublished (1999)

[183] C. Reichhardt, C.J. Olson and F. Nori, cond-mat/9703021 (1997)

[184] M. Benkraouda and J.R. Clem, *Phys. Rev. B* **53** (1996) 5716

Study of Many Interfaces and Inhomogenous Systems by Meams of the Surface Green Function Matching and the Transfer Matrix Methods

V. R. Velasco *

Abstract

We show that the joint use of the surface Green function matching method and the full transfer matrix technique simplifies greatly the study of continuous media having many non-equivalent interfaces and of inhomogeneous media. We study, as applications, the transverse elastic waves in finite Fibonacci superlattices and in wells (A-B-A systems) having a parabolic variation of their elastic properties.

1. Introduction

Problems involving significant inhomogeneities are very interesting and common. These may arise, for instance, in epitaxially grown structures like graded wells, quantum wells with parabolic or other z-dependent profiles, n-i-p-i superlattices, z-dependent image potential wells, etc. These man-made structures are complemented by some natural layered structures like the ocean and the athmosphere, with a marked horizontal stratification[1]. Many of these problems can be envisaged as an $A - B - C$ configuration where the part B is the "difficult" or inhomogeneous one, A and C being the homogeneous ones. Different methods can be used to study these situations[1,2], but on all of them the matching at different interfaces must be considered. The surface Green function matching (SGFM) method[3], that involves the Green functions of the constituent media, is a very efficient method to deal with homogeneous systems with many interfaces[4,5]. It has been shown that the combination of the SGFM method with the transfer matrix technique as presented in[6] is a very efficient way of studying inhomogenous systems [7–10]. These applications have been concerned with the electronic properties of inhomogeneous systems, described by means of simple models.

The vibrational properties of inhomogeneous systems have been also studied by using different methods[1,11]. In all these cases is quite common to solve the problem by dividing the medium into a number of layers in which the material properties are assumed to be constant. In this way it is possible to get a very reasonable approximation to practically any real situation. The price to be paid is the increase in the number of interfaces to be considered.

Another field of research in which systems with many interfaces are present is that of aperiodic low-dimensional systems[12,13]. These aperiodic systems, having a structural order described by means of deterministic sequences, can be seen as an intermediate case

*Instituto de Ciencia de Materiales, CSIC Cantoblanco, 28049 Madrid, Spain

between periodic and disordered one-dimensional (1D) solids. They have exotic properties not present in crystalline and amorphous solids. Among the different quasi-periodic models the Fibonacci system is the 1D version of the quasi-crystals, and it has been the subject of several theoretical studies[14–18] as a model of the quasi-crystals. Fibonacci superlattices have been grown[13,19] by employing GaAs and AlAs slabs as generating blocks. The Raman spectra of the longitudinal acoustic modes of these superlattices were obtained[13] and some studies of the vibrational properties of the 1D systems were presented in[14–18]. Some studies on the elastic waves in Fibonacci superlattices have been done in the last years[19–23]. Quite recently the study of the sagittal waves in Fibonacci superlattices[24] and of the purely transverse elastic waves propagating along symmetry directions of piezoelectric Fibonacci superlattices[25] have been presented. These studies[24,25] have been done by using an extension of the SGFM method[3] to deal with structures having $N > 2$ non-equivalent interfaces[4]. When the systems studied have a high number of interfaces the dimension of the system Green function is also high and this poses strong requirements on the computer time and memory needed to solve the problem.

The transfer matrix technique[6], when combined with the SGFM method can reduce these problems. We shall present in Section 2 a brief description of the SGFM and the transfer matrix methods for the case of $N > 2$ non-equivalent interfaces. In Section 3 we shall present some results for the case of the transverse elastic waves in Fibonacci superlattices. In Section 4 we shall consider the case of transverse elastic waves in inhomogeeous systems, and conclusions will be presented in Section 5.

2. The SGFM method and the full transfer matrix technique

In the simplest heterostructures one has to solve a matching problem at one (single heterojunction) or two (quantum well or superlattices) physically distinct interfaces. In the case of most complicated systems of physical interest one can be faced with the matching at a larger number N of different interfaces.

In the SGFM method the extension from one to two interfaces is made by defining the simultaneous bi-projection on the two-surface domain[3], which is the entire interface domain when matching is effected. For a given bulk medium the Green function is a matrix of order n, depending on the model or the physical problem under study. Then the bi-projection is of order $2n$, i.e. a 2×2 supermatrix in which each element is a $n\times n$ matrix. The generalization to an arbitrary number N of interfaces for the continuum case has been given in[4] and we shall present here only the elements needed for our problem.

The unit projectors on the different interfaces are denoted by i_j. Each constituent medium has an extended *bulk* Green function G_j. We stress that each G_j is the Green function of an *extended pseudomedium*[3,7] which satisfies the same differential equations in domain j and arbitrary boundary conditions at its interfaces. In simple cases these can be just the G_j of the regular bulk media j, but in some cases it may be convenient in practice to use the flexibility provided by the arbitrary choice of boundary conditions in order to effect some practical simplifications[3,7–10]; we stress that all the G_j are *bulk* Green functions. The purpose of the analysis is to match all these Green functions at the N interfaces and thus obtain the Green function G_s of the full system in terms of the given G_j. We must note that in general we shall have on the left a semi-infinite medium L with bulk Green function G_L and on the right a semi-infinite medium R with with bulk Green function G_R. Then we have N interfaces and $N+1$ physically different media, at least in principle. It is convenient, as in the case of the quantum well[3], to define formally domain 1 to consist of

L/R on the left/right. Then medium 1 is, by definition, medium L/R on the left/right and G_1 is G_L/G_R in L/R. We then have formally N interfaces and N media.

A generic domain j is bounded by the interfaces $\imath_{j-1}$ and $\imath_j$. We define the partial interface projector

$$\mathcal{I}_j = \imath_{j-1} + \imath_j, \tag{1}$$

which spans the space of 2×2 supermatrices. Then we define, out of G_j, the projection

$$\tilde{G}_j = \mathcal{I}_j G_j \mathcal{I}_j. \tag{2}$$

The total interface projector is given by

$$\mathcal{I} = \sum_{j=1}^{N} \imath_j. \tag{3}$$

In the matrix representation $\mathcal{I}$ is the $N \times N$ unit supermatrix matrix, i.e., all diagonal elements are equal to 'unity', which simply means the number 1 if $n = 1$ (a scalar problem, like the transverse elastic waves equation) or the $n \times n$ unit matrix if $n > 1$. We shall henceforth denote this as 'unity' without need for further explanation. Thus, in the supermatrix format of $\mathcal{I}$, the partial projector $\imath_j$ has 'unity' in the j-th diagonal element and zero elsewhere and $\mathcal{I}_j$ has 'unity' in the $(j-1)$-th and j-th diagonal elements and zero elsewhere. In the space of $\mathcal{I}_j$ the projection $\tilde{G}_j$ is the 2×2 supermatrix

$$\tilde{G}_j = \left\| \begin{matrix} < \imath_{j-1}|G_j|\imath_{j-1} > & < \imath_{j-1}|G_j|\imath_j > \\ < \imath_j|G_j|\imath_{j-1} > & < \imath_j|G_j|\imath_j > \end{matrix} \right\| \equiv \mathcal{I}_j \tilde{G}_j \mathcal{I}_j. \tag{4}$$

When this is represented in the $N \times N$ supermatrix format of $\mathcal{I}$, the four nonvanishing elements of $\mathcal{I}\tilde{G}_j\mathcal{I}$ are those of (4) in the positions labelled $(j-1, j-1), (j-1, j), (j, j-1)$ and (j, j) as row and column indices of the $N \times N$ supermatrix. We note that with this notation $\tilde{G}_1$ is a particular case. In the space of $\mathcal{I}_1$ (which, by definition, is $\imath_1 + \imath_N$) this is the 2×2 diagonal supermatrix

$$\tilde{G}_1 = \left\| \begin{matrix} < \imath_1|G_L|\imath_1 > & 0 \\ 0 & < \imath_N|G_R|\imath_N > \end{matrix} \right\| \equiv \mathcal{I}_1 \tilde{G}_1 \mathcal{I}_1. \tag{5}$$

Any G_j for $j > 1$ is a propagator which can propagate from $\imath_{j-1}$ to $\imath_j$, or vice versa, through a medium j. The topology of domain 1 is different, as it consists of two disconnected subdomains. The block diagonal form of (5) corresponds to the fact that there is no propagation from, say, $\imath_1$ to $\imath_N$ with propagator G_L or vice versa. Thus when $\tilde{G}_1$ is represented in the $N \times N$ supermatrix format of $\mathcal{I}$, it has the nonvanishing elements (1,1) and (N, N) being zero all the rest.

Now in the final results all intervening elements are put together in the large $N \times N$ supermatrix format, but many algebraic steps, *including matrix inversions*, are carried out in the smaller space of 2×2 supermatrices. Obviously the large matrix $\mathcal{I}\tilde{G}_j\mathcal{I}$ cannot be inverted, but the small matrix $\mathcal{I}_j\tilde{G}_j\mathcal{I}_j$ can, in the space of $\mathcal{I}_j$ and, after inversion, the result can be put in its place in the large $N \times N$ supermatrix format. Symbols like $\tilde{G}_j^{-1}$ will be often used in this sense. Strictly speaking this is to be understood as $\mathcal{I}_j\tilde{G}_j^{-1}\mathcal{I}_j$ so that, by definition,

$$\mathcal{I}_j\tilde{G}_j^{-1}\mathcal{I}_j \cdot \mathcal{I}_j\tilde{G}_j\mathcal{I}_j = \mathcal{I}_j\tilde{G}_j^{-1} \cdot \tilde{G}_j\mathcal{I}_j = \mathcal{I}_j \tag{6}$$

which can be abbreviated as

$$\tilde{G}_j^{-1} \cdot \tilde{G}_j = \mathcal{I}_j \tag{7}$$

It is understood that there is everywhere a dependence on the eigenvalue variable and on the 2D wavevector $\boldsymbol{\kappa}$ introduced by the 2D Fourier transform, so that only the dependence on the position variable z is explicitly displayed. A symbol like $< i_{j-1}|G_j|i_j >$ denotes the value of $< z|G_j|z' >$ when z is at the position corresponding to the $(j-1)$-th interface and z' at that of the j-th interface. We shall indicate by z_j that z is in the domain j, so that $< z_j|G|z_k >$

when $j \neq k$ is zero if G is any of the G_m, but is just what we want to obtain if G is G_s. Again, $< z_1|G_1|z_1' >$ requires special consideration, depending on whether 1 means L or R. Thus

$$\begin{aligned} < z_L|G_1|z_L' > &= < z_L|G_L|z_L' >; \\ < z_L|G_1|z_R' > &= 0 = < z_R|G_1|z_L' >; \\ < z_R|G_1|z_R' > &= < z_R|G_R|z_R' > . \end{aligned} \tag{8}$$

Then we define row N-supervectors in the following way:

$$< z_j|G_j|\mathcal{I}_j >= [0,0,\ldots,0,< z_j|G_j|i_{j-1} >,< z_j|G_j|i_j >,0,\ldots,0] \tag{9}$$

It is necessary to take into account the following: (i) The former equation is $< z_j|G_j|\mathcal{I}_j >$ when put in the large format of N-supervectors. In the space of $\mathcal{I}_j$ it would be only a 2-supervector. (ii) In the same way we define the column N-supervector ,$< \mathcal{I}_j|G_j|z_j' >$ with the two nonvanishing components ,$< i_{j-1}|G_j|z_j' >$ and ,$< i_j|G_j|z_j' >$. As particular cases we have again

$$\begin{aligned} < z_L|G_1|\mathcal{I}_1 >=< z_L|G_L|\mathcal{I}_1 > &= [< z_L|G_1|i_1 >,0,\ldots,0,0]; \\ < z_R|G_1|\mathcal{I}_1 >=< z_R|G_R|\mathcal{I}_1 > &= [0,0,\ldots,0,< z_R|G_R|i_N >]. \end{aligned} \tag{10}$$

Performing the SGFM analysis we arrive finally to the expressions:

$$< z_j|G_s|z_j' >=< z_j|G_j|z_j' > + < z_j|G_s|\mathcal{I}_j' > \cdot\tilde{G}_j^{-1} \cdot (\tilde{G}_s - \tilde{G}_j) \cdot \tilde{G}_j^{-1} \cdot < \mathcal{I}_j|G_j|z_j' >, \tag{11}$$

and

$$< z_j|\tilde{G}|z_k' >=< z_j|G_j|\mathcal{I}_j > \cdot\tilde{G}_j^{-1} \cdot < \mathcal{I}_j|\tilde{G}_s|\mathcal{I}_k > \cdot\tilde{G}_k^{-1} \cdot < \mathcal{I}_k|\tilde{G}_k|z_k' > . \tag{12}$$

The results (11) and (12) are the formal extension to the N-interface case of the standard SGFM results for $N = 2^3$. If we can obtain $\tilde{G}_s$, then (11) and (12) provide all the $< z|G_s|z' >$ elements for z and z' anywhere. In order to do this it is necessary to impose the matching conditions at the interfaces.

The details of the matching formula for $\tilde{G}_s$ depend on the matching conditions to be imposed, which in turn depend on the physical problem under study. In the general case the matching conditions require the continuity of a linear differential form[3,4]

$$\mathcal{A} = -\tilde{S}^{-1} \cdot' G + \tilde{P} \cdot \tilde{G}, \tag{13}$$

where $\tilde{S}$ and $\tilde{P}$ depend on the specific problem under study. The detailed SGFM analysis is given in[4] and will be omitted here for brevity's sake. After this analysis we obtain the general matching formula as

$$\tilde{G}_s^{-1} = -\mathcal{I}\left(\mathcal{A}_1^{-1} \cdot \tilde{G}_1^{-1} - \sum_{j=2}^{N} \tilde{\mathcal{A}}_j^{-1} \cdot \tilde{G}_j^{-1}\right)\mathcal{I}. \tag{14}$$

The case of a polytype superlattice can be equally studied by combining this analysis with the introduction of the Bloch periodicity as in the binary superlattice[3]. In the former case of a sandwich structure, the extremes of the system are open and L and R are semi-infinite, and therefore there are no cross terms between L and R. The off-diagonal elements, or blocks, of $'\tilde{G}_1$, or $\tilde{\mathcal{A}}_1$, vanish and these supermatrices are block-diagonal. In the superperiodic structure[4] and due to the Bloch periodicity the interfaces i_N and i_0 are physically identical, as are i_{N+1} and i_1. Then the amplitudes at i_N/i_{N+1} are equal to those at i_0/i_1 except for the phase factor $f = \exp(iqd)$, where q is the superwavevector associated with the superperiod d. Then domains L and R appear as mathematically linked with the appropriate phase factor and the off diagonal elements of the supermatrices $\tilde{G}_1$, $'\tilde{G}_1$ and $\tilde{\mathcal{A}}_1$ no longer vanish.

In fact they become

$$\begin{aligned} \tilde{G}_1 &= \left\| \begin{array}{cc} <i_1|\mathcal{G}_1|i_1> & f^{-1}<i_1|G_1|i_0> \\ f<i_N|G_1|i_{N+1}> & <i_N|\mathcal{G}_1|i_N> \end{array} \right\|; \\ '\tilde{G}_1 &= \left\| \begin{array}{cc} <i_1|'\mathcal{G}_1^{(+)}|i_1> & f^{-1}<i_1|'G_1|i_0> \\ -f<i_N|'G_1|i_{N+1}> & -<i_N|'\mathcal{G}_1^{(-)}|i_N> \end{array} \right\|; \\ \tilde{\mathcal{A}}_1 &= \left\| \begin{array}{cc} <i_1|\mathcal{A}_1^{(+)}|i_1> & f^{-1}<i_1|\mathcal{A}_1|i_0> \\ -f<i_N|\mathcal{A}_1|i_{N+1}> & -<i_N|\mathcal{A}_1^{(-)}|i_N> \end{array} \right\|. \end{aligned} \tag{15}$$

The rest of the analysis would proceed in the same way as in the former case, and the matching formula is the same as that of (14) with the only difference that $'\tilde{G}_1$ or $\tilde{\mathcal{A}}_1$ are now given by (15).

It is then clear that in the case of many interfaces (N), $\tilde{G}_S$ will be a $Nn \times Nn$ matrix, n being the size of the G_j matrices, and it depends on the model we are studying. If Nn is a big number then our study will pose heavy requirements on computer memory and time to handle the inversion of big matrices, that must be performed many times.

Another possible approach to these problems is provided by the concept of the *transfer matrix*. This term is given different names in the literature. If one writes down the form of the function as a combination of some basis functions, often two plane waves of the same energy or frequency running in opposite directions, then the transfer matrix is defined[26,27] so that it transfers the coefficients of this combination from one point to another. There is a different concept of the transfer matrix which transfers the function itself and its derivative with respect to some coordinate from one point to another. A preliminary account of this transfer matrix was given in[28] for the case of constant potential. The mathematical properties of this transfer matrix were studied in detail in[6] with emphasis on the Schrödinger equation, although the formal theory is equally valid for any linear second-order differential equation or else for a set of coupled linear first-order differential equations.

The Green function and the transfer matrix approaches appeared to be unrelated. In fact the two concepts have an inherent difference. The transfer matrix makes no reference to boundary conditions while a Green function is not completely defined until these are specified.

In[7–10] it was proved that there was a relationship between these two approaches and that it was possible to turn it into a practical scheme for doing calculations based on a hybrid approach which uses the advantages of either side as convenience advises. This analysis can be found in a very complete way in[3] and it will not be repeated here. We shall present here only some expressions needed to understand the notation.

The analysis is centered on an $A - B - C$ configuration where B is the difficult part, meaning that there is no easy way to evaluate the corresponding Green function. This situation does correspond to many cases of physical interest in which B is some transition region between two perfect media A and C.

The problems of interest will involve n ordinary, linear, differential equations of arbitrary order. Here we shall consider, specifically, second order differential systems. Let $\mathbf{\Psi}(z)$, an n-component vector with components $\psi_1(z), \ldots, \psi_n(z)$ be one solution of the system under study for a given value of the eigenvalue variable. The basis of this system consists of $2n$ linearly independent solutions labelled $j = 1, 2, \ldots, 2n$. From the j-th solution we define the $2n$-component vector

$$\mathbf{\Phi}_j(z) = \begin{bmatrix} \phi_{1,j}(z) \\ \vdots \\ \phi_{n,j}(z) \\ \phi_{n+1,j}(z) \\ \vdots \\ \phi_{2n,j}(z) \end{bmatrix} = \begin{bmatrix} \psi_{1,j}(z) \\ \vdots \\ \psi_{n,j}(z) \\ \psi'_{1,j}(z) \\ \vdots \\ \psi'_{n,j}(z) \end{bmatrix} . \tag{16}$$

The $2n \times 2n$ transfer matrix $\mathbf{M}(z, z_0)$, which transfers amplitudes and derivatives from z_0 to z is then defined by

$$\mathbf{\Phi}(z) = \mathbf{M}(z, z_0) \cdot \mathbf{\Phi}(z_0). \tag{17}$$

It is convenient to define the n component vectors $\mathbf{\Phi}_{Aj}(z)$, which represents $\mathbf{\Psi}_j(z)$ and $\mathbf{\Phi}_{Dj}(z)$, which represents $\mathbf{\Psi}'_j(z)$, so that

$$\mathbf{\Phi}_j(z) = \begin{bmatrix} \mathbf{\Phi}_{Aj}(z) \\ \mathbf{\Phi}_{Dj}(z) \end{bmatrix}, \tag{18}$$

where A/D denotes amplitude/derivative. The $2n \times 2n$ matrix $\mathbf{M}$ is consequently partitioned in four $n \times n$ matrices $\mathbf{M}_{AA}, \mathbf{M}_{AD}, \mathbf{M}_{DA}, \mathbf{M}_{DD}$, so (17) is cast as

$$\mathbf{\Phi}_\alpha(z) = \sum_{\beta=A,D} \mathbf{M}_{\alpha\beta}(z, z_0) \cdot \mathbf{\Phi}_\beta(z_0) \quad (\alpha = A, D). \tag{19}$$

The matrix $\mathbf{M}$ has the obvious properties

$$M_{\alpha\beta}(z_0, z_0) = \delta_{\alpha\beta} \; , \quad \frac{dM_{A\alpha}(z, z_0)}{dz} = M_{D\alpha}(z, z_0). \tag{20}$$

It can also be proved that

$$det\|\mathbf{M}(z, z_0)\| = 1 \tag{21}$$

and it is obvious that the inverse of $\mathbf{M}(z, z_0)$ is $\mathbf{M}(z_0, z)$.

For a general differential system of the form

$$\left[\frac{d}{dz}(-\mathbf{S}^{-1}\frac{d}{dz} + \mathbf{P}) + \mathbf{Y}\frac{d}{dz} + \mathbf{W}\right] \cdot \mathbf{\Psi} = 0 \tag{22}$$

it can be shown[3] that

$$\tilde{\mathbf{G}}_B = \begin{Vmatrix} \mathbf{M}_{DA}^{-1} \cdot \mathbf{M}_{DD} & \mathbf{M}_{DA}^{-1} \\ -\mathbf{N}_{DA}^{-1} & \mathbf{M}_{AA} \cdot \mathbf{M}_{DA}^{-1} \end{Vmatrix}_B \cdot \tilde{\mathbf{S}}_B \ ;$$

$$\tilde{\mathbf{G}}_B^{-1} = \tilde{\mathbf{S}}_B^{-1} \cdot \begin{Vmatrix} \mathbf{M}_{AD}^{-1} \cdot \mathbf{M}_{AA} & -\mathbf{M}_{AD}^{-1} \\ \mathbf{N}_{AD}^{-1} & \mathbf{M}_{DD} \cdot \mathbf{M}_{AD}^{-1} \end{Vmatrix}_B \ , \tag{23}$$

with

$$\mathbf{N}_{DA}^{-1} = \mathbf{M}_{AD} - \mathbf{M}_{AA} \cdot \mathbf{M}_{DA}^{-1} \cdot \mathbf{M}_{DD} \ ,$$

$$\mathbf{N}_{AD}^{-1} = \mathbf{M}_{DA} - \mathbf{M}_{DD} \cdot \mathbf{M}_{AD}^{-1} \cdot \mathbf{M}_{AA} \ . \tag{24}$$

For the scalar case (23) simplifies to

$$\tilde{\mathbf{G}}_B = S_B \begin{Vmatrix} \frac{M_{DD}}{M_{DA}} & \frac{1}{M_{DA}} \\ \frac{1}{M_{DA}} & \frac{M_{AA}}{M_{DA}} \end{Vmatrix}_B \ ,$$

$$\tilde{\mathbf{G}}_B^{-1} = S_B^{-1} \begin{Vmatrix} \frac{M_{AA}}{M_{AD}} & -\frac{1}{M_{AD}} \\ -\frac{1}{M_{AD}} & \frac{M_{DD}}{M_{AD}} \end{Vmatrix}_B \ . \tag{25}$$

If now we use the freedom to choose extended pseudomedia we can choose a $\mathbf{G}_B$ for which

$$'\mathcal{G}_{Bl}^{(+)} = 0 \ , \ '\mathcal{G}_{Bl}^{(-)} = -\mathbf{S}_B \ , \ '\mathcal{G}_{Br}^{(+)} = \mathbf{S}_B \ , \ '\mathcal{G}_{Br}^{(-)} = 0 \ , \tag{26}$$

And then the matching formula is

$$\tilde{\mathbf{G}}_S^{-1} = -\tilde{\mathbf{A}}_e \cdot \tilde{\mathbf{G}}_e + \tilde{\mathbf{G}}_i^{-1} + \tilde{\mathcal{P}}_i \ , \tag{27}$$

where e indicates the external part including the semi-infinite media A and C.

After having obtained $\tilde{\mathbf{G}}_S$ it is possible to obtain all the $\mathbf{G}_S$ elements by using the SGFM formulas[3].

The advantage of the transfer matrix method for an inhomogeneous case or for a system with many interfaces is that any given physical property represented by a function $V(z)$ can be approximated to any desired accuracy in the (z, z') interval by taking a succession of N steps from $z' = z_0$ to z_1, to z_2, ..., to z_{N-1}, to $z = z_N$, with $V(z_n) = V_n$=constant for every n-th interval from z_{n-1} to z_n. Within each interval the transfer matrix $\mathbf{M}$ can be used with the local value of V_n. The transfer matrix $\mathbf{M}(z, z')$ can then be obtained to any desired accuracy by using the multiplicative property

$$\mathbf{M}(z, z') = \mathbf{M}(z, z_{N-1}) \cdot \mathbf{M}(z_{N-1}, z_{N-2}) \cdots \mathbf{M}(z_2, z_1) \cdot \mathbf{M}(z_1, z') \ . \tag{28}$$

This expression can be used when no discontinuity is present at the interfaces separating the different intervals. In general this is not true and discontinuities are present so we must modify equation (28) to take into account these discontinuities.

For the general case of a differential system represented by (22) it is clear that we shall have

$$\begin{bmatrix} \mathbf{\Psi} \\ -\mathbf{S}^{-1}\frac{d}{dz}\mathbf{\Psi} + \mathbf{P}\cdot\mathbf{\Psi} \end{bmatrix}_{z+} = \begin{bmatrix} \mathbf{\Psi} \\ -\mathbf{S}^{-1}\frac{d}{dz}\mathbf{\Psi} + \mathbf{P}\cdot\mathbf{\Psi} \end{bmatrix}_{z-} , \tag{29}$$

which can be put as

$$\begin{bmatrix} \mathbf{I} & \mathbf{0} \\ \mathbf{P} & -\mathbf{S}^{-1} \end{bmatrix}_{z+} \cdot \begin{bmatrix} \mathbf{\Psi} \\ \frac{d}{dz}\mathbf{\Psi} \end{bmatrix}_{z+} = \begin{bmatrix} \mathbf{I} & \mathbf{0} \\ \mathbf{P} & -\mathbf{S}^{-1} \end{bmatrix}_{z-} \cdot \begin{bmatrix} \mathbf{\Psi} \\ \frac{d}{dz}\mathbf{\Psi} \end{bmatrix}_{z-} . \tag{30}$$

From here it is easy to obtain:

$$\begin{bmatrix} \mathbf{\Psi} \\ \frac{d}{dz}\mathbf{\Psi} \end{bmatrix}_{z+} = \begin{bmatrix} \mathbf{I} & \mathbf{0} \\ \mathbf{P} & -\mathbf{S}^{-1} \end{bmatrix}_{z+}^{-1} \cdot \begin{bmatrix} \mathbf{I} & \mathbf{0} \\ \mathbf{P} & -\mathbf{S}^{-1} \end{bmatrix}_{z-} \cdot \begin{bmatrix} \mathbf{\Psi} \\ \frac{d}{dz}\mathbf{\Psi} \end{bmatrix}_{z-} . \tag{31}$$

It is then clear that at each interface the transfer matrix of the former interval must be multiplied by

$$\begin{bmatrix} \mathbf{I} & \mathbf{0} \\ \mathbf{P} & -\mathbf{S}^{-1} \end{bmatrix}_{z+}^{-1} \cdot \begin{bmatrix} \mathbf{I} & \mathbf{0} \\ \mathbf{P} & -\mathbf{S}^{-1} \end{bmatrix}_{z-} , \tag{32}$$

to get the continuity of the physical properties involved.

It can be seen that using the relationship between the Green function and the full transfer matrix we are able to use (27) instead of (14) and we are then considering a problem with two interfaces instead of a problem with N interfaces. This reduces substantially the memory and computer time requirements of the problem. Instead of inverting many times $Nn \times Nn$ matrices, we multiply many times N matrices of size $n \times n$.

After having presented here a resume of the formal aspects we shall study now some practical and real cases.

3. Transverse elastic waves in Fibonacci superlattices

This is a case in which we have a system formed by different homogeneous media and many interfaces. The Fibonacci superlattices grown experimentally[13,19] were obtained by stacking recursively along the x_2 direction with two generators, blocks A and B (formed by GaAs and AlAs slabs), mapping the mathematical rule in the Fibonacci sequence

$$S_1 = \{A\} \ , \ S_2 = \{AB\} \ , \ S_3 = \{ABA\} \ , \cdots, \ S_n = S_{n-1}S_{n-2} \ . \tag{33}$$

The sagittal elastic waves in Fibonacci superlattices were studied in[24], while the transverse elastic waves in Fibonacci superlattices were studied in[25]. In both cases the SGFM method for $N > 2$ interfaces was employed

We shall discuss now the simplification introduced in the calculations by the use of the transfer matrix. We shall concentrate, for simplicity's sake, in the transverse elastic waves.

As in[25] our systems will be formed by hexagonal crystals of the *6mm* class with the geometry considered in[29–31]. In this case the C-axis is parallel to the x_3 direction and we shall take the direction x_2 normal to the interface, being x_1 the propagation direction. The wave-vector is then $(k_1,k_2,k_3)=(\kappa,q,0)$. In this case there is a decoupling between motion

Table 1: Mass densities and elastic coefficients employed in our calculations for the finite Fibonacci superlattices.

	ρ (kg/m^3)	C_{44} (10^{10} N/m^2)
ZnO	5676	4.6
CdS	4819	1.458

along the x_1 and x_2 directions, this case corresponding to the sagittal waves studied in[24], and the motion along the x_3 direction, due to the symmetry[31]. We shall consider here the transverse waves associated to the elastic displacement u_3. The Fibonacci superlattices studied here are obtained by stacking recursively along the x_2 direction with two generators, blocks A and B, following the Fibonacci sequence as given in (33). A schematic representation of the AB stacking, together with the reference frame described above is presented in figure 1.

We shall choose as our constituent materials CdS and ZnO. In our calculations A consists of $(CdS)_i/(ZnO)_j$ layers and B of $(CdS)_i/(ZnO)_k$ layers. The thicknesses of the different layers were d(CdS)=17 Å and d(ZnO)=42 Å for block A and d(CdS)=17 Å and d(ZnO)=20 Å for block B. This is a structure similar to the experimentally grown Fibonacci superlattices[13,19]. We shall concentrate on finite structures formed by a given Fibonacci sequence, with stress free bounding surfaces. In this case a slight modification in the general formulas[4] is needed. The interface projected Green function of the system is then given by

$$\tilde{G}_s^{-1} = -\mathcal{I}\left(0_1 - \sum_{j=2}^{N} \tilde{\mathcal{A}}_j^{-1} \cdot \tilde{G}_j^{-1}\right)\mathcal{I} \ , \tag{34}$$

where $\mathcal{I}$ is the total interface projector (3). In the matrix representation $\mathcal{I}$ is the $N \times N$ unit supermatrix matrix. In this format 0_1 is simply a 2×2 zero supermatrix corresponding to the domain 1.

The dispersion relation is obtained from the peaks of the projected density of states

$$-\frac{1}{\pi}\, Im\, Tr\, \tilde{\mathbf{G}}_S(z, z; \omega) \ . \tag{35}$$

In order to obtain these states, corresponding to δ functions, in the computer, an imaginary part of 0.001 GHz was added to the frequency. A sampling in ω from 0 to 80 GHz in steps of 0.005 GHz was performed. In this way it is possible to obtain a sufficiently dense mesh of points to get an accurate description of the dispersion relation.

The material data employed in our calculations are given in Table 1. We have considered Fibonacci generations ranging from the second to the fifteenth and including up to 1974 slabs of constituent materials in the case of the fifteenth generation.

It is clear now that by using the SGFM method for several interfaces[4] it would be necessary to invert a 1974×1974 matrix 16,000 times, if we consider the sampling given above, and this is quite demanding on computer resources.

On the other hand if we use the transfer matrix technique the equation (32) reduces in the case of the transverse elastic waves to

$$\mathbf{C}(z+, z-) = \begin{bmatrix} 1 & 0 \\ 0 & C_{44}(z+) \end{bmatrix}^{-1} \cdot \begin{bmatrix} 1 & 0 \\ 0 & C_{44}(z-) \end{bmatrix} = \begin{bmatrix} 1 & 0 \\ 0 & \frac{C_{44}(z-)}{C_{44}(z+)} \end{bmatrix} , \tag{36}$$

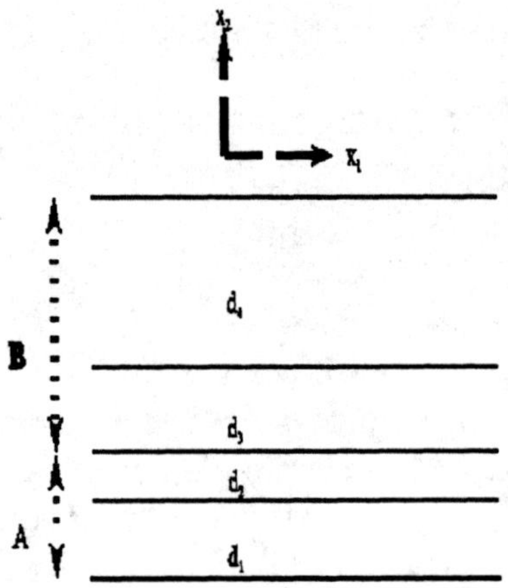

Figure 1: Schematic view of the AB blocks entering the Fibonacci sequence, together with the reference frame used in the calculations. The x_3 axis is perpendicular to the plane of the figure.

and it is clear now that (28) transforms into

$$\begin{aligned}\mathbf{M}(z, z') &= \mathbf{M}(z, z_{N-1}) \cdot \mathbf{C}((z+)_{N-1}, (z-)_{N-1}) \cdot \mathbf{M}(z_{N-1}, z_{N-2}) \cdots \mathbf{M}(z_2, z_1) \cdot \\ &\quad \mathbf{C}((z+)_1, (z-)_1) \cdot \mathbf{M}(z_1, z')\end{aligned} \tag{37}$$

being

$$\mathbf{M}(z_{n+1}, z_n) = \begin{bmatrix} \cosh \beta_t (z_{n+1} - z_n) & \frac{\sinh \beta_t (z_{n+1} - z_n)}{\beta_t} \\ \beta_t \sinh \beta_t (z_{n+1} - z_n) & \cosh \beta_t (z_{n+1} - z_n) \end{bmatrix}, \tag{38}$$

and

$$\beta_t = \sqrt{\kappa^2 - \frac{\rho \omega^2}{C_{44}}} \ . \tag{39}$$

It is then clear that it will be fastest to solve (27) by using equations (36) through (39) than to solve (14), but in the case of the Fibonacci superlattices it is possible to perform a further simplification. This is due to the recursion (33). Instead of multiplying all the matrices involved in the R generation it is possible to obtain the transfer matrices corresponding to the previous generations giving the R generation and to obtain the transfer matrix of the system by multiplying the matrices of the intermediate generations. This reduces considerably the computer time required to solve the problem.

The tenth generation involving 178 interfaces would be solved in a PENTIUM II 266 MHz computer after 11.7 hours, when using the Green function for N interfaces approach. On the other hand the solution would only take 70 seconds or 20 seconds, when solved by using the transfer matrix technique in the two approaches explained above.

We have considered, with both the SGFM and the transfer matrix methods, Fibonacci generations ranging from the second to the thirteenth and including up to 754 slabs of constituent materials in the case of the thirteenth generation.

In Fig.2 we present a range of frequency values versus its ordering number for increasing frequencies for the finite seventh (42 slabs), tenth (178 slabs) and thirteenth (754 slabs) finite

Fibonacci generations having stress free boundary conditions. It can be seen in the figure the existence of the Fibonacci structure reflected in the presence of a series of primary and secondary gaps together with a kind of self-similar structure for all the generations considered, and corresponding to the κ=(0,0) point. The isolated points seen in the primary gaps correspond to localized states. This is evident in Figure 3 where we present the local density of states (LDOS) in arbitrary units of the 14.37 GHz mode of the finite tenth generation Fibonacci superlattice versus the length of the system, that comprises 55 A blocks and 34 B blocks, thus totalling a lenght of 4,503 Å . The rapid attenuation of the mode when going away of the left surface is evident in the figure. Localized states also appear in the secondary gaps, but they do not show in the figure because of the scale.

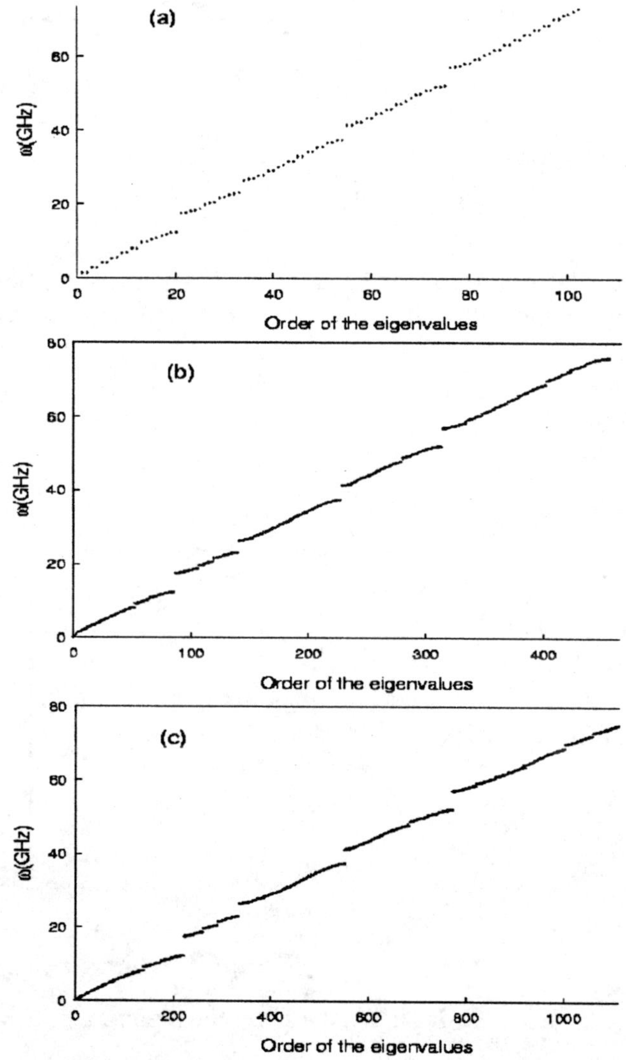

Figure 2: Frequencies ordered by increasing values versus the ordering number of the modes at the κ=(0,0) point of the following finite Fibonacci multilayers with stress-free surfaces: (a) the seventh generation; (b) the tenth generation; (c) the thirteenth generation.

In order to study higher Fibonacci generations the SGFM method for N interfaces

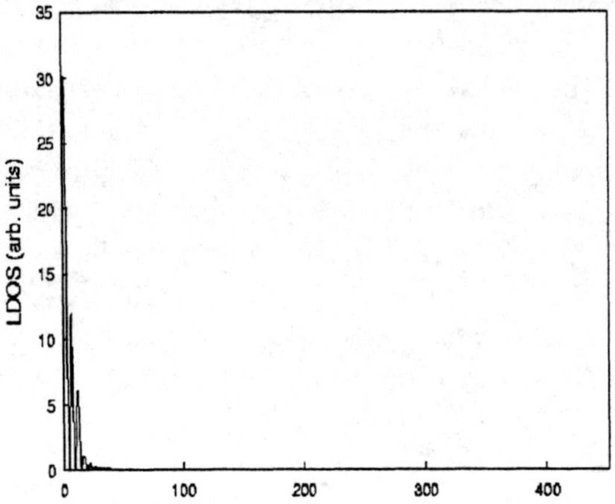

Figure 3: The LDOS in arbitrary units of the 14.37 GHz mode of the finite tenth generation Fibonacci multilayer versus the length of the multilayer.

becomes quite expensive on computer resources. On the other hand the combination of the SGFM method with the full transfer matrix technique allows us to study these higher generations.

Figure 4 gives the same information than Fig.2, but for a finite fifteenth Fibonacci generation, obtained by using the SGFM method and the full transfer matrix technique. In the present case the system has 1974 slabs, and the solution took 30 seconds in a PENTIUM II 266 MHz.

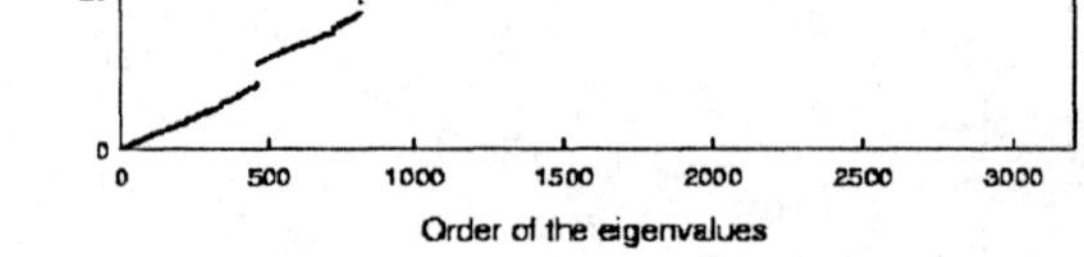

Figure 4: As in figure 2 for the finite fifteenth Fibonacci generation with stress-free surfaces.

It is clear that no new features are present in this case, and there is no further need to study even higher Fibonacci generations. It is then clear that in the case of the elastic transverse waves the Fibonacci spectrum can be obtained even for low Fibonacci generations. This situation is different to that found for sagittal waves in Fibonacci superlattices[24], where the Fibonacci structure of the spectrum was not clearly seen due to the mixing of the longitudinal and transverse waves entering the sagittal wave. This entailed only the study up to the seventh Fibonacci generation, when using the SGFM method for N interfaces,

Table 2: Mass densities and elastic coefficients employed in our calculations for the parabolic wells.

	ρ (kg/m^3)	C_{44} (10^{10} N/m^2)
GaAs	5316.9	5.94
AlAs	3721.8	5.42

due to the rapidly increasing computer resources needed for the solution of the problem. The use of the transfer matrix technique in this case would allow to study higher Fibonacci generations with a reasonable cost on computer resources.

4. Transverse elastic waves in inhomogeneous media

The problems associated to the elastic waves in inhomogeneous systems have been studied intensely in the past[32]. Several problems associated to the propagation of sound in the ocean where there exists a layer in which the temperature, and thus the sound velocity, decreases with depth, were considered[33,34]. Several general questions of wave propagation in a semi-infinite solid of varying elasticity were presented in[35]. In all these cases different profiles for the variation of the elasticity of the system were considered, and different approximations were used.

Besides this kind of problems new inhomogeneous systems have been created recently[36] with the advent of new growth techniques, like the molecular-beam epitaxy (MBE).

In that way it was possible to mimic an ideal inverse parabolic quantum well. This well is composed of thick $Al_{0.36}Ga_{0.64}As$ barriers and an $Al_xGa_{1-x}As$ well where the composition x varies parabolically from 0.02 at the edges to 0.12 at the center. To obtain experimentally this ideal system in the well region, an Al composition grading was made in two ways: (i) an $Al_{0.36}Ga_{0.64}As$/GaAs variable superlattice structure having a constant period of 10 Å and (ii) an analog structure with a multiple step x variation of the $Al_xGa_{1-x}As$. The x variation for the analog well was made in 23 steps (two monolayers each) across the 130 Å well. The x value in each step was controlled by the Al source temperature.

We shall study this kind of system with the x variation given by

$$x = \frac{4(x_{min} - x_{max})}{d^2} z^2 + \frac{4(x_{max} - x_{min})}{d} z + x_{min} \ , \tag{40}$$

being d the thickness of the well, x_{min} the minimum value of x in the well, x_{max} the maximum value of x in the well and z the coordinate along the growth direction. The well is confined to the region $0 \leq z \leq d$.

The elastic constants and mass densities employed in our calculations are given in Table 2. In order to obtain the elastic properties for the ternary compound we shall use a linear interpolation in the spirit of the virtual crystal approximation, and thus we obtain

$$\begin{aligned} C_{44}(ternary) &= x\, C_{44}(AlAs) + (1-x)\, C_{44}(GaAs) \ , \\ \rho(ternary) &= x\, \rho(AlAs) + (1-x)\, \rho(GaAS) \ . \end{aligned} \tag{41}$$

By using the SGFM method and the transfer matrix technique it is now possible to study the transverse elastic waves of a well with elastic properies having the variation given by (40) and (41).

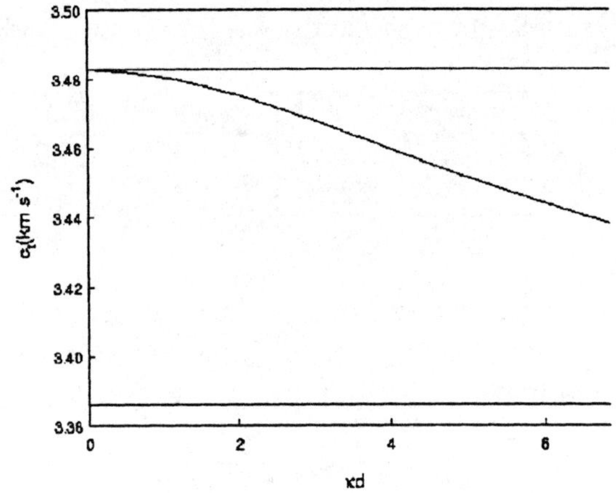

Figure 5: Velocity of the transverse elastic waves in the system having the parabolic variation of the elastic properties given in (40), as a function of κd.

In Fig.5 we represent the variation of the velocity of the transverse elastic waves as a function of κd for a parabolic well grown in 10 steps.

From (41) it is easy to see that the velocity of the transverse elastic waves $c_t = \sqrt{\frac{C_{44}}{\rho}}$ is higher for the ternary in the semi-infinite parts than for any possible composition in the well. Thus the velocity of the transverse wave in the system decreases from the value at the barrier for κd=0 towards the values in the well region for increasing values of κd.

This can be understood if we take into account that for a simple $A - B - A$ structure, where A and B are homogeneous media having C_{44}, ρ and C'_{44}, ρ', respectively, the dispersion relation for the transverse elastic waves (if $\sqrt{\frac{C'_{44}}{\rho'}} < \sqrt{\frac{C_{44}}{\rho}}$) is given by

$$C'_{44}\beta'_t \tanh \beta'_t d = C_{44}\beta_t \ , \tag{42}$$

where

$$\beta_t = \sqrt{\kappa^2 - \frac{\rho\omega^2}{C_{44}}} \ , \ \beta'_t = \sqrt{\kappa^2 - \frac{\rho'\omega^2}{C'_{44}}} \ . \tag{43}$$

From (42) we can see that in the limit $d \to 0$ the only solution is $\omega = \sqrt{\frac{C_{44}}{\rho}}\,\kappa$, and thus the velocity of the wave is that corresponding to the medium A. On the other hand for $d \to \infty$, the solution would correspond to $\omega = \sqrt{\frac{C'_{44}}{\rho'}}\,\kappa$. This is the kind of general behaviour for a well system, and is reflected in figure 5.

In Fig.6 we see the LDOS corresponding to the wave with c_t=3.48208 km s^{-1}, as a function of the distance for, $-d \leq z \leq 2d$. The well region corresponds to $0 \leq z \leq d$, and is indicated by the vertical lines in the figure. It can be seen how the wave is localized in the well region, and the rapid decay in the external region.

Instead of the inverse well corresponding to the x variation given by (40) we can consider a direct well with the x variation given by

$$x = \frac{4(x_{max} - x_{min})}{d^2} z^2 + \frac{4(x_{min} - x_{max})}{d} z + x_{max} \ . \tag{44}$$

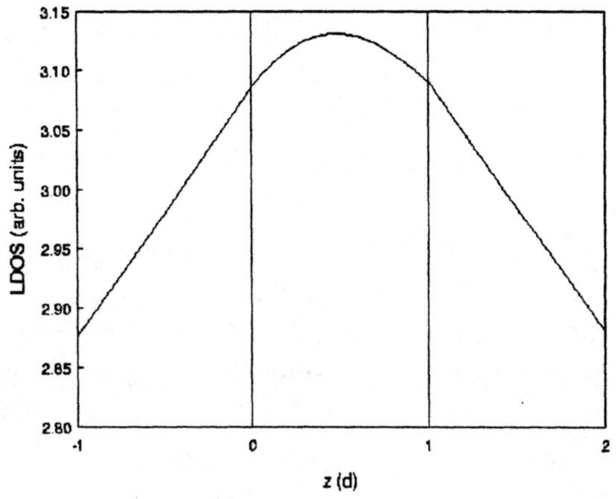

Figure 6: The LDOS in arbitrary units of the c_t=3.4808 km s^{-1} mode of the inverse parabolic well wersus z, for $-d \leq z \leq 2d$.

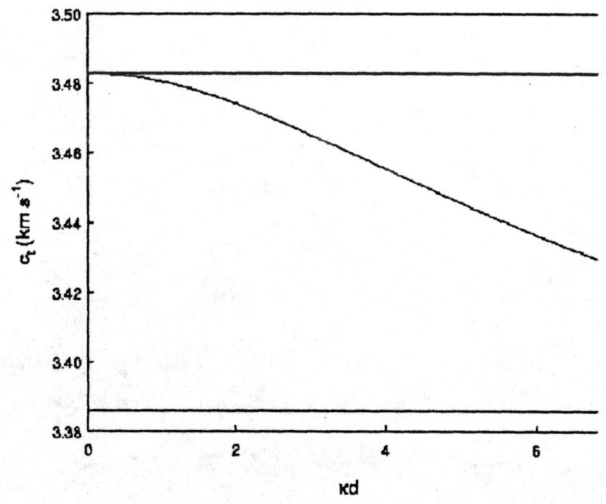

Figure 7: As in figure 5 for the system having the parabolic variations of the elastic properties given in (44).

The variation of the velocity of the transverse elastic waves , as a function of κd, for this parabolic well, grown in 10 steps, is represented in Fig.7. We observe the same behaviour than that previously found in figure 5, because of the reasons given before. The differences in the values of the velocity of the transverse elastic waves between both wells are very small, for the variations of x considered here. Nevertheless the possibility is open to study different kinds of systems with other variations of x and different profiles having linear, exponential, ... etc., variations of the elastic properties.

5. Conclusions

We have seen that the combination of the SGFM method and the full transfer matrix technique opens many interesting possibilities for the study of the physical properties of continuous media with many non-equivalent interfaces, or exhibiting inhomogeneities in their properties. The transfer matrix technique allows us to reduce this kind of problems to the study of $A-B-C$ systems, where B is the difficult one. In this way we reduce the size of the matrices to be considered, and consequently the computer resources needed to solve the problem. This has been illustrated by means of the study of the transverse elastic waves in high generation Fibonacci superlattices, and in well regions having a parabolic variation of the elastic properties. The road is now open for the study of other inhomogeneous systems and physical properties.

Acknowledgements

I am grateful to J. E. Zárate, Dr. L. Fernández-Alvarez, Dr. H. Rodriguez-Coppola and to Profs. F. García-Moliner and R. Pérez-Alvarez for useful discussions. This work was partially supported by the Spanish Dirección General de Enseñanza Superior through grant No. PB96-0916.

References

[1] Brekhovskikh, L. M. and O. A. Godin. 1998.*Acoustics of Layered Media I*, Springer, Berlin.

[2] Smith, D. L. and C. Mailhiot. 1990. *Rev. Mod. Phys.*, **62**, 173-234.

[3] García-Moliner, F. and V. R. Velasco. 1992. *Theory of Single and Multiple Interfaces. The Surface Green Function Matching Method*, World Scientific, Singapore.

[4] Pérez-Alvarez, R., F. García-Moliner and V. R. Velasco. 1995.*J. Phys.: Condens. Matter*,**7**, 2037-2049.

[5] Fernández-Alvarez, L., G. Monsivais, S. Vlaev and V. R. Velasco. 1996. *Surf. Sci.*, **369**, 367-378.

[6] Mora, M. E., R. Pérez and Ch. B. Sommers. 1985. *J. Physique*, **46**, 1021-1026.

[7] Pérez-Alvarez, R., H. Rodriguez-Coppola, V. R. Velasco and F. García-Moliner. 1988. *J. Phys. C*, **21**, 2197-2206.

[8] Velasco, V. R., F. García-Moliner, H. Rodriguez-Coppola and R. Pérez-Alvarez. 1990. *Phys. Scr.*,**41**, 375-382.

[9] García-Moliner, F., R. Pérez-Alvarez, H. Rodriguez-Coppola and V. R. Velasco. 1990. *J. Phys. A*, **23**, 1405-1420.

[10] Rodriguez-Coppola, H., V. R. Velasco, R. Pérez-Alvarez and F. García-Moliner. 1996. *Phys. Scr.*, **53**, 377-381.

[11] Rayleigh, J. W. S. 1945. *The Theory of Sound, Vol. 2*, Dover, New York.

[12] Levine, D. and P. J. Steinhardt. 1984. *Phys. Rev. Lett*, **53**, 2477-2480.

[13] Merlin, R., K. Bajema, R. Clarke, F.-Y. Juang and P. K. Bhattacharya. 1985. *Phys. Rev. Lett.*, **55**, 1768-1770.

[14] Nori, F. and J. P. Rodriguez. 1986. *Phys. Rev. B*, **34**, 2207-2211.

[15] Kohmoto, M., L. P. Kadanoff and C. Tang. 1983. *Phys. Rev. Lett.*, **50**, 1870-1872.

[16] Niu, Q. and F. Nori. 1986. *Phys. Rev. Lett.*, **57**, 2057-2060.

[17] Niu, Q. and F. Nori. 1990. *Phys. Rev. B*, **42**, 10329-10341.

[18] Liu, Y. and R. Riklund. 1987. *Phys. Rev. B*, **35**, 6034-6042.

[19] Todd, J., R. Merlin, R. Clarke, K. M. Mohanty and J. D. Ax. 1986. *Phys. Rev. Lett.*, **57**, 1157-1160.

[20] Bajema, K. and R. Merlin. 1987. *Phys. Rev. B*, **36**, 4555-4557.

[21] Lockwood,D. J., A. H. Macdonald, G. C. Aers, M. W. C. Dharma-Wardana, R. L. S. Devine and W. T. Moore. 1987. *Phys. Rev. B*, **36**, 9286-9289.

[22] Dharma-Wardana, M. W. C., A. H. Macdonald, D. J. Lockwood, J. M. Baribeau and D. C. Houghton. 1987. *Phys. Rev. Lett.*, **58**, 1761-1764.

[23] Tamura, S. and J. P. Wolfe. 1987. *Phys. Rev. B*, **36**, 3491-3494.

[24] Fernández-Alvarez, L. and V. R. Velasco. 1998. *Phys. Rev. B*, **57**, 14141-14147.

[25] Zárate, J. E., L. Fernández-Alvarez and V. R. Velasco. 1999. *Superlatt. Microstruct.*, **25**, 519-526.

[26] James, H. M. 1949. *Phys. Rev.*, **76**, 1602.

[27] Strandberg, M. W. P. 1982. *Phys. Rev. B*, **25**, 5147-5149.

[28] Smith, R. A. 1961. *Wave Mechanics of Crystalline Solids*, Wiley, New York.

[29] Bleustein, J. L. 1968. *Appl. Phys. Lett.*, **13**, 412-413.

[30] Gulyaev, Yu. V. 1969. *Soviet Phys. JETP Lett.*, **9**, 63-65.

[31] Dieulesaint, E. and D. Royer. 1980. *Elastic Waves in Solids*, J. Wiley & Sons, Chichester.

[32] Ewing, W. M., W. S. Jardetzky and F. Press. 1957. *Elastic Waves in Layered Media, Ch. 7*, Mc Graw-Hill, New York.

[33] Pekeris, C. L. 1946. *J. Acoust. Soc. Amer.*, **18**, 295-315.

[34] Morse, R. W. 1950. *J. Acoust. Soc. Amer.*, **22**, 857-860.

[35] Sezawa, K. 1931. *Bull. Earthquake Research Inst. (Tokyo)*, **9**, 310-315.

[36] Chen, W. Q., S. M. Wang, T. G. Andersson and T. G. Thordson. 1993. *Phys. Rev. B*, **48**, 14264-14268.

LOOK, SIR: ROBINSON INSISTS HE'S NOT GOING TO GET ON BOARD UNLESS WE CHANGE THIS ONE FOR SOME SORT OF A "FERRIMAGNETIC CHAIN"... DO WE HAVE ONE OF THOSE?

Quantum Heisenberg Ferrimagnetic Chains

N. B. Ivanov * U. Schollwöck †

Abstract

We present recent progress in understanding the physics of a novel class of one-dimensional ferrimagnetic Heisenberg systems composed of different antiferromagnetically coupled quantum spins in the elementary cell. Generically, quantum mixed-spin systems sustain macroscopically ordered ferrimagnetic ground states and, simultaneously, strong quantum spin-fluctuation effects. Various specific ground-state and thermodynamic properties combining the behavior of uniform ferromagnetic and antiferromagnetic one-dimensional systems are reviewed.

1. Introduction

Quantum spin chains have for a long time been attracting the attention of both theorists and experimentalists. One of the main reasons for this continuing fascination is the dominant role played by quantum fluctuations leading to a rich variety of observed physical phenomena in these systems [1]. In 1983 Haldane put forward his famous conjecture according to which half-odd-integer antiferomagnetic Heisenberg spin chains should have a gapless spectrum, and hence be critical at zero temperature with algebraic decay of correlations, while integer spin chains should be gapped with exponential decay of correlations [2, 3]. By now, this conjecture has been well established both theoretically, based on precise numerical calculations (see, e.g., Ref. [4] and references therein]), and experimentally, the first experimental evidence for the Haldane gap being neutron-scattering experiments on the spin $S = 1$ system $CsNiCl_3$ [5]. Recent important developments in the field are connected with numerous studies of quantum spin chains and ladders in external magnetic fields. In particular, in the presence of a uniform magnetic field a characteristic sequence of plateaus in the magnetization curve has been predicted [6, 7] and experimentally observed [8, 9]. Contrary to the case of spin chains, the Haldane gap and the dispersion of elementary excitations in spin ladders are controlled by two different energy scales, namely the transverse and longitudinal exchange couplings: for a review on spin ladders, see Ref. [10]. Ladder systems are prime candidates to study quantum phase transitions where the Haldane gap is destroyed by external magnetic fields and they offer the possibility of an extremely rich quantum behavior unsuspected in more conventional spin systems [11].

In view of the recent achievements for uniform quasi-one dimensional (1D) systems, ferrimagnetic chains and ladders open an interesting new direction. The point is that besides some computational simplifications characteristic of 1D systems, the presence of two or more

*Theoretische Physik II, Universität Augsburg, D-86135 Augsburg, Germany. Permanent address: Institute of Solid State Physics, Bulgarian Academy of Sciences, Tzarigradsko chaussee-72, 1784 Sofia, Bulgaria

†Sektion Physik, Ludwig-Maximilians-Universität München Theresienstraße 37/III, D-80333 München

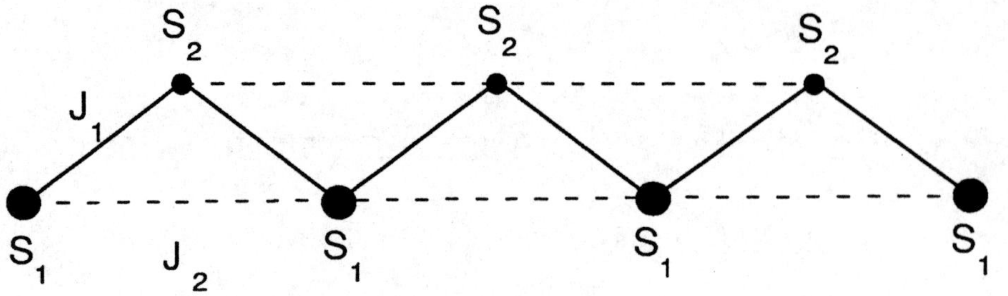

Fig. 1. Structure of the quasi-one-dimensional bimetallic compound $NiCu(pba)(H_2O)_3{\cdot}2H_2O$ with alternating site spins $S_1 = S_{Ni} = 1$ and $S_2 = S_{Cu} = \frac{1}{2}$.

different quantum spins in the unit cell considerably increases the number of situations of interest: the topology of spin arrangements plays an essential role in the structure of the ground state and low-energy excitations. Typically, ferrimagnetic chains exhibit long-range ordered ferrimagnetic ground states that may be referred to as *quantized ferromagnetic states*: these are macroscopic magnetically ordered states characterized both by a ferromagnetic order parameter, which is quantized in integral (or half-integral) multiples of the number of unit cells, and staggered magnetizations which are not quantized. What makes such phases and the related phase transitions interesting is the fact that the ferromagnetic order parameter is a conserved quantity: in Heisenberg systems this is expected to imply strong constraints on the critical behavior [12].

Turning to the ferrimagnetic compound engineering, in the last decade it has become possible to synthesize a large variety of quasi-1D mixed-spin materials [13, 14]. MnCu $(dto)_2(H_2O)_3{\cdot}4.5H_2O$ is the first structurally characterized ferrimagnetic chain [15, 16], where dto = dithiooxalato . Two families of ferrimagnetic chains are described by ACu(pba) $(H_2O)_3{\cdot}nH_2O$ and ACu(pbaOH)$(H_2O)_3{\cdot}nH_2O$ [17, 14], where pba = 1,3-propylenebis (oxamato), pbaOH = 2-hydroxo-1,3-propylenebis (oxamato), and A=Ni, Fe, Co, and Mn. These are quasi-1D bimetallic molecular magnets containing two different transition metal ions per unit cell alternatively distributed on the lattice (see Fig. 1 from Ref. [18]).

Published experimental work implies that the magnetic properties of these mixed-spin compounds are basically described by a Heisenberg spin model with antiferromagnetically coupled nearest-neighbor localized spins (see Fig. 2):

$$\mathcal{H} = J\sum_{n=1}^{N}\left[\mathbf{S}_1(n) + \mathbf{S}_1(n+1)\right]\cdot\mathbf{S}_2(n) - \mu_B H\sum_{n=1}^{N}\left[g_1 S_1^z(n) + g_2 S_2^z(n)\right], \; J > 0 . \quad (1)$$

Here the integers n number the unit cells containing two kinds of site spin operators with quantum spin numbers $S_1 > S_2$. g_1 and g_2 are the g-factors of the respective magnetic moments of the spins S_1 and S_2, μ_B is the Bohr magneton, and H is the external uniform magnetic field applied along the z direction. As an example, the following parameters of (1) have been extracted from magnetic measurements on the recently synthesized quasi-1D bimetallic compound $NiCu(pba)(D_2O)_3{\cdot}2D_2O$ [18] with a structure which is similar to that shown in Fig. 1: $(S_1, S_2) \equiv (S_{Ni}, S_{Cu}) = (1, \frac{1}{2})$, $J/k_B = 121K$, $g_1 \equiv g_{Ni} = 2.22$, $g_2 \equiv g_{Cu} = 2.09$.

The diamond lattice (Fig. 2) is a simple example of another class of Heisenberg ferrimagnetic chains constructed of one kind of antiferromagnetically coupled site spins S but with a different number of lattice sites in the $\mathcal{A}$ and $\mathcal{B}$ sublattices [19-24]. Moreover, there is

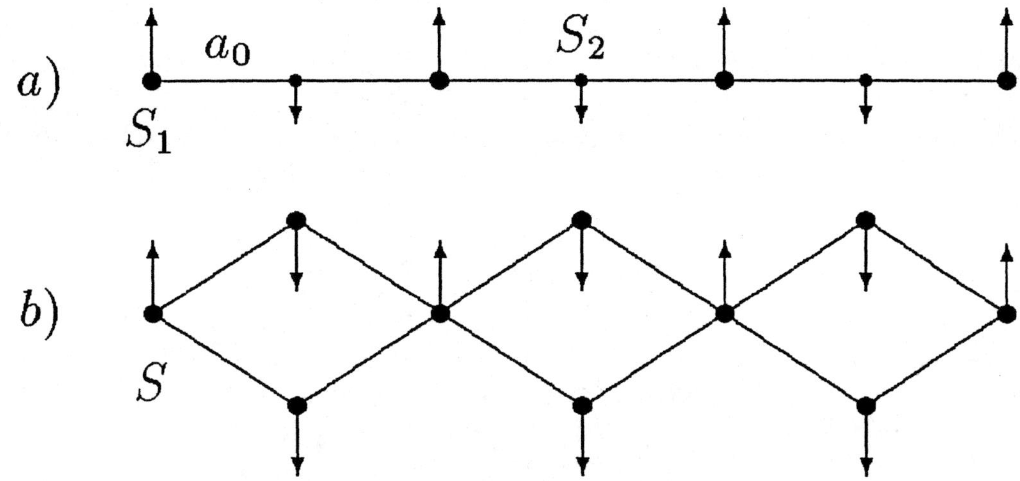

Fig. 2. Classical spin states of two generic types of bipartite ferrimagnetic chains: a) mixed-spin alternating chain with site spins S_1 and S_2 ($S_1 > S_2$); b) uniform-spin diamond chain with a site spin S

a number of experimental works on organic and inorganic magnetic materials of this kind, where ferrimagnetic behavior has been observed [25]. Since interactions in the alternating and diamond structures set up bipartite lattices, the Lieb-Mattis theorem [26] is applicable: it predicts ferrimagnetic ground states with a total spin $S_g = (S_1 - S_2)N$ for the mixed-spin system and $S_g = SN$ for the diamond system, where N is the number of unit cells in the periodic chain and S is the value of the site spin in the diamond model. Various more complicated hypothetical 1D periodic structures with ferrimagnetic properties may easily be constructed by changing the lattice topology and values of the site spins [27, 28, 29]: it is necessary to ensure different maximum total spins of the $\mathcal{A}$ and $\mathcal{B}$ sublattices of the bipartite structures. Generically, one may expect that all these structures show similar low-enegry properties controlled entirely by symmetries of the Heisenberg Hamiltonian (1) and the ferrimagnetic ground state.

It is remarkable that only in the recent few years interest in the physical properties of 1D quantum ferrimagnets has considerably increased. Most of the early efforts have been concentrated on the synthesis of molecular ferromagnets, and a relatively small amount of work has been devoted to theoretical studies [30-34]. Drillon et al. [30] performed exact-diagonalization results for relatively short mixed-spin chains, while Seiden [31] and Georges et al. [32] presented a theory for the static magnetic properties of ferrimagnetic chains containing alternating quantum-classical site spins. De Vega and Woynarovich [33] solved a class of integrable quantum chain models containing different kind of spins, whereas Martins [34] explored the effect of a magnetic field in an integrable Heisenberg chain with mixed spins.

A goal of the present work is to present recent progress in understanding the ground-state and thermodynamic properties of 1D ferrimagnetic Heisenberg systems. The discussion is basically focussed on experimentally important mixed-spin ferrimagnetic systems with two kinds of alternating site spins. In Section 1, using spin-wave theory (SWT) as a tool, properties of the ground state and low-energy excitations are considered. When taken to an appropriate order in $1/S$, SWT is shown to give both a very clear physical picture and precisely enough quantitative results. Here we also discuss the role of frustration in 1D ferrimagnetic chains by the use of a generic mixed-spin Heisenberg model with next-nearest-neighbor antiferromagnetic couplings. Section 2 is basically devoted to thermodynamic properties of 1D ferrimagnets both in zero and finite external magnetic fields. Results of various theoretical and numerical approaches [such as the density matrix renormalization group (DMRG), quantum Monte Carlo (QMC), and the modified SWT] are compared and analyzed.

2. Ground-state properties and elementary excitations

For the Heisenberg model (1) in a zero magnetic field, the Lieb-Mattis theorem [26] predicts the total spin $S_g = (S_1 - S_2)N$ of the ground state and the following ordering of energy levels: If $E(S_t)$ is the lowest energy eigenvalue belonging to total spin S_t, for finite N the lowest energy eigenvalues in each total spin sector order monotonically if $S_t \geq S_g$:

$$E(S_t + 1) > E(S_t) \quad (S_t \geq S_g)\,. \tag{2}$$

For all total spin sectors with $S_t < S_g$, the following inequality will of course hold:

$$E(S_t) > E(S_g) \quad (S_t < S_g)\,. \tag{3}$$

The existence of a magnetic ground state characterized by the magnetization density $M_0 = (S_1 - S_2)/2a_0$ (a_0 is the lattice constant) sets the main properties of 1D ferrimagnets in the low-energy region. The Heisenberg Hamiltonian (1) for $H = 0$ is invariant under a simultaneous rotation of the spin variables (symmmetry group $SO(3)$), whereas the ferrimagnetic ground state beaks this symmetry spontaneously down to $SO(2)$ (the magnetization points in one specific direction). Consequences of this symmetry breaking for the structure of the excitation spectrum are dictated by the nonrelativistic version of Goldstone's theorem [35]: in the absence of long-range forces, a spontaneous symmetry breaking $SO(3) \to SO(2)$ leads to low-energy excitations whose energy tends to zero for wave-vectors $k \to 0$. In contrast to the relativistic version (applicable for antiferromagnets), the theorem does neither specify the exact form of the dispersion relation for small k, nor does it determine the number of different Goldstone modes: these features are not fixed by symmetry considerations alone – rather, they depend on the specific properties of the system. In a hydrodynamic description, the quadratic form of the magnon energies E_k in ferromagnets

$$E_k = \frac{\rho_s}{M_0} k^2 + \mathcal{O}(k^4)\,, \tag{4}$$

where M_0 is the magnetization density, and ρ_s is the spin-stiffness constant, results entirely from the symmetry of the ferromagnetic state and the fact that the order parameter is itself a constant of the motion [36]. This form of the Goldstone modes, which is just the Landau-Lifshitz formula, may be rigorously obtained by simple sum-rule arguments [37]. Similar arguments are applicable for the Heisenberg ferrimagnet as well [38, 39]. However, due to the doubling of the unit cell in the ferrimagnetic state, there appears a second spin-wave branch in the ferrimagnetic case with finite energy gap Δ at $k = 0$: $\Delta \sim M_0/\chi_{||}$, where $\chi_{||}$ is the magnetic susceptibility parallel to $\mathbf{M}_0$ [40].

The finite magnetization of the ground state opens also an interesting and rare opportunity to apply spin-wave theory to a low-dimensional quantum spin system. Moreover, as demonstrated below, SWT confirms the above symmetry considerations concerning the structure of the low-energy excitation spectra, and (if taken to an appropriate order in $1/S$)

gives results for the ground-state parameters which are close to the most precise numerical estimates.

2.1. *Spin-wave theory for 1D Heisenberg ferrimagnets*

To develop a spin-wave theory (see, e.g., Refs. [41, 42] and references therein), one should firstly transform the original spin Hamiltonian (1) to an equivalent boson Hamiltonian by use of an appropriate boson representation of spin operators. We adopt the following Dyson-Maleev representation of spin operators:

$$S_1^+(i) = \sqrt{2S_1}\left(a_i - \frac{1}{2S_1}a_i^\dagger a_i a_i\right),\ S_1^-(i) = \sqrt{2S_1}a_i^\dagger,\ S_1^z(i) = S_1 - a_i^\dagger a_i\,,$$

$$S_2^+(j) = \sqrt{2S_2}\left(b_j^\dagger - \frac{1}{2S_2}b_j^\dagger b_j^\dagger b_j\right),\ S_2^-(j) = \sqrt{2S_2}b_j,\quad S_2^z(j) = b_j^\dagger b_j - S_2\,, \tag{5}$$

where $S_\alpha^\pm = S_\alpha^x \pm \imath S_\alpha^y$, $\alpha = 1, 2$. a_i and b_j are boson operators defined on the lattice sites $i \in \mathcal{A}$ and $j \in \mathcal{B}$, respectively.

Substituting the latter expressions in Eq. (1), one finds the following boson representation of $\mathcal{H}$ in terms of the Fourier transforms a_k and b_k of a_i and b_j:

$$\mathcal{H}_B = -2\sigma S^2 NJ + \mathcal{H}_0' + V_{DM}', \tag{6}$$

where

$$\mathcal{H}_0' = 2SJ\sum_k\left[a_k^\dagger a_k + \sigma b_k^\dagger b_k + \sqrt{\sigma}\gamma_k\left(a_k^\dagger b_k^\dagger + a_k b_k\right)\right], \tag{7}$$

and

$$V_{DM}' = -\frac{J}{N}\sum_{1-4}\delta_{12}^{34}\left(2\gamma_{1-4}a_3^\dagger a_2 b_1^\dagger b_4 + \sqrt{\sigma}\gamma_{1+2-4}a_3^\dagger b_2^\dagger b_1^\dagger b_4 + \frac{1}{\sqrt{\sigma}}\gamma_4 a_3^\dagger a_2 a_1 b_4\right). \tag{8}$$

$\gamma_k = \cos(ka_0)$ is the lattice structure factor and $\delta_{12}^{34} \equiv \delta(1+2-3-4)$ is the Kronecker δ-function. We have adopted the convention $(k_1, k_2, k_3, k_4) \equiv (1, 2, 3, 4)$ and the notations $\sigma \equiv S_1/S_2 > 1$ and $S_2 \equiv S$. Here the external magnetic field is $H = 0$.

In the above expressions $\mathcal{H}_0' = \mathcal{O}(S)$ is the quadratic boson Hamiltonian of the standard linear spin-wave theory (LSWT) and $V_{DM}' = \mathcal{O}(1)$ is the quartic Dyson-Maleev boson interaction. Omitting completely the boson interaction V_{DM}', Pati et al. [43, 44] and Brehmer et al. [45] have recently used the quadratic LSWT Hamiltonian $\mathcal{H}_0'$ for a qualitative description of the 1D ferrimagnetic model (1). $\mathcal{H}_0'$ is easily diagonalized by use of the following Bogoliubov transformation to the quasiparticle boson operators α_k and β_k:

$$\begin{aligned} a_k &= u_k(\alpha_k - x_k\beta_k^\dagger),\quad b_k = u_k(\beta_k - x_k\alpha_k^\dagger), \\ u_k &= \sqrt{\frac{1+\varepsilon_k}{2\varepsilon_k}},\quad x_k = \frac{\eta_k}{1+\varepsilon_k}, \end{aligned} \tag{9}$$

where $\varepsilon_k = \sqrt{1-\eta_k^2}$ and $\eta_k = 2\sqrt{\sigma}\gamma_k/(1+\sigma)$.

As a matter of fact, for the 1D ferrimagnetic model (1) the quadratic Hamiltonian (7) is not the most appropriate choice of a starting zeroth-order approximation for $1/S$ perturbation series. Since we are mostly interested in systems in the extreme quantum limit with small site spins, it is more appropriate from the very beginning to redefine $\mathcal{H}_0'$

by adding the first-order $1/S$ corrections to magnon dispersions. The latter corrections come entirely from a normal ordering of the quasiparticle boson operators α_k and β_k in the quartic interaction $V_{DM}^{'}$. It is important that these corrections (similar to Oguchi's corrections in antiferromagnets [46]) renormalize the magnon excitation spectra (and the ground-state energy) without changing their basic structure, i.e., the number of Goldstone modes. Unlike antiferromagnets, where Oguchi's corrections are numerically small even for site spins $S = \frac{1}{2}$, in extreme quantum 1D ferrimagnets the magnon dispersions are significantly renormalized [47] (see below). As a result of the normal-ordering procedure, the boson Hamiltonian (6) is recast to the following basic form:

$$\mathcal{H}_B = E_0 + \mathcal{H}_0 + V\,, \qquad V = V_2 + V_{DM}\,. \tag{10}$$

The ground-state energy per lattice site calculated to first order in $1/S$ reads

$$\frac{E_0}{2NJ} = -\sigma S^2 + S\left[2\sqrt{\sigma}a_1 + (\sigma+1)a_2\right] - a_1^2 - a_2^2 - \frac{\sigma+1}{\sqrt{\sigma}}a_1 a_2\,, \tag{11}$$

where

$$a_1 = -\frac{\sqrt{\sigma}}{2(\sigma+1)}\frac{1}{N}\sum_k \frac{\gamma_k^2}{\varepsilon_k}, \qquad a_2 = -\frac{1}{2} + \frac{1}{2N}\sum_k \frac{1}{\varepsilon_k}. \tag{12}$$

When expressed in terms of quasiparticle operators, the corrected LSWT Hamiltonian reads

$$\mathcal{H}_0 = 2SJ\sum_k \left[\omega_k^{(\alpha)}\alpha_k^\dagger\alpha_k + \omega_k^{(\beta)}\beta_k^\dagger\beta_k\right]. \tag{13}$$

The normal-ordered quasiparticle interaction V contains the quadratic term

$$V_2 = J\sum_k \left[V_k^{(+)}\alpha_k^\dagger\beta_k^\dagger + V_k^{(-)}\alpha_k\beta_k\right], \tag{14}$$

where

$$V_k^{(+,-)} = -\frac{a_2(\sigma-1)^2}{\sqrt{\sigma}(\sigma+1)}\left(1 \pm \frac{\sigma+1}{\sigma-1}\varepsilon_k\right)\frac{\gamma_k}{\varepsilon_k}, \tag{15}$$

and, in addition, the normal-ordered quartic Dyson-Maleev interaction V_{DM} containing nine vertex functions: $V^{(i)} = V^{(i)}_{12;34}$, $i = 1, \ldots, 9$. Explicit expressions for V_{DM} and the related vertex functions are presented in Appendix A.

2.1.1. *Noninteracting magnons*

The quadratic Hamiltonian (13) describes two branches of spin-wave excitations (magnons) with the dispersions

$$E_k^{(\alpha,\beta)} \equiv 2SJ\omega_k^{(\alpha,\beta)} = 2SJ\left[\left(1 - \frac{a_1}{S\sqrt{\sigma}}\right)\left(\frac{\sigma+1}{2}\varepsilon_k \mp \frac{\sigma-1}{2}\right) - a_2\frac{1-\gamma_k^2}{\varepsilon_k S}\right]. \tag{16}$$

The α excitations are gapless acoustical magnons in the subspace with $S^z_{tot} = (S_1 - S_2)N - 1$, whereas the β excitations are gapful optical magnons in the subspace $S^z_{tot} = (S_1 - S_2)N + 1$: $\mathbf{S}_{tot} = \sum_{n=1}^N [\mathbf{S}_1(n) + \mathbf{S}_2(n)]$. It is easy to show that the effect of an external magnetic field on the dispersions $E_k^{(\alpha,\beta)}$ reduces to the simple substitution $E_k^{(\alpha,\beta)} \longmapsto E_k^{(\alpha,\beta)} \pm g\mu_B H$ provided that $g_1 = g_2 \equiv g$.

The spin stiffness constant ρ_s related to the acoustical magnons plays a basic role [together with the magnetization density $M_0 = (S_1 - S_2)/2a_0$] in the low-temperature thermodynamics of the model (1) and may be obtained from the Landau-Lifshitz relation (4) and Eq. (16):

$$\rho_s^{(0)} = Ja_0 S_1 S_2 \left[1 - \frac{a_1}{S\sqrt{\sigma}} - a_2\frac{\sigma+1}{S\sigma}\right]. \tag{17}$$

In the same free-quasiparticle approximation, the spectral gap of optical magnons at $k = 0$ reads

$$\Delta_0 = 2J(S_1 - S_2)\left(1 - \frac{a_1}{S\sqrt{\sigma}}\right). \tag{18}$$

In the case $(S_1, S_2) = (1, \frac{1}{2})$ Eqs.(17) and (18) give $\rho_s^{(0)}/Ja_0S_1S_2 = 0.761$ and $\Delta_0 = 1.676J$. On the other hand, the LSWT Hamiltonian $\mathcal{H}_0'$ produces the results $\rho_s^{(0)}/Ja_0S_1S_2 = 1$ and $\Delta_0 = J$, which are also the parameters of the related classical system. The numerical estimate for the gap $\Delta = 1.759J$ [48] clearly demonstrates the importance of the $1/S$ corrections in Eq. (16).

Next, let us consider the macroscopic sublattice magnetizations $m_1 = \langle S_1^z(i)\rangle = S_1 - \langle a_i^+ a_i\rangle$ and $m_2 = \langle S_2^z(j)\rangle = -S_2 + \langle b_j^+ b_j\rangle$ which are finite in the ferrimagnetic state. In the case of small site spins, these are strongly reduced as compared to their classical values (respectively, S_1 and $-S_2$):

$$m_1 = S_1 - a_2, \qquad m_1 + m_2 = S_1 - S_2. \tag{19}$$

As an example, for $(S_1, S_2) = (1, \frac{1}{2})$ Eq. (19) predicts a 42% reduction of the small spin $S_2 = \frac{1}{2}$. However, the off-diagonal quadratic interaction V_2 in Eq. (10) produces important first-order corrections for the sublattice magnetizations m_1 and m_2. Thus, to first order in $1/S$ Eq. (19) should be replaced by the more precise expression

$$m_1 = S_1 - a_2 - \frac{\sqrt{\sigma}}{2S(\sigma+1)^2}\frac{1}{N}\sum_k \left(V_k^+ + V_k^-\right)\frac{\gamma_k}{\varepsilon_k^2}. \tag{20}$$

Now for the ferrimagnetic chain $(1, \frac{1}{2})$ the last expression gives the result $m_1 = 0.816$, which differs only 3% from the DMRG numerical estimate 0.79248 [44].

Summarizing, it may be stated that the free-quasiparticle approximation based on the Hamiltonian (13) gives a good qualitative description of the ground-state properties of the model (1). Further improvement of the SWT results may be achieved by considering the role of quasiparticle interactions [49]. Since the first-order $1/S$ corrections for the ground state energy E_0 and the dispersions $E_k^{(\alpha,\beta)}$ have already been taken into account by the normal-ordering procedure, corrections to Eqs. (11) and (16) arise only up from the second order of the perturbation series in V. It is demonstrated below that already the second-order SWT produces results which are close to the the most precise numerical estimates.

2.1.2. *Role of the magnon-magnon interaction V*

As an example, we consider the second-order corrections for the magnon dispersions $\omega_k^{(\alpha,\beta)}$, Eq. (16) [50]. Second-order corrections for the dispersion of acoustical magnons $\omega_k^{(\alpha)}$ are connected with the self-energy diagrams in Fig. 3. The respective analytic expressions read

$$\delta\omega_k^{(\alpha)}(a) = -\frac{1}{(2S)^2}\frac{V_k^{(+)}V_k^{(-)}}{\omega_k^{(\alpha)} + \omega_k^{(\beta)}}, \tag{21}$$

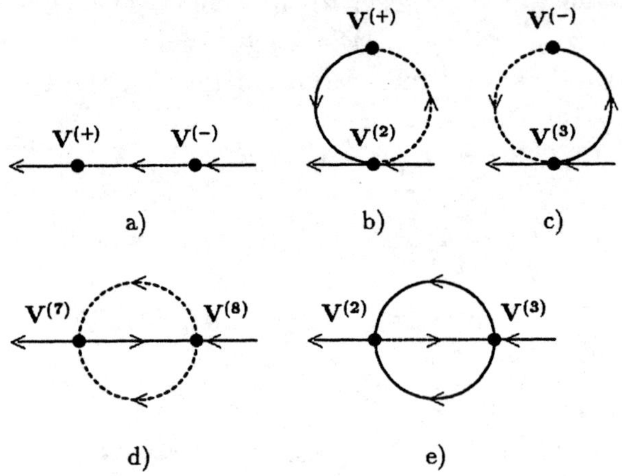

Fig. 3. Second-order self-energy diagrams giving the corrections for the dispersion of acoustical magnons $\omega_k^{(\alpha)}$. Solid and dashed lines represent, respectively, the propagators of α and β magnons. The Dyson-Maleev vertex functions $V^{(i)}$, $i = 1, \ldots, 9$ are defined in Appendix A.

$$\delta\omega_k^{(\alpha)}(bc) = \frac{1}{(2S)^2}\frac{2}{N}\sum_p \frac{V_p^{(+)}V_{kp;pk}^{(2)} + V_p^{(-)}V_{kp;pk}^{(3)}}{\omega_p^{(\alpha)} + \omega_p^{(\beta)}}, \tag{22}$$

$$\delta\omega_k^{(\alpha)}(d) = -\frac{1}{(2S)^2}\frac{2}{N^2}\sum_{2-4}\delta_{k2}^{34}\frac{V_{43;2k}^{(8)}V_{k2;34}^{(7)}}{\omega_k^{(\alpha)} + \omega_2^{(\alpha)} + \omega_3^{(\beta)} + \omega_4^{(\beta)}}, \tag{23}$$

$$\delta\omega_k^{(\alpha)}(e) = -\frac{1}{(2S)^2}\frac{2}{N^2}\sum_{2-4}\delta_{k2}^{34}\frac{V_{43;2k}^{(3)}V_{k2;34}^{(2)}}{-\omega_k^{(\alpha)} + \omega_2^{(\beta)} + \omega_3^{(\alpha)} + \omega_4^{(\alpha)}}. \tag{24}$$

Since the vertex functions $V_k^{(-)}$, $V_{kp;pk}^{(2)}$, $V_{kp;pk}^{(3)}$, $V_{43;2k}^{(8)}$, and $V_{43;2k}^{(3)}$ vanish at $k = 0$ (see Appendix A), the gapless structure of $\omega_k^{(\alpha)}$ is preserved by each of the above corrections. Using this feature of the Dyson-Maleev representation, one can easily find, in particular, the second-order term in the series for the spin-stiffness constant ρ_s:

$$\frac{\rho_s}{Ja_0S_1S_2} = 1 + \frac{r_1}{2S} + \frac{r_2}{(2S)^2} + \mathcal{O}\left(\frac{1}{S^3}\right), \tag{25}$$

where $r_1 = -2a_1/\sqrt{\sigma} - 2a_2(\sigma+1)/\sigma$ is obtained from Eq. (17). To generate a standard $1/S$ series, instead of the renormalized dispersions $\omega_k^{(\alpha,\beta)}$, Eq. (16), one has to substitute in Eqs. (21) to (24) the bare excitation energies (without the $1/S$ corrections). However, as we are mostly interested in the extreme quantum systems with small S_1 and S_2, it seems more appropriate (especially for the dispersions) to consider the quasiparticle interaction V as a small perturbation (dilute bose gas), i.e., to use the renormalized dispersions $\omega_k^{(\alpha,\beta)}$ in Eqs. (21) to (24). Comparison of SWT results for the excitation gap Δ with available numerical estimates confirms such a point of view: the series with the renormalized dispersions $\omega_k^{(\alpha,\beta)}$ give somewhat better results in the extreme quantum systems $(1, \frac{1}{2})$ and $(\frac{3}{2}, 1)$.

Second-order corrections for the dispersion of optical magnons $\omega_k^{(\beta)}$ are connected with similar diagrams. One of the corrections coincides with $\delta\omega_k^{(\alpha)}(a)$, whereas the other four

Table 1. Spin-wave results for the ground-state energy per cell $e_0 = E_0/JN$, the on-site magnetization m_1 of the sublattice A, the reduced spin-stiffness constant $r_s = \rho_s/Ja_0S_1S_2$, and the reduced optical magnon gap $\delta = \Delta/2J(S_1-S_2)$ of the model Hamiltonian (1). The results for E_0 and m_1 are obtained by using standard second-order spin-wave expansions [49]. The results for ρ_s and Δ correspond, respectively, to second- and third-order series in the boson interaction V [50]. e_0', m_1', and δ' indicate the numerical DMRG estimates for e_0, m_1 [44], and δ [52], respectively. The stars mark DMRG estimates from Ref. [52].

(S_1,S_2)	e_0	e_0'	m_1	m_1'	δ	δ'	r_s
$(1,\frac{1}{2})$	−1.4543	−1.4541	0.7939	0.7925	1.7744	1.76	0.7892
$(\frac{3}{2},\frac{1}{2})$	−1.9670	−1.9672	1.3567	1.3574	1.4217	1.42	0.9201
$(2,\frac{1}{2})$	−2.4741	−2.474*	1.8898		1.2938	1.29	0.9566
$(\frac{3}{2},1)$	−3.8632	−3.8619	1.1462	1.1443	1.6381	1.63	0.7592
$(2,1)$	−4.8940	−4.894*	1.7501		1.3685	1.37	0.8902

diagrams give the following contributions:

$$\delta\omega_k^{(\beta)}(bc) = \frac{1}{(2S)^2}\frac{2}{N}\sum_p \frac{V_p^{(+)}V_{kp;pk}^{(5)} + V_p^{(-)}V_{kp;pk}^{(6)}}{\omega_p^{(\alpha)} + \omega_p^{(\beta)}}, \tag{26}$$

$$\delta\omega_k^{(\beta)}(d) = -\frac{1}{(2S)^2}\frac{2}{N^2}\sum_{2-4}\delta_{k2}^{34}\frac{V_{43;2k}^{(7)}V_{k2;34}^{(8)}}{\omega_k^{(\beta)} + \omega_2^{(\beta)} + \omega_3^{(\alpha)} + \omega_4^{(\alpha)}} \tag{27}$$

$$\delta\omega_k^{(\beta)}(e) = -\frac{1}{(2S)^2}\frac{2}{N^2}\sum_{2-4}\delta_{k2}^{34}\frac{V_{43;2k}^{(5)}V_{k2;34}^{(6)}}{-\omega_k^{(\beta)} + \omega_2^{(\alpha)} + \omega_3^{(\beta)} + \omega_4^{(\beta)}}. \tag{28}$$

Comparing the corrected dispersion $\omega_k^{(\beta)} + \delta\omega_k^{(\beta)}$ with available exact-diagonalization results for the $(S_1,S_2) = (1,\frac{1}{2})$ system [51], one finds an excellent agreement in the whole Brillouin zone, the most pronounced deviations (less than 1%) being near the zone center [50].

Results for the ground-state energy E_0, the on-site magnetization m_1, the spin-stiffness constant ρ_s, and the optical magnon gap Δ are summarized in Table 1. The results for E_0 and m_1 are obtained in the framework of standard second-order spin-wave series [49]. The results for ρ_s and Δ come, respectively, from second- and third-order series in V [50]. As a whole, there is an excellent agreement with the available DMRG results. The most pronounced deviations are found for the systems $(1,\frac{1}{2})$ and $(\frac{3}{2},1)$ where the parameter $S_1 - S_2$ is minimal. In the extreme quantum case $(S_1,S_2) = (1,\frac{1}{2})$, the SWT results for E_0, m_1, and Δ differ by 0.014%, 0.18%, and 0.88%, respectively, from the DMRG estimates.

2.2. *Frustrated ferrimagnetic chains*

Over the last years a great deal of interest has been concentrated on the physics of quantum Heisenberg spin chains with competing exchange couplings both for half-integer and integer site spins (see, e.g., Refs.[53-57] and the references therein). In general, frustration reduces the antiferromagnetic correlations and in some cases may produce various exotic quantum ground states such as the dimerized state, the spin-nematic state, or some sort of spin-liquid states. At the same time, little is known about the role of frustration in the mixed spin chains which typically have ground states with an extensive magnetic moment . An important question is what happens when magnetic order disappears due to frustration. What makes such transitions interesting is that the order parameter is also a conserved

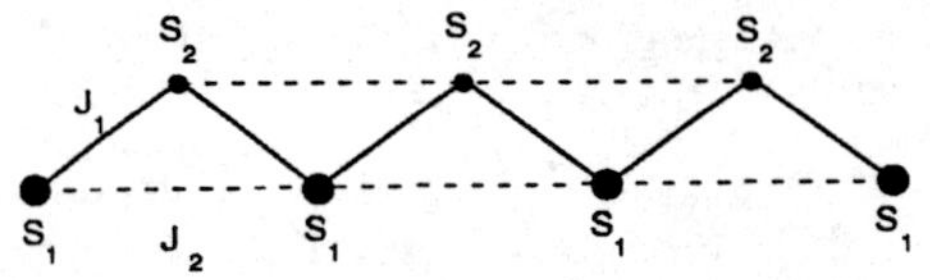

Fig. 4. $J_1 - J_2$ frustrated mixed-spin chain: $J_1,\ J_2 > 0$.

quantity; in systems with a Heisenberg $SO(3)$ symmetry this is expected to lead to strong constraints on the critical field theories [12, 58]. Theoretical models of quantum ferrimagnetic systems with competing interactions have already been discussed in the literature [22, 33, 59, 60]. However, these models consider complicated multiple spin-spin interactions which are far from the real experiment.

Heisenberg mixed-spin chains containing two different alternating site spins $S_1 > S_2$ per unit cell and interacting via competing antiferromagnetic neighbor and next-neighbor couplings, $J_1 > 0$ and $J_2 > 0$ give a simple realistic example of a frustrated 1D ferrimagnetic system [47] (see Fig. 4). The Hamiltonian of this system reads

$$\mathcal{H}_F = J_1 \sum_{n=1}^{N} [\mathbf{S}_1(n) + \mathbf{S}_1(n+1)] \cdot \mathbf{S}_2(n) + J_2 \sum_{n=1}^{N} [\mathbf{S}_1(n) \cdot \mathbf{S}_1(n+1) + \mathbf{S}_2(n) \cdot \mathbf{S}_2(n+1)] \ . \tag{29}$$

In this Subsection we set $J_1 \equiv 1$ and $\alpha \equiv J_2/J_1$.

In the classical limit of two vector spins of relative length σ, the ground state of Eq. (29) can be described by the ansatz

$$\mathbf{S}_{in} = S_i \left[\hat{\mathbf{u}} \cos(Qn) + \hat{\mathbf{v}} \sin(Qn)\right], \quad i = 1, 2, \tag{30}$$

where $\hat{\mathbf{u}} \perp \hat{\mathbf{v}}$ are unit vectors in the spin space. The classical commensurate ferrimagnetic state with a pitch angle between neighboring spins $\theta \equiv Q/2 = \pi$ is stable up to the phase transition point

$$\alpha_C = \frac{\sigma}{2(\sigma^2 + 1)}. \tag{31}$$

For larger values of frustration, $\alpha > \alpha_C$, the stable state is a spiral state with an ordering wave vector given by

$$\cos\frac{Q}{2} = -\frac{\sigma}{2\alpha(\sigma^2 + 1)}. \tag{32}$$

In the limit $\sigma = 1$, Eq. (32) reduces to the well-known result for homogenous frustrated antiferromagnetic chains, and α_C reaches its maximum value 1/4. In the limit $\alpha \to \infty$, $\theta = \pi/2$ and the system is composed of two decoupled antiferromagnetic chains with site spins S_1 and S_2.

Turning now to the *quantum* case, one might expect the classical result to be a guide to possible forms of order or, if order is destroyed by quantum fluctuations, to the nature of short-ranged correlations in the quantum ferrimagnet. Thus, a special feature of the discussed model, Eq. (1), is the existence of an order-disorder quantum phase transition for sufficiently strong frustration: in the limit $\alpha \to \infty$, long-range Néel order is excluded, as decoupled antiferromagnetic chains emerge. The character of the latter transition should crucially depend on the values of quantum site spins S_1 and S_2.

2.2.1. *Spin-wave analysis*

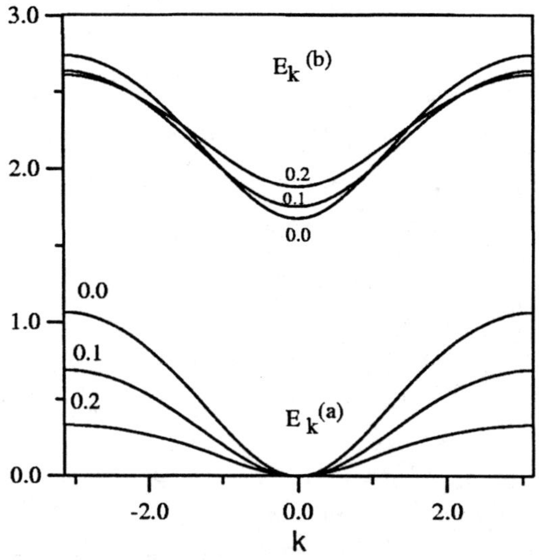

Fig. 5. $J_1 - J_2$ ferrimagnet: Magnon dispersions $E_k^{(\alpha,\beta)}$ in a first-order $1/S$ spin-wave approximation for different values of the frustration parameter $\alpha = J_2/J_1$: $\alpha = 0$, 0.1, and 0.2. $(S_1, S_2) = (1, 1/2)$.

It is naturally to suppose that in the weakly frustrated region SWT still gives a realistic picture of the low-energy physics of the system. As to the Lieb-Mattis theorem, it may not necessarily work in the frustrated system, but due to the continuity principle one may expect that the order of states, Eqs. (2) and (3), survives up to some finite α [61]. This suggestion is also confirmed by exact-diagonalization studies of small clusters [62]. On the other hand, strong frustration will cause a substantial rearrangement of levels, as is shown numerically below.

It is instructive to trace the changes in the excitation spectra, Eq. (16), predicted by spin-wave analysis (see Fig. 5 from Ref. [47]) Near the classical transition point α_C the acoustical branch is strongly flattened, whereas the changes in the optical branch are modest (Fig.5). There is a smooth increase of the gap Δ up to the classical transition point. This means that the optical magnons do not play any important role in the mechanism of the transition. Exactly at the classical transition point $\alpha = \alpha_C$, the quadratic term in the small k expansion of $E_k^{(\alpha)}$ vanishes, so that it takes the form $E_k^{(\alpha)} = ck^4 + O(k^6)$, $\quad c > 0$, $\alpha = \alpha_C$. For $\alpha > \alpha_C$, the classical ground state is the spiral state described above, Eq. (32). Now zero modes appear at $k = 0$ and at $k = \pm(2\pi - Q)$. However, the classically broken SO(3) symmetry in the spiral state is generally expected to be restored by quantum fluctuations in one-dimensional systems [63], so that a crucial role of the magnon-magnon interactions should be expected near and beyond the classical transition point.

SWT also qualitatively captures the asymptotic behavior of the finite-range spin-spin correlations Let us consider the spin-spin correlator $K_{12}^{\perp}(r) = \langle S_1^+(n) S_2^-(n+r) \rangle$. In a LSWT one finds

$$K_{12}^{\perp}(r) = -\frac{2S\sqrt{\sigma}}{N} \sum_k \frac{\eta_k'}{\epsilon_k'} \cos(kr), \tag{33}$$

where $\eta_k' = 2\sqrt{\sigma}\cos(ka_0)/(c_k + d_k)$, $c_k = 1 - \alpha[1 - \cos(2ka_0)]$, $d_k = \sigma - \alpha[1 - \cos(2ka_0)]$, and $\epsilon_k' = \sqrt{1 - \eta_k'^2}$.

Table 2. DMRG results (Ref. [47]) for the transverse spin-spin correlation length ξ in the systems $(S_1, S_2) = (\frac{3}{2}, 1)$, $(1, \frac{1}{2})$, and $(\frac{3}{2}, \frac{1}{2})$ for different values of the frustration parameter $\alpha \equiv J_2/J_1$.

α	0.00	0.02	0.04	0.06	0.08	0.10	0.12	0.14	0.16	0.18
$(\frac{3}{2},1)$	1.74	1.67	1.60	1.53	1.45	1.37	1.29	1.20	1.11	1.01
$(1,\frac{1}{2})$	1.01	0.97	0.93	0.89	0.84	0.80	0.74	0.69	0.61	0.46
$(\frac{3}{2},\frac{1}{2})$	0.75	0.72	0.67	0.64	0.56	0.53	0.45			

The asymptotic behavior of $K_{12}^{\perp}(r)$ for long distances r depends on the analytic structure of the denominator in Eq. (33). It is easy to check that at the point $\alpha = 1/2(\sigma+1)$ (called in what follows a disorder point α_D) the poles in Eq. (33) change from being purely imaginary (for $0 \leq \alpha < \alpha_D$) to being mixed real-imaginary (for $\alpha > \alpha_D$). Let us first focus on the case $0 \leq \alpha < \alpha_D$. Turning Eq. (33) into an integral, one obtains the following 2D Ornstein-Zernike type asymptotic expression in the region $0 \leq \alpha < \alpha_D$, where the correlation length is given by the absolute value of the imaginary part of the poles:

$$K_{12}^{\perp}(r) \sim \frac{e^{-r/\xi}}{\sqrt{r}}, \qquad r \to \infty, \qquad \alpha < \alpha_D. \tag{34}$$

This form may be generically expected for a one-dimensional quantum problem from field-theoretical arguments. ξ is the correlation length of the short-range transverse spin fluctuations. In the limit $\alpha = 0$, $\xi = 1/\ln\sigma$ [43]. The function $\xi(\alpha)$, $0 \leq \alpha < \alpha_D$, compared to the DMRG results, is shown in Fig. 6 from Ref. [47]. A well-pronounced effect of frustration is the reduction of the correlation length.

In the region $\alpha > \alpha_D$ the poles in Eq. (33) move away from the imaginary axis. Due to the emergence of a real part, incommensurate real-space spin-spin correlations appear: in Eq. (34), an additional factor $\cos qr$ enters, where $q \neq 0$ is given by the real part of the poles. This analysis, which is in complete analogy to the standard analysis of disorder points in classical physics [64], allows one to identify – in a quantum system – the special point α_D as a disorder point of the first kind as introduced in classical thermodynamics, marked by the onset of incommensurate finite-range spin-spin correlations in the chain.

2.2.2. *DMRG and exact-diagonalization analysis*

DMRG is particularly adapted to the problem, for it allows treating large enough frustrated systems. DMRG is also free of the negative sign problems characteristic of Monte Carlo methods when applied to frustrated systems. An analysis of the correlations calculated by DMRG for large values of frustration is difficult due to the incommensurability of spin-spin correlations already predicted by spin-wave theory. In the small frustration region, however, the behavior of the transverse spin-spin correlations can clearly be extracted for distances r up to 20 (the limit is due to the very small correlation length ξ, leading to a very fast decay to numerical noise). Fits clearly reveal that the transverse correlations obey the 2D Ornstein-Zernike form, Eq. (34), in accordance with the qualitative SWT analysis. The fit can be done over many orders of magnitude. A fit to a purely exponential law is much worse. The correlation lengths given in Table 2 are extracted from these fits. As an example, in Fig. 6 we show the correlation length ξ vs. α in the case $(1, \frac{1}{2})$.

Recently, it has been argued by one of the authors [55], that in quantum spin chains at $T = 0$ the presence of a phase transition from a commensurate to an incommensurate phase in the classical limit should be generically reflected in the presence of so-called disorder

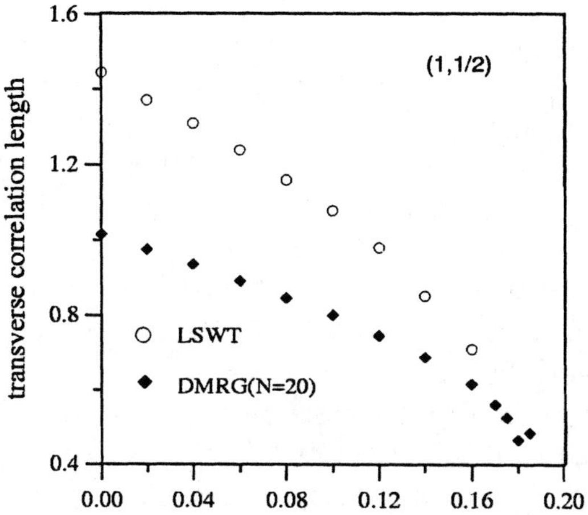

Fig. 6. The transverse spin-spin correlation length in the frustrated $(1, 1/2)$ system vs. $\alpha = J_2/J_1$ obtained from the LSWT and the DMRG methods. The disorder point marking the onset of incommensurate real-space finite-range spin-spin correlations is located at $J_2/J_1 \approx 0.177$.

Table 3. DMRG results (Ref. [47]) for the disorder point α_D and the first-order phase transition point α_T in the systems $(S_1, S_2) = (\frac{3}{2}, 1)$, $(1, \frac{1}{2})$, and $(\frac{3}{2}, \frac{1}{2})$. $\alpha_C = \sigma/2(\sigma^2+1)$ is the classical commensurate-incommensurate transition point. $\alpha_D' = 1/2(\sigma+1)$ is the LSWT result for the disorder point.

(S_1, S_2)	α_D'	α_D	α_C	α_T
$(\frac{3}{2}, 1)$	1/5	0.235(5)	3/13	0.280(1)
$(1, \frac{1}{2})$	1/6	0.177(2)	1/5	0.231(1)
$(\frac{3}{2}, \frac{1}{2})$	1/8	0.14(2)	3/20	0.161(1)

points, a concept of classical statistical mechanics. Indeed, such disorder points have already been identified in a variety of cases, such as frustrated spin chains, as well as in the AKLT model [55]. Theory predicts that a disorder point α_D is characterized by the onset of incommensurate real-space correlations, by an infinite derivative of the correlation length $d\xi/d\alpha$ at $\alpha = \alpha_D - 0$ (i.e. on the commensurate side of the disorder point), as well as by a dimensional reduction: at the disorder point, correlations follow an Ornstein-Zernike behavior typical of a system with lower than the true physical dimension. One recognizes this typical behavior for the function $\xi(\alpha)$ (Fig. 6), showing the existence of a disorder point, at $\alpha \approx 0.177$; analogous observations are possible for other spin values (Table 3). Generically, it can be expected that the disorder point shifts to smaller values of frustration for larger σ.

Finally, let us consider the structure of the low-energy levels around the classical transition as obtained from the DMRG and ED simulations. Both methods seem to be effective tools revealing a complicated picture of low-energy level crossings. At the phase transition point $\alpha_T > \alpha_C$ (see Table 3) a first-order transition (level crossing) takes place from the ferrimagnetic ground state with $S_g = (S_1 - S_2)N$ to a singlet $S_g = 0$ ground state. DMRG shows that at least from $\alpha = 0.25$ onwards the ground state has $S_g = 0$.

Summarizing, the above analysis shows that frustration produces a strong reduction of the finite-range spin-spin correlation length: such a phenomenon is typical for antiferromagnetic chains. In the $J_1 - J_2$ 1D ferrimagnet the quantum spin fluctuations destroy the classical continuous transition and generate a discontinuous transition from the ferrimagnetic $S_g = (S_1 - S_2)N$ state to a singlet state. For smaller values of frustration, $\alpha < \alpha_D$, the transverse short-range spin-spin correlations show typical $2D$ Ornstein-Zernike behavior. A number of open issues can be indicated. One of the important questions concerns the nature of the $S_g = 0$ ground states established beyond the ordered ferrimagnetic phase. As mentioned above, the classically broken SO(3) symmetry in the spiral state is generally expected to be restored by quantum fluctuations in one-dimensional systems. Thus, one is enforced to look for possible magnetically disordered states. A valuable information can be obtained from the Lieb-Schultz-Mattis theorem [65] adapted to mixed-chain spin models [27]. The theorem is applicable to the systems $(S_1, S_2) = (1, \frac{1}{2})$ and $(\frac{3}{2}, 1)$, and says that the model (29) either has gapless excitations or else has degenerate ground states. Thus, one should look for phases with presumably some discrete symmetry broken. An analysis of possible quantum spin phases has already been implemented in the context of the model of frustrated ferromagnetic chains ($J_1 < 0$, $J_2 > 0$) [66, 67]. In particular, two possible phases characterized by nematic type long-range correlations and additional symmetry breaking of reflections about a bond or about a site have been suggested. In principle, the instability of ferrimagnetic state may be expected to show the same main features as the ferromagnetic instability appearing in the frustrated ferromagnetic chain. However, the variety of possible nonmagnetic quantum phases beyond α_T may be considerably enlarged due to the existence of two kinds of site spins, S_1 and S_2, in the mixed chain model. The complicated picture of low-energy level crossings points towards a rich phase diagram of the frustrated $J_1 - J_2$ ferrimagnetic model.

3. Thermodynamics of ferrimagnets

Of great theoretical and experimental interest are the thermodynamic properties of one-dimensional ferrimagnets, both in zero external magnetic field and in finite magnetic field. In the following, we will review key properties.

Numerous experiments have been carried out in the past on the susceptibility of ferrimagnetic chains [68]. Mostly, these experiments suffer from one spin being rather big (typically $S = \frac{5}{2}$), making the ferrimagnet rather classical. Quantum effects are most pronounced in the $(S_1, S_2) = (1, \frac{1}{2})$ case. As mantioned above, Hagiwara et al. [18, 69] were able to carry out measurements on such a ferrimagnet, $NiCu(pba)(D_2O)_3 \cdot 2D_2O$; see Fig. 1. In this compound, the $S = 1$ is carried by the Ni atoms, the $S = \frac{1}{2}$ by the Cu atoms. This substance has an exchange interaction of about 121 K and shows a three dimensional ordering transition at 7 K, obscuring the very low temperature features of the 1D ferrimagnet somewhat. Still, low-field results are in good agreement with the following theoretical results. Regrettably, the magnetic fields necessary to reach experimentally the theoretically most interesting Luttinger phase of the ferrimagnet (see below) would be for this compound well beyond 100 T, making the search for a compound with weaker interactions (and lower associated characteristic field strengths) an interesting challenge for experimentalists.

3.1. *Ferrimagnets in zero external field*

In zero external magnetic field, both susceptibility and specific heat are expected to be characterized by a crossover from ferromagnetic to antiferromagnetic behavior: the low-temperature thermodynamics reflects the dispersion relation of the lower gapless ferromagnetic branch, whereas the intermediate-temperature behaviour is dominated by that of the upper gapped antiferromagnetic branch. One therefore expects for the specific heat C_v to find for $T \to 0$ a ferromagnetic $T^{1/2}$ dependence, for intermediate temperatures an antiferromagnetic Schottky-like peak, and for high temperatures a paramagnetic T^{-2} decay. For the susceptibility χ, one expects first a T^{-2} divergence for $T \to 0$ as in a ferromagnet and a Curie law T^{-1} decay at high temperatures. In a paramagnet, Curie's law implies $\chi T = const$ at all temperatures; ferromagnetic interactions enhance magnetization at low temperatures, leading to a monotonic decay of χT with T; the inverse is true for antiferromagnetic interactions [70]. Thus one expects a crossover from a monotonic decay to a monotonic growth of χT with T.

Using several numerical and analytic tools the specific heat and the magnetic susceptibility at zero field have been calculated. Spin-wave [43, 45, 44] and perturbation [48, 71] calculations suggest that the low-temperature properties of the model are qualitatively the same regardless of the values of S_1 and S_2. Alcaraz and Malvezzi [22] performed finite-size calculations combined with a scaling analysis in the cases of $(S_1, S_2) = (1, \frac{1}{2})$ and $(S_1, S_2) = (\frac{3}{2}, \frac{1}{2})$ and indeed concluded that the appearance of quadratic dispersion relations should be expected for arbitrary isotropic mixed-spin chains showing ferrimagnetism instead of antiferromagnetism. Thus large-scale numerical investigations as reported below have been restricted to the case of $(S_1, S_2) = (1, \frac{1}{2})$, which is the quantum limit of ferrimagnets. Let us set the framework for numerical results by an analytic discussion.

3.1.1. *Spin wave thermodynamics*

Although it is already a rather special feature of the ferrimagnetic chains to be amenable to spin wave analysis at zero temperature, a conventional SWT approach to thermodynamics [41] is also plagued by the usual divergence of the zero-field magnetization in low-dimensional magnets; for the specific heat one obtains merely the leading term at lowest temperature. In the case of low-dimensional ferromagnets, Takahashi [72, 73] was able to describe the low-temperature thermodynamics by imposing a constraint on the total magnetization in the spin-wave description. This approach has been successfully extended to quantum antiferromagnets [74-78], and has been followed up for ferrimagnets by Yamamoto et al. [79, 80]. The alternative Schwinger-boson spin-wave theory has also been recently applied to 1D ferrimagnets at low temperature [81]. As LSWT considerably underestimates the antiferromagnetic gap Δ, one has to go to quartic order in the operators and treat it at some approximative level to obtain quantitative results for the dispersion before evaluating thermodynamics.

In their approach, Yamamoto et al. [79, 80] neglected the quasiparticle interaction V in Eq. (10) and used the quadratic quasiparticle Hamiltonian $\mathcal{H}_0$, Eq. (13) supplemented by the term $\mu \sum_{n=1}^{N} [S_1^z(n) + S_2^z(n)]$ insuring Takahashi's zero magnetization constraint at finite temperature. For ferrimagnets the constraint reads

$$\langle S_1^z + S_2^z \rangle = (S_1 - S_2)N - \sum_k (n_k^{(\alpha)} - n_k^{(\beta)}) = 0, \tag{35}$$

where $n_k^{(\alpha)}$ and $n_k^{(\beta)}$ are the population numbers for the α and β boson quasiparticle excitations:

$$n_k^{(\alpha,\beta)} = \frac{1}{\exp\left(\frac{E_k^{(\alpha,\beta)}}{T} \mp \frac{\mu}{T}\right) - 1}, \tag{36}$$

with the dispersions $E_k^{(\alpha,\beta)}$ from Eq. (16). Eq. (35) is an equation for the chemical potential

$\mu = \mu(T)$.

Then the free energy and the magnetic susceptibility at thermal equilibrium are found to be

$$F = E_0 + \mu(S_1 - S_2)N - T\sum_k \ln\left[(1+n_k^{(\alpha)})(1+n_k^{(\beta)})\right] , \tag{37}$$

$$\chi = \frac{1}{3T}\sum_k \left[n_k^{(\alpha)}(1+n_k^{(\alpha)}) + n_k^{(\beta)}(1+n_k^{(\beta)})\right] . \tag{38}$$

Here and in what follows we set $k_B \equiv 1$ and $g\mu_B \equiv 1$. The ground-state energy E_0 is defined by Eq. (11). It is instructive to represent the related expansions in powers of $T^{1/2}$ for the specific heat and magnetic susceptibility per cell in terms of the dimensionless temperature $t = T/\rho_s^{(0)} M_0$:

$$\frac{C_v}{M_0} = \frac{3}{8}\frac{\zeta(\frac{3}{2})}{\sqrt{\pi}} t^{\frac{1}{2}} - \frac{t}{2} + \left[const(S_1,S_2) - \frac{15\zeta(\frac{1}{2})}{32\sqrt{\pi}}\right] t^{\frac{3}{2}} + \mathcal{O}\left(t^2\right) , \tag{39}$$

$$\frac{\chi T}{M_0} = \frac{2}{3}t^{-1} - \frac{\zeta(\frac{1}{2})}{\sqrt{\pi}} t^{-\frac{1}{2}} + \frac{\zeta^2(\frac{1}{2})}{2\pi} + \mathcal{O}\left(t^{\frac{1}{2}}\right) , \tag{40}$$

where $M_0 = (S_1 - S_2)/2a_0$ is the magnetization density of the ferrimagnetic ground state, a_0 is the lattice spacing, $\rho_s^{(0)}$ is the spin-stiffness constant defined with Eq. (17), $\zeta(z)$ is Riemann's zeta function, and $const(S_1, S_2) = 15(S_1^2 + S_1S_2 + S_2^2)\zeta(\frac{5}{2})/128\sqrt{\pi}$. As may be expected, the above expansions coincide with Takahashi's original expansions for the Heisenberg ferromagnetic chain (with $M_0 = S/a_0$ and $\rho_s = JS^2 a_0$). Apart from the coefficient of $t^{3/2}$ in the expansion for C_v, the above expressions were shown to reproduce very well the thermodynamic Bethe-ansatz calculations for the spin-$\frac{1}{2}$ ferromagnet [82, 83]. It is interesting to note that without the factor $const(S_1, S_2)$, Eqs. (39) and (40) fulfill the general hypothesis according to which in 1D Heisenberg ferromagnets all observables should be universal functions of the bare couplings M_0, ρ_s, and H, realizing a *no-scale-factor universality* [84, 85]. In particular, this means that the free-energy density has the generic form

$$\frac{F}{N} = TM_0\Phi_F\left(\frac{\rho_s M_0}{T}, \frac{H}{T}\right) , \tag{41}$$

where $\Phi_F(x,y)$ is a universal scaling function with no arbitrary scale factors: for example, the site spin S enters only indirectly, through ρ_s and M_0. The appearence of the factor $const(S_1, S_2)$ in Eq. (39) (violating the above scaling hypothesis) is probably an artifact of Takahashi's spin-wave theory: this is also confirmed by the observation that without the latter coefficient the last term in Eq. (39) better approaches Bethe's ansatz result.

At high temperatures, Takahashi's constraint allows the number of bosons of each mode to diverge, as the magnetizations carried by the two types of quasiparticles counteract and the total magnetization constraint is still satisfied. The specific heat becomes a monotonically increasing function. To overcome this difficulty, Yamamoto et al. have proposed a new, alternative constraint [79], namely that the *staggered* magnetization should be zero:

$$\langle : S_1^z - S_2^z : \rangle = (S_1 + S_2)N - \sum_k \frac{n_k^{(\alpha)} + n_k^{(\beta)}}{\varepsilon_k} = 0 . \tag{42}$$

The normal ordering is taken with respect to both quasiparticle boson operators α and β. Normal ordering was introduced, because it was found that under normal ordering the

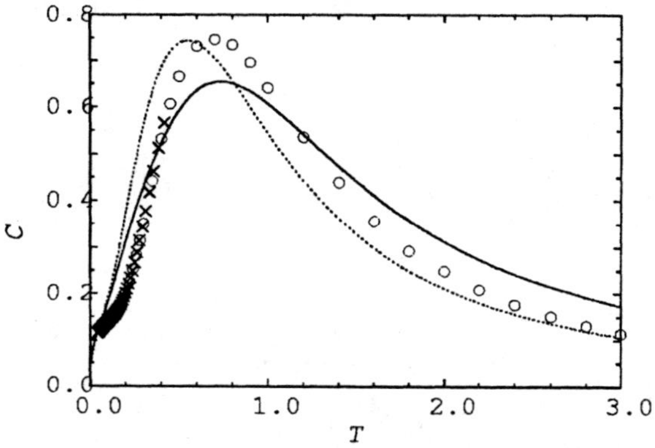

Fig. 7. Temperature dependence of the specific heat per unit cell: Spin wave calculations with noninteracting spin waves (dotted lines) and with the improved dispersion relations (solid lines) are confronted with numerical findings [QMC (○) and DMRG(×).

condition (42) has the appealing feature to reproduce the low-temperature series expansions (39) and (40). The self-consistency equations now read

$$F = E_0 + \mu(S_1 + S_2)N - T\sum_k \ln\left[(1+n_k^{(\alpha)})(1+n_k^{(\beta)})\right] , \qquad (43)$$

$$\chi = \frac{1}{3T}\sum_k \left[n_k^{(\alpha)}(1+n_k^{(\alpha)}) + n_k^{(\beta)}(1+n_k^{(\beta)})\right] , \qquad (44)$$

with

$$n_k^{(\alpha,\beta)} = \frac{1}{\exp\left(\frac{E_k^{(\alpha,\beta)}}{T} - \frac{\mu}{\varepsilon_k T}\right) - 1} , \qquad (45)$$

where the chemical potential μ is now due to the condition (42). This set of equations has no closed analytic solution. In the case of $(S_1, S_2) = (1, \frac{1}{2})$, Yamamoto et al. [80] have numerically evaluated (43) and (44) in the thermodynamic limit (Fig. 7 and Fig. 8 from Ref. [80]). The solid curves represent the calculations starting from the improved dispersion relations $E_k^{(\alpha,\beta)}$, Eq. (16), while the dashed lines show the calculations with the bare dispertions [without the $1/S$ corrections in Eq. (16) [79]. Anticipating numerical results shown in the figures, one sees that the agreement is not perfect, but all relevant features are quantitatively rather well reproduced over a very large temperature range, a remarkable success for spin wave theory in a one dimensional system.

3.1.2. *Numerical results*

Numerically, the major complications in studying ferrimagnets in zero field come from the high ground state degeneracy, the gapless dispersion generating very long range fluctuations with strong finite size effects, limiting the use of exact diagonalization, and from the fact that some particular interesting physics is at very low temperatures, hard to access e.g. by Monte Carlo or transfer matrix methods.

Yamamoto et al. employ a QMC method based on a Suzuki-Trotter decomposition [86] of the checkerboard type [87], accelerating and refining the thermodynamic calculations by using an improved algorithm [88]. Their QMC results are on rather large systems; QMC can in any case not reach such low temperatures that the divergence of the ferromagnetic fluctuation correlation length would exceed these system sizes. Further numerical progress at very low temperatures can be made by introducing the transfer matrix DMRG. Based

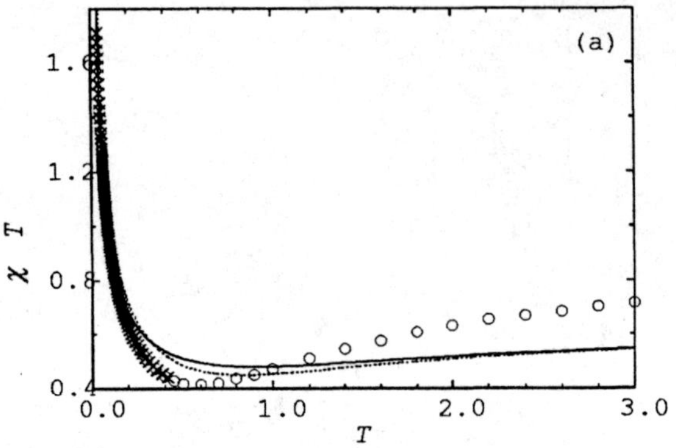

Fig. 8. Temperature dependence of the magnetic susceptibility times temperature per unit cell: Spin wave calculations with noninteracting spin waves (dotted lines) and with the improved dispersion relations (solid lines) are confronted with numerical findings [QMC (○) and DMRG(×)].

on White's original density matrix renormalization group idea [89, 90] developed at $T = 0$, it is applied to the renormalization of (quantum) transfer matrices [91, 92] instead of Hamiltonians. This means that system sizes are automatically infinite, while the power of the original DMRG to treat very large systems in real space translates itself into a very fine Trotter decomposition allowing access to very low temperatures. This extension of the DMRG method has recently been applied very successfully to the low-temperature thermodynamics of several low-dimensional strongly correlated systems [93, 94, 95]. All QMC and DMRG results presented in the following are for all practical purposes quasiexact.

Fig. 7 shows the numerically obtained temperature dependence of the specific heat of the $(S_1, S_2) = (1, \frac{1}{2})$ Heisenberg ferrimagnetic spin chain. Fig. 7 shows a good agreement between the QMC and DMRG calculations. At intermediate temperatures, the specific heat indeed exhibits a Schottky-like anomaly typical of antiferromagnets. It has been found by comparisons with a dimer model of ferrimagnets that the temperature dependence of the specific heat in that temperature range straightforwardly reflects the antiferromagnetic gap and therefore the model can be treated well as a two-level system unless it is at too low temperatures $T \ll \Delta \simeq 1.759J$.

Unlike QMC, DMRG results allow to approach the $T^{1/2}$ asymptotic behavior characteristic of ferromagnets, as DMRG can treat (due to its special features) the ferromagnetic excitations, which are Goldstone modes with diverging correlation length. Calculations of the specific heat on finite systems indeed show that all antiferromagnetic features can be reproduced in rather short systems, while the ferromagnetic ones emerge only slowly with growing system size.

Fig. 8 shows the temperature dependence of the magnetic susceptibility times the temperature, χT, of the $(S_1, S_2) = (1, \frac{1}{2})$ Heisenberg ferrimagnetic spin chain with QMC and DMRG calculations again in excellent agreement. The product χT diverges as T^{-1} at low temperatures as in a ferromagnet [82, 83], while it approaches $[S_1(S_1+1)+S_2(S_2+1)]/3 = 11/12$ at high temperatures. χT shows the expected minimum while shifting from the quantum ferromagnetic behavior to the classical antiferromagnetic behavior.

Summarizing, it can be stated that combining numerical at $T \gtrsim 0.05J$ and analytic findings at $T \lesssim 0.05J$, a quantitative picture of the thermodynamics of ferrimagnets at all finite temperatures emerges. Analytic approximations moreover give a clear understanding of the observed features.

3.2. *Ferrimagnets in finite magnetic fields*

In finite field all dispersion relations are shifted and the ground state degeneracy is lifted, such that experimentally easily observable changes in the thermodynamic quantities should occur.

As a guide for the eye, one can check what happens to simple linear spin wave theory at finite magnetic fields. The introduction of a Zeeman field term $H \sum_i S_i^z$ in the Hamiltonian leaves all eigenstates invariant, while shifting the eigenenergies by HS^z_{tot}, where S^z_{tot} is the total magnetization of the eigenstate. Spin wave theory gives eigenstates of S^z_{tot}; thus the only effect of an external field on a one-particle excitation carrying magnetization ± 1 is the Zeeman term introducing a gap $\Delta^{FM}(H) = H$ to the FM excitations $E_k^{(\alpha)}$, while the AFM branch $E_k^{(\beta)}$ has a reduced gap: $\Delta^{AFM}(H) = \Delta^{AFM}(0) - H$. In linear spin wave theory, m-particle excitations are simply linear superpositions of m one-particle excitations. As the field acts only through the Zeeman term, it is sufficient in this approximation to consider a one-particle excitation to study field effects. This has been done by Maisinger et al. [96].

Let us start by increasing the field from zero. As long as AFM fluctuations are gapped, we expect Néel order at $T = 0$, making previous spin wave analyses applicable. At $H = \Delta^{AFM}(0)$, AFM excitations become gapless, allowing spin-up flips at no energy cost, such that the simple result $M(T = 0) = 0.5$, if we specialise in the following on the case $(S_1, S_2) = (1, \frac{1}{2})$, breaks down. One concludes that there is a so-called lower critical field $H_{c1} = \Delta(H = 0)$, above which $M(T = 0) > 0.5$, and our spin wave analysis breaks down. Using the numerical value for the gap, $\Delta = 1.76J$, one predicts $H_{c1} = 1.76J$.

Incidentally, for a field $H = \Delta(H = 0)/2$ (i.e. $H_{c1}/2$), magnetization $M(T)$ should be constant with T, while it is monotonically decreasing with T for $H < \Delta/2$, and monotonically increasing with T for $H > \Delta/2$. This is because the dispersion relations of the harmonic spin waves increasing and decreasing magnetization become identical at this particular field value.

Starting from very strong fields, linear spin wave analysis based on the fully polarized state yields for $(S_1, S_2) = (1, \frac{1}{2})$ a dispersion relation

$$E_k^{(\alpha,\beta)} = H - J\left[\frac{3}{2} \pm \frac{1}{2}\sqrt{5 + 4\cos(2ka_0)}\right] = H - J\omega_1^{F\mp}(k), \tag{46}$$

where $\omega_1^{F\pm}(k)$ is the two-branched exact ferromagnetic one-particle excitation: both excitation branches reduce magnetization. Below a so-called upper critical field $H_{c2} = 3J$, one dispersion branch acquires negative energy, such that a breakdown of the simple result for magnetization per cell $M(T = 0) = 3/2$ is predicted for $H < H_{c2}$.

The existence of two critical fields H_{c1} (up to which the ground state magnetization persists) and H_{c2} (which marks the beginning of full polarization) in the ferrimagnet is analogous to the finite field behaviour of the $S = 1$ (Haldane) chain [97-101] and gapped spin-$\frac{1}{2}$ chains and spin ladders [11, 102, 103]. In all these systems the characteristic presence of gapped excitations for both $0 < H < H_{c1}$ and for $H > H_{c2}$ breaks down at the critical fields; in SWT this is indicated by an instability of an arbitrary number of spin waves (spin-wave condensation). In all these systems, we expect the occurrence of a critical (Luttinger liquid, XY) phase in the intermediate regime $H_{c1} < H < H_{c2}$. This phase should resemble a (critical) anisotropic XXZ Heisenberg chain with the anisotropy effect induced by the magnetic field; power law correlation functions as well as a linear behaviour of the specific

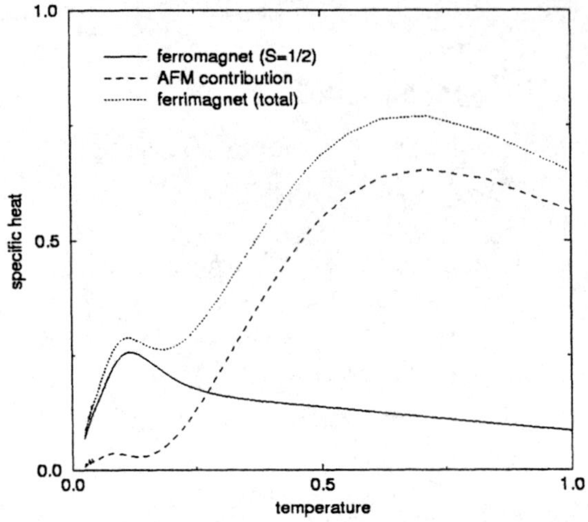

Fig. 9. C at $H = 0.05J$ for a ferrimagnet (dotted, DMRG) and a $S = \frac{1}{2}$ ferromagnet (solid, DMRG). To a very good approximation, the difference in specific heat (dashed) can be identified as AFM contribution.

heat $C_v = \gamma T$ (with γ related to H) are expected. The ferrimagnetic chain, however, having two spins per unit cell, is richer than the $S = 1$ chain and the interplay of two elementary excitations leads to new crossover phenomena between high and low field behaviour in both specific heat and magnetization. The fact that this also holds for spin ladders will be pursued below.

3.2.1. *Low magnetic fields*

Let us first consider the case of low magnetic fields which is still essentially dominated by the zero field behavior. One expects [96] the specific heat $C_v(T)$ to acquire for low fields $H < H_{c1}$ a generic double-peak structure, because two gapped excitation modes exist: both gapped antiferromagnets as well as ferromagnets in external fields (see Fig. 9 from Ref. [96]) show exponential activation of C_v with a pronounced peak, whose position is related to the gap or field size. Peak positions should thus shift with H to higher (for the FM contribution) and lower (AFM contribution) temperatures because of the Zeeman term. At the critical fields, where the dispersion of gapless excitations is proportional to k^2, $C_v \propto \sqrt{T}$ is expected for $T \to 0$.

Another approach to the understanding of C_v for low fields can be obtained by studying the decoupled-dimer limit, useful in the $H = 0$ case [48]: every second interaction is switched off, yielding a flat dispersion, but the FM and AFM elementary excitations are still gapless and gapped, respectively. This very naive analysis also gives a double-peak structure for C_v. With increasing field, the double-peak structure is smeared out and the two peaks merge into one.

To obtain quantitatively reliable values for the specific heat and the magnetization, the transfer matrix DMRG [91, 92, 93, 94] was applied to the problem [96]. DMRG shows that in the specific heat per elementary cell a two-peak structure evolves for small fields, with the high-temperature AFM peak moving to smaller temperatures and the low-temperature FM peak moving to higher temperatures (see Fig. 10 from Ref. [96]). Using for a first rough estimate $C_v \propto x^2 \cosh^{-2} x \quad x = J\Delta/2T$, leading to $\Delta \approx 2.4T_{peak}$, one finds a zero-field gap $\Delta_{cv} \approx 1.70J$ in reasonable agreement with the numerically observed gap $\Delta \approx 1.76J$; it shifts linearly and almost proportionally with H to lower T. As Fig. 9 shows, the low-T

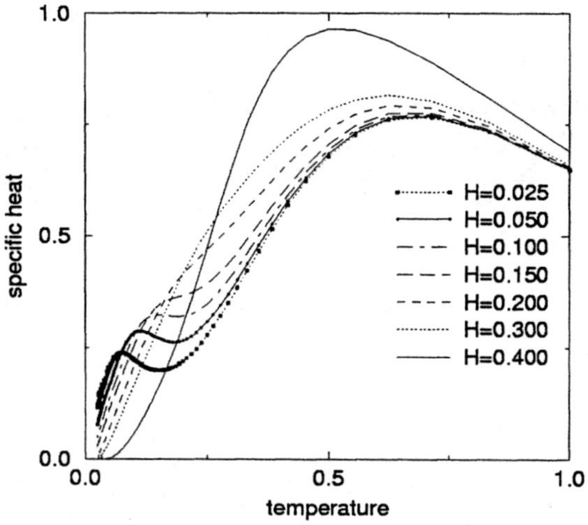

Fig. 10. Specific heat C vs. T for fields $H \leq 0.4J$.

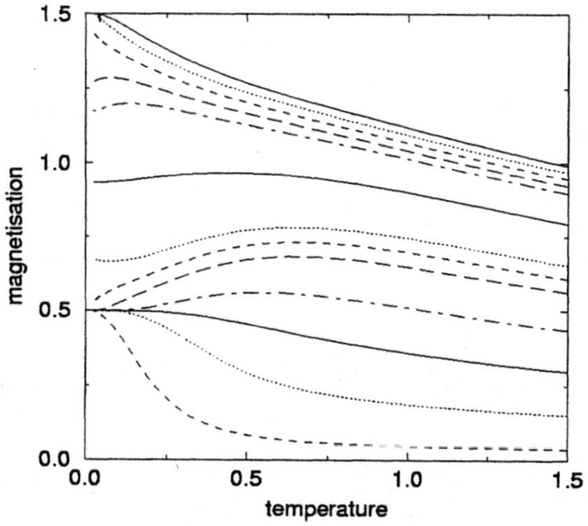

Fig. 11. Magnetisation vs. T for fields $H = 0.1$, 0.4, 0.8, 1.2, 1.6, 1.75, 1.9, 2.4, 2.8, 2.9, 3.0, 3.1, $3.2J$ (from the lowest to the highest curve). Note the rather rich non-monotonic behaviour.

peak is almost identical with that of a $S = \frac{1}{2}$ ferromagnet; the "remaining" specific heat can be attributed to AFM excitations (due to renormalizations, the naive comparison is of course not perfect). For $H > 0.2J$, the two peaks merge into one. The rather rich structure of the specific heat at experimentally accessible low fields calls for experimental verification.

Studying the magnetization per elementary cell by DMRG, Maisinger et al. [96] find for $H < H_{c1}$ that $M = 1/2$ at $T = 0$. How this magnetization is reached for fixed field, is due to a competing effect of magnetizing interactions and thermal fluctuations. For high temperature, thermal fluctuations will always suppress magnetization. For fields below $H = 0.8J$, demagnetizing FM fluctuations cost less in energy and also suppress magnetization. For higher fields, magnetizing AFM fluctuations will increase magnetization. One observes therefore an intermediate magnetization peak before thermal fluctuations suppress again magnetization (see Fig. 11 from Ref. [96]). An almost flat magnetization curve is observed up to $T \approx 0.2J$ at $H \approx 0.88J$, in excellent accordance with the analytic

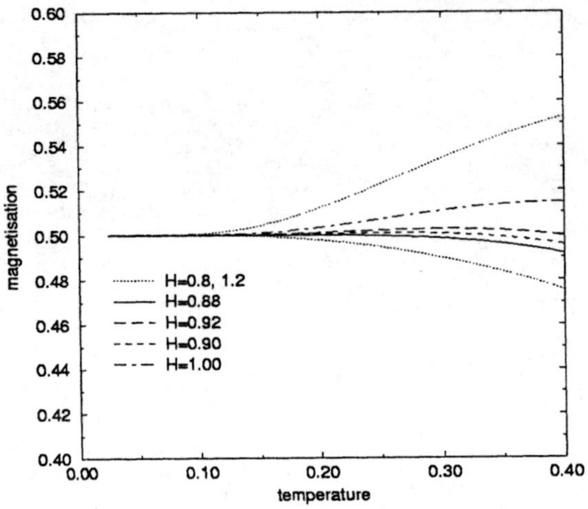

Fig. 12. T-independence of the low-T magnetisation for magnetic fields close to half the critical field $H_{c1} = 1.76(1)$.

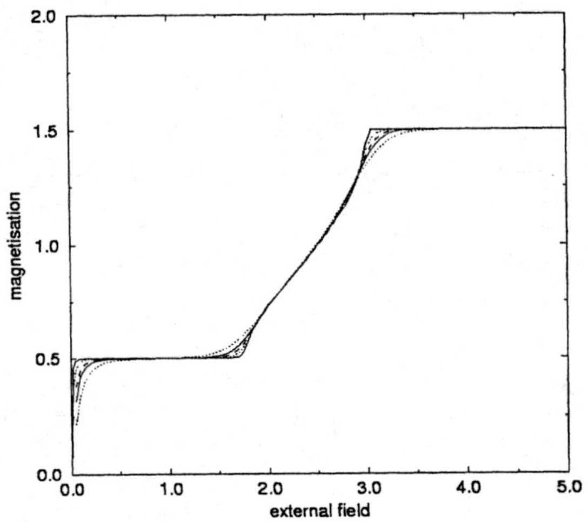

Fig. 13. Magnetisation vs. H for $T = 0.025$, 0.05 (DMRG), $T = 0.08$, 0.1, $0.15J$ (quantum MC, to show consistency of methods). Curves approach singular square-root behaviour for $T \to 0$.

prediction (Fig. 12 from Ref. [96]).

Magnetization for $T \to 0$ allows to locate the two critical fields. The lower critical field (Fig. 13 from Ref. [96]) is found at $H_{c1} = 1.76(1)J$, the ferrimagnetic magnetization at $T = 0$ is broken up, with $M(H) - M(H_{c1}) \propto \sqrt{H - H_{c1}}$ (cf. equivalent observations in Haldane magnets [97- 101]).

Numerically, one locates the upper critical field at $H_{c2} = 3.00(1)J$, in excellent agreement with the analytic result. Again, a square root behaviour in the approach to the maximum magnetization can be observed.

3.2.2. *Luttinger phase in high magnetic fields*

Maisinger et al. [96] have studied the thermodynamics of the critical phase using the transfer matrix DMRG. In the critical phase the fact that the method works in the thermodynamic limit is very useful. The main observation is that the behavior of the specific heat is rather symmetrical about a field $(H_{c1} + H_{c2})/2$: From a shoulderlike structure (and the usual mean field peak at higher temperatures), an additional low-temperature peak structure "pops up", being most pronounced at that particular field values. Away from the symmetry field, the peak is weaker and flatter, and starts showing a weak differentiation into two peaks.

From an analytic point of view, the Luttinger-like critical phase needs many-body effects for understanding, as the transition is understood in terms of a magnon condensation, leading to a very intricate quantum ground state. Kolezhuk et al. [51] have attempted a description of the Hamiltonian by an effective model appropriate for the critical phase.

The general idea of a mapping to an effective model is to reduce the Hilbert space of the problem. For the $(1, \frac{1}{2})$ ferrimagnet the set of the cell wave functions ψ_{jm} consists of a doublet ($j = \frac{1}{2}$) and a quartet ($j = \frac{3}{2}$). One would naively expect that in a strong field it is necessary to keep three states with $(j, m) = (\frac{1}{2}, \frac{1}{2})$, $(\frac{3}{2}, \frac{1}{2})$, $(\frac{3}{2}, \frac{3}{2})$; it is clear, however, that there is only *one* ground state and *one* excitation (the optical magnon) becoming gapless at $H = H_{c1}$, so that the interplay should be effectively between *two* cell wavefunctions being "proper" linear combinations of ψ_{jm}.

To identify these combinations, Kolezhuk et al. use a variational matrix product (MP) approach which has proved to be very successful for the $(1, \frac{1}{2})$ ferrimagnet without external field. [104, 105-109]. The following (non-normalized) variational ground state wave function was proposed [104]:

$$\Psi_0 = \mathrm{tr}\,(g_1 g_2 \cdots g_N), \tag{47}$$

where the cell wavefunction is given by

$$g_n = uM^{0,\frac{1}{2}} + vM^{1,\frac{1}{2}} + M^{1,\frac{3}{2}},$$

$$M^{0,\frac{1}{2}} = \mathrm{Id}\,\psi_{\frac{1}{2},\frac{1}{2}},$$

$$M^{1,\frac{1}{2}} = -\frac{1}{\sqrt{3}}\sigma^0\psi_{\frac{1}{2},\frac{1}{2}} + \sqrt{\frac{2}{3}}\sigma^{+1}\psi_{\frac{1}{2},-\frac{1}{2}},$$

$$M^{1,\frac{3}{2}} = \frac{1}{\sqrt{2}}\sigma^{-1}\psi_{\frac{3}{2},\frac{3}{2}} + \frac{1}{\sqrt{6}}\sigma^{+1}\psi_{\frac{3}{2},-\frac{1}{2}} - \frac{1}{\sqrt{3}}\sigma^0\psi_{\frac{3}{2},\frac{1}{2}}.$$

Here $\sigma^0 = \sigma_z$ $\sigma^{\pm1} = \mp\frac{1}{\sqrt{2}}(\sigma_x + i\sigma_y)$ are the Pauli matrices, Id is the 2×2 unit matrix, ψ_{jm} denote spin states of the n-th cell, and u, v are the variational parameters whose optimal values minimizing the energy of the Hamiltonian (1) at $H = 0$ were found [104] to be $u = 1.3026$, $v = 1.0788$, with the variational ground state energy per unit cell $E_0^{\mathrm{var}} = -1.449\,J$ (to compare with the numerical value $-1.455\,J$). It was shown that Ψ_0 possesses the correct quantum numbers of $S_{\mathrm{tot}} = S^z_{\mathrm{tot}} = N/2$ and provides an excellent description of the ground state correlations. The MP approach is well suited to this problem since the quantum fluctuations are extremely short-ranged [43, 45, 44, 104].

The simplest trial wave function for the optical magnon with momentum k is then given by:

$$|k\rangle = \sum_{n=1}^{N} e^{ikn}|n\rangle, \quad |n\rangle = \mathrm{tr}\,(g_1 \cdots g_{n-1}\widetilde{g}_n g_{n+1} \cdots g_N), \tag{48}$$

where the excited cell wavefunction is given by

$$\widetilde{g}_n = 1\,\psi_{\frac{3}{2},\frac{3}{2}} + w\,\sigma^{+1}\,\psi_{\frac{1}{2},\frac{1}{2}} + \sqrt{12/5}\,f\,Q^{1,\frac{3}{2}}, \tag{49}$$

$$Q^{1,\frac{3}{2}} = \sqrt{3/5}\,\sigma^0\,\psi_{\frac{3}{2},\frac{3}{2}} - \sqrt{2/5}\,\sigma^{+1}\,\psi_{\frac{3}{2},\frac{1}{2}}.$$

The form of $\widetilde{g}_n$ in (48) is dictated by the requirement that $|n\rangle$ is the state with $S_{\rm tot} = S^z_{\rm tot} = N/2+1$, therefore [104] $\widetilde{g}_n$ should carry the "hyperspin" quantum numbers $(\frac{3}{2},\frac{3}{2})$, while each of the g_n carries $(\frac{1}{2},\frac{1}{2})$. Here f, w are still free (variational) parameters. In general the states $|n\rangle$ are not mutually orthogonal, merely to Ψ_0. Further $\widetilde{g}_n$ can be represented in the form

$$\widetilde{g}_n = \frac{f-1}{\sqrt{2}}\, g_n\, \sigma^{+1} - \frac{f+1}{\sqrt{2}}\, \sigma^{+1}\, g_n + \widetilde{w}\, \sigma^{+1}\, \psi_{\frac{1}{2},\frac{1}{2}}\,,$$

so that $|k\rangle$ in fact depends only on $\widetilde{w} = \{w + \sqrt{2}(u + \frac{f}{\sqrt{3}})\}$. Thus, one parameter in (48) is redundant. For the choice

$$f = \frac{5z}{6(2+z)} + \frac{5}{\sqrt{6}}\, w\, \frac{v - u\sqrt{3}}{2+z}\,, \qquad (50)$$
$$z \equiv 2\sqrt{3}\, uv + \{3 + 12v^2(u^2+1)\}^{1/2}\,,$$

the set of one-magnon states $\{|n\rangle\}$ becomes mutually *orthogonal*, considerably simplifying further calculations. Moreover, one can show that arbitrary N'-magnon states $|n_1,\ldots n_{N'}\rangle$ become orthogonal; however, they are not normalized, and, generally, $\langle n, n+m|n, n+m\rangle$ does not coincide with $\langle n|n\rangle^2$, though it tends very fast to the latter value as m increases. The one-magnon norm and the matrix elements of the ferrimagnetic Hamiltonian are given by

$$\begin{aligned}
\mathcal{N}_0 &\equiv \langle n|n\rangle = 0.1173w^2 - 0.0337w + 0.1432\,,\\
\langle n|\widetilde{H}|n'\rangle &= Jx^{|n-n'|}(A + B\delta_{|n-n'|,1} + C\delta_{nn'})\,,\\
x &= 0.2147, \quad A = -0.0483w^2 + 0.0808w - 0.0284\,,\\
B &= -0.0839w^2 + 0.0379w + 0.0010\,, \qquad (51)\\
C &= 0.3359w^2 - 0.0495w + 0.5860\,,
\end{aligned}$$

where $\widetilde{H} \equiv \widehat{H} - NE_0^{\rm var}$. Now the dispersion is obtained by minimizing the excitation energy $\varepsilon(k) = \langle k|\widetilde{H}|k\rangle / \langle k|k\rangle$,

$$\varepsilon(k) = \frac{J}{\mathcal{N}_0}\Big\{C + 2Bx\cos k + A\frac{1-x^2}{1+x^2-2x\cos k}\Big\}\,. \qquad (52)$$

Just minimizing the gap $\Delta_{\rm opt} = \varepsilon(k=0)$ yields $w = w_0 = -3.8605$, with $\Delta^{\rm var}_{\rm opt} \simeq 1.754\,J$, and the resulting dispersion for $w(k) = w_0$ is in excellent agreement with exact diagonalization data.

Now the problem is mapped to an anisotropic $S = \frac{1}{2}$ Heisenberg spin chain by introducing effective spin-$\frac{1}{2}$ states $|\alpha_n = \pm\frac{1}{2}\rangle$ at each cell, and identifying $|\alpha_1\alpha_2\cdots\alpha_N\rangle = {\rm tr}(R_1R_2\cdots R_N)$, where $R_n = \widetilde{g}_n,\, g_n$ for $\alpha_n = \frac{1}{2}, -\frac{1}{2}$, respectively. Introducing the spin-$\frac{1}{2}$ operators $s_n^{z,\pm}$, and restricting all effective interactions to nearest neighbors only, one obtains the effective Hamiltonian for the critical phase of the $(1,\frac{1}{2})$ ferrimagnet:

$$\widehat{H}_e = \sum_n \frac{J_{xy}}{2}(s_n^+ s_{n+1}^- + s_n^- s_{n+1}^+) + J_z s_n^z s_{n+1}^z - 2h_e s_n^z, \qquad (53)$$

where

$$J_{xy} = -2t_1, \quad J_z = U_1, \quad h_e = (H - t_0 - U_1)/2, \qquad (54)$$
$$t_m \equiv \frac{\langle n|\widetilde{H}|n+m\rangle}{\mathcal{N}_0}, \quad U_m \equiv \frac{\langle n, n+m|\widetilde{H}|n, n+m\rangle}{\langle n, n+m|n, n+m\rangle} - 2t_0.$$

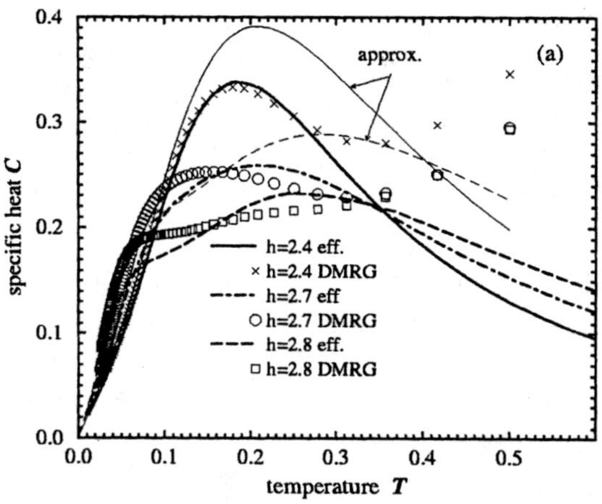

Fig. 14. Specific heat in the Luttinger regime: analytical vs. numerical results in the low-T regime.

From (51) one can see that the hopping amplitudes $t_m \propto x^m$ are very small for $m \geq 2$, $t_2 = -0.0242$, $t_3 = -0.0052$, ..., which justifies keeping only nearest-neighbor interaction terms in (53). The same reasoning holds for neglecting more complicated interaction terms.

The numerical values for the parameters t_0, t_1, and U_1 are $t_0 = 2.3370$, $t_1 = -0.2608$, and $U_1 = 0.1187$, so that the spin-$\frac{1}{2}$ chain (53) is in a gapless XY phase with $\Delta = J_z/J_{xy} \simeq 0.227J$. At the critical value [110] of the field $h_e = h_{e,c} = \frac{1}{2}J_{xy}(J + \Delta)$ the spin-$\frac{1}{2}$ chain undergoes a transition into the saturated ferromagnetic phase. The obvious symmetry $h_e \mapsto -h_e$ corresponds for the ferrimagnet to the symmetry around $H_0 = \frac{1}{2}(H_{c1} + H_{c2})$. Such a symmetry was indeed observed [96] for the behavior of the specific heat, though for the $H < H_0$ side it is somewhat plagued by the low-T contribution of the acoustical magnon branch. The critical fields for the model (53) are slightly different from real ones, $h_e = \pm h_{e,c}$ corresponds to $H = H'_{c1} \simeq 1.81J$ and $H = H'_{c2} \simeq 3.09J$, but the agreement is still remarkably good. For numerical comparisons the field interval is thus rescaled from (H'_{c1}, H'_{c2}) onto (H_{c1}, H_{c2}).

The temperature dependence of the specific heat for the anisotropic spin-$\frac{1}{2}$ chain has been calculated [51] using Klümper's version [111, 112] of the thermodynamic Bethe ansatz; this amounts to solving numerically the following system of two coupled nonlinear integral equations:

$$\begin{aligned} y_1 &= \frac{\pi\beta}{\gamma\cosh\frac{\pi x}{\gamma}} - \frac{\pi\beta\widetilde{h}}{2(\pi-\gamma)} - \int_{-\infty}^{+\infty} dx' \Big\{ K(x-x')L(y_1) \\ &- K(x-x'-i\gamma+i\delta)L(y_2) \Big\}, \end{aligned} \tag{55}$$

the other equation being of the same form with $i \to -i$, $y_1 \leftrightarrow y_2$, and $\widetilde{h} \to -\widetilde{h}$. Here $L(y) = \ln(1+e^{-y})$, $\Delta = \cos\gamma$, δ is an infinitesimal positive number, $\widetilde{h} = 4h_e/\sin\gamma$, $\beta = 2T/(J_{xy}\sin\gamma)$, and the kernel $K(x)$ is defined as

$$K(x) = \frac{1}{4\pi}\int_{-\infty}^{+\infty} dq \frac{\sinh[(\pi/2-\gamma)q]\,\cos qx}{\cosh(\gamma q/2)\sinh[(\pi-\gamma)q/2]}\,. \tag{56}$$

Then the specific heat $C_v(T)$ can be found as

$$C_v = \beta^2 \frac{\partial^2 f}{\partial \beta^2}, \quad f = \frac{1}{2\gamma} \int_{-\infty}^{+\infty} dx \, \frac{L(y_1) + L(y_2)}{\cosh(\pi x/\gamma)} \tag{57}$$

In Fig. 14 (from Ref. [51]) the results given by (57) are compared with the numerical transfer matrix DMRG data for the low-temperature part of $C_v(T)$ for the ferrimagnet in the thermodynamic limit. The agreement is generally satisfactory, and becomes very good near $H = H_0$. The "pop-up" peak behavior is clearly reproduced.

Kolezhuk et al. [51] also give a rather intuitive understanding of the behavior of $C_v(T)$ by mapping the XXZ chain to spinless fermions via a Jordan-Wigner transformation. The resulting Hamiltonian is studied in first-order perturbation theory for the quartic part in Δ. A simple expression for the fermion dispersion $\epsilon(k)$ results:

$$\begin{aligned} \epsilon(k) &= \Delta - \lambda - \cos k \\ &- (2\Delta/\pi)\, \theta(1-\lambda)\{ \arccos \lambda + (1-\lambda^2)^{1/2} \cos k \} . \end{aligned} \tag{58}$$

Here $\epsilon(k)$ is measured in J_{xy} units, $\lambda = 2h_e/J_{xy}$, and θ is the Heaviside function. The specific heat can now be trivially calculated; in Fig. 14, two curves obtained within the above approximation are shown, revealing that this simple approach nevertheless succeeds to capture the important physics. In zero field the contributions into the specific heat from particles and holes are equal; with increasing field, the hole bandwidth grows up, while the particle bandwidth decreases, and the average band energies do not coincide. This leads to the presence of two low-temperature peaks in $C_v(T)$: holes yield a strong, round peak moving towards higher temperatures with increasing the field, and the other peak (due to the particles) is weak, sharp, and moves to zero when h_e tends to $h_{e,c}$. In the spinon language, one has essentially the same picture for two bands of spinons with opposite spin. Remarkably this feature is also found in the numerical data, but it is not very pronounced.

4. Concluding remarks

Quasi-1D quantum Heisenberg ferrimagnets open an intriguing new direction in the field of low-dimensional magnetism. The appearence of quantized ferromagnetic phases with a conserved ferromagnetic order parameter makes these phases and the related phase transitions interesting, since in systems with a Heisenberg $SO(3)$ symmetry this is expected to lead to strong constraints on the critical field theories. Thus quasi-1D quantum Heisenberg ferrimagnetic systems with competing interactions (capable to destroy the macroscopic ferrimagnetic state) constitute an interesting direction for future studies.

Ferrimagnetic spin ladders set up another interesting direction for future studies. In particular, we would like to emphasize that the features of the critical phase discussed above are *general* and should manifest themselves in the critical regime of essentially *any* gapped 1D spin system in a sufficiently strong magnetic field, as far as it is possible to construct the mapping to a spin-$\frac{1}{2}$ chain. Spin-$\frac{1}{2}$ ladder compounds which often show a gap in zero field are natural candidates for such theoretical mappings and for experiments, although in these systems the physical origin of the gap (Haldane gap) is quite different from the case of the ferrimagnets. A particularly suitable candidate for experiment is $Cu_2(C_5H_{12}N_2)_2Cl_4$ [11, 102, 113], where the lower and upper critical fields are of the order of only 7 T and 13 T respectively. This compound however is plagued by a three-dimensional transition at very low temperature in the critical phase, obscuring the peculiar low-temperature features of the critical regime somewhat. Hagiwara et al. [114] were however able to separate clearly the (round) additional low temperature peak from the (sharp) transition peak, giving experimental proof of the features above. Comparing to theory, there are however some deviations. First, while the general story of the low-temperature peak is confirmed, the additional weak splitting into two is not observed. But this is a rather subtle feature anyways. More troubling is that features are correctly located, but at low temperatures a rather strong quantitative difference between experiment and theory emerges for the specific heat, which is linked to the transition, affecting the entropy balance of the compound. Much more theoretical work is needed here to come to a quantitatively satisfying picture.

Acknowledgments

The authors thank J. Richter, H.-J. Mikeska, A. K. Kolezhuk, S. Yamamoto, T. Fukui, S. Brehmer, K. Maisinger and M. Hagiwara for their various collaborations in this field. This work was supported by the Deutsche Forschungsgemeinschaft (Grant No. 436BUL 113/106/0, Scho 621/2) and the Bulgarian Science Foundation (Grant No. F812/98).

Appendix A Dyson-Maleev vertices for the ferrimagnetic model

The normal-ordered Dyson-Maleev quasiparticle interaction V_{DM} in Eq. (10) reads

$$\begin{aligned} V_{DM} &= -\frac{zJ}{4N}\sum_{1-4}\delta_{12}^{34}\Big[V^{(1)}_{12;34}\alpha_1^\dagger\alpha_2^\dagger\alpha_3\alpha_4 + 2V^{(2)}_{12;34}\alpha_1^\dagger\beta_2\alpha_3\alpha_4 + 2V^{(3)}_{12;34}\alpha_1^\dagger\alpha_2^\dagger\beta_3^\dagger\alpha_4 \\ &+ 4V^{(4)}_{12;34}\alpha_1^\dagger\alpha_3\beta_4^\dagger\beta_2 + 2V^{(5)}_{12;34}\beta_4^\dagger\alpha_3\beta_2\beta_1 + 2V^{(6)}_{12;34}\beta_4^\dagger\beta_3^\dagger\alpha_2^\dagger\beta_1 \\ &+ V^{(7)}_{12;34}\alpha_1^\dagger\alpha_2^\dagger\beta_3^\dagger\beta_4^\dagger + V^{(8)}_{12;34}\beta_1\beta_2\alpha_3\alpha_4 + V^{(9)}_{12;34}\beta_4^\dagger\beta_3^\dagger\beta_2\beta_1\Big], \end{aligned} \tag{A.1}$$

where $(k_1, k_2, k_3, k_4) \equiv (1, 2, 3, 4,)$, $\delta_{12}^{34} \equiv \delta(1 + 2 - 3 - 4)$ is the Kronecker δ-function, and z is the lattice coordination number. The explicit form of the symmetric vertex functions for the ferrimagnetic Heisenberg model (1) read

$$V^{(i)}_{12;34} = u_1u_2u_3u_4\bar{V}^{(i)}_{12;34}, \qquad i = 1, 2\cdots, 9, \tag{A.2}$$

where

$$\begin{aligned} \bar{V}^{(1)}_{12;34} &= + \;\; \gamma_{1-3}x_1x_3 + \gamma_{1-4}x_1x_4 + \gamma_{2-3}x_2x_3 + \gamma_{2-4}x_2x_4 \\ &\;\;\; - \;\; \sigma^{1/2}(\gamma_{1-3-4}x_1x_3x_4 + \gamma_{2-3-4}x_2x_3x_4) - \sigma^{-1/2}(\gamma_1x_1 + \gamma_2x_2), \\ \bar{V}^{(2)}_{12;34} &= - \;\; \gamma_{1-3}x_1x_2x_3 - \gamma_{1-4}x_1x_2x_4 - \gamma_{2-3}x_3 - \gamma_{2-4}x_4 \\ &\;\;\; + \;\; \sigma^{1/2}(\gamma_{1-3-4}x_1x_2x_3x_4 + \gamma_{2-3-4}x_3x_4) + \sigma^{-1/2}(\gamma_1x_1x_2 + \gamma_2), \\ \bar{V}^{(3)}_{12;34} &= - \;\; \gamma_{1-3}x_1 - \gamma_{1-4}x_1x_3x_4 - \gamma_{2-3}x_2 - \gamma_{2-4}x_2x_3x_4 \\ &\;\;\; + \;\; \sigma^{1/2}(\gamma_{1-3-4}x_1x_4 + \gamma_{2-3-4}x_2x_4) + \sigma^{-1/2}(\gamma_1x_1x_3 + \gamma_2x_2x_3), \\ \bar{V}^{(4)}_{12;34} &= + \;\; \gamma_{1-3}x_1x_2x_3x_4 + \gamma_{1-4}x_1x_2 + \gamma_{2-3}x_3x_4 + \gamma_{2-4} \\ &\;\;\; - \;\; \sigma^{1/2}(\gamma_{1-3-4}x_1x_2x_3 + \gamma_{2-3-4}x_3) - \sigma^{-1/2}(\gamma_1x_1x_2x_4 + \gamma_2x_4), \\ \bar{V}^{(5)}_{12;34} &= - \;\; \gamma_{1-3}x_2x_3x_4 - \gamma_{1-4}x_2 - \gamma_{2-3}x_1x_3x_4 - \gamma_{2-4}x_1 \\ &\;\;\; + \;\; \sigma^{1/2}(\gamma_{1-3-4}x_2x_3 + \gamma_{2-3-4}x_1x_3) + \sigma^{-1/2}(\gamma_1x_2x_4 + \gamma_2x_1x_4), \\ \bar{V}^{(6)}_{12;34} &= - \;\; \gamma_{1-3}x_4 - \gamma_{1-4}x_3 - \gamma_{2-3}x_1x_2x_4 - \gamma_{2-4}x_1x_2x_3 \\ &\;\;\; + \;\; \sigma^{1/2}(\gamma_{1-3-4} + \gamma_{2-3-4}x_1x_2) + \sigma^{-1/2}(\gamma_1x_3x_4 + \gamma_2x_1x_2x_3x_4), \end{aligned}$$

$$\bar{V}^{(7)}_{12;34} = + \gamma_{1-3}x_1x_4 + \gamma_{1-4}x_1x_3 + \gamma_{2-3}x_2x_4 + \gamma_{2-4}x_2x_3,$$
$$- \sigma^{1/2}(\gamma_{1-3-4}x_1 + \gamma_{2-3-4}x_2) - \sigma^{-1/2}(\gamma_1x_1x_3x_4 + \gamma_2x_2x_3x_4)$$
$$\bar{V}^{(8)}_{12;34} = + \gamma_{1-3}x_2x_3 + \gamma_{1-4}x_2x_4 + \gamma_{2-3}x_1x_3 + \gamma_{2-4}x_1x_4$$
$$- \sigma^{1/2}(\gamma_{1-3-4}x_2x_3x_4 + \gamma_{2-3-4}x_1x_3x_4) - \sigma^{-1/2}(\gamma_1x_2 + \gamma_2x_1),$$
$$\bar{V}^{(9)}_{12;34} = + \gamma_{1-3}x_2x_4 + \gamma_{1-4}x_2x_3 + \gamma_{2-3}x_1x_4 + \gamma_{2-4}x_1x_3$$
$$- \sigma^{1/2}(\gamma_{1-3-4}x_2 + \gamma_{2-3-4}x_1) - \sigma^{-1/2}(\gamma_1x_2x_3x_4 + \gamma_2x_1x_3x_4).$$

Here u_k and x_k are the coefficients of the Bogoliubov transformation, Eq. (9), $\gamma_k = \cos(ka_0)$, and $\sigma = S_1/S_2$.

References

[1] Affleck, I. 1989. *J. Phys.: Condens. Matter*, **1**, 3047-3072.

[2] Haldane. F. D. M. 1983. *Phys. Lett. A*, **93**, 464-468.

[3] Haldane. F. D. M. 1983. *Phys. Rev. Lett.*, **50**, 1153-1156.

[4] Kashurnikov, V. A., N. V. Prokof'ev, B. V. Svistunov and M. Troyer. 1999. *Phys. Rev. B*, **59**, 1162-1167.

[5] Buyers, W. J. L., R. M. Morra, R. L. Armstrong, M. J. Hogan, P. Gerlach and K. Hirakawa. 1986. *Phys. Rev. Lett.*, **56**, 371-374.

[6] Hida, K. 1994. *J. Phys. Soc. Jpn.*, **63**, 2359-2364.

[7] Fledderjohann, A., C. Gerhardt, M. Karbach, K.-H. Mütter and R. Wießner. 1999. *Phys. Rev. B*, **59**, 991-999.

[8] Narumi, Y., M. Hagiwara, R. Sato, K. Kindo, H. Nakano and M. Takahashi. 1998. *Physica B*, **246-247**, 509-512.

[9] Shiramura, W., K. Takatsu, B. Kurniawan, H. Tanaka, H. Uekusa, Y. Ohashi, K. Takizawa, H. Mitamura and T. Goto. 1998. *J. Phys. Soc. Jpn.*, **67**, 1548-1551.

[10] Dagotto, E. and T. M. Rice. 1996. *Science*, **271**, 618.

[11] Giamarchi, T. and A. M. Tsvelik. 1999. *Phys. Rev. B*, **59**, 11389-11407.

[12] Sachdev, S. and T. Senthil. 1996. *Annals of Physics*, **251**, 76.

[13] Kahn, O., Y. Pei and Y. Journaux. 1992. in: *Inorganic Materials*, John Wiley & Sons Ltd, New York, 59-114.

[14] Kahn, O. 1993. *Molecular Magnetism*, VCH, New York.

[15] Verdaguer, M., A. Gleizes, J. P. Renard and J. Seiden. 1984. *Phys. Rev. B*, **29**, 5144-5155.

[16] Gleizes, A. and M. Verdaguer. 1984. *J. Am. Chem. Soc.*, **106**, 3727.

[17] Pei, Y., M. Verdaguer, O. Kahn, J. Sletten and J. P. Renard. 1987. *Inorg. Chem.*, **26**, 138.

[18] Hagiwara, M., K. Minami, Y. Narumi, K. Tatani and K. Kindo. 1998. *J. Phys. Soc. Jpn.*, **67**, 2209-2211.

[19] Macêdo, A. M. S., M. C. dos Santos, M. D. Coutinho-Filho and C. A. Macêdo. 1995. *Phys. Rev. Lett.*, **74**, 1851-1854.

[20] Raposo, E. P. and M. D. Coutinho-Filho. 1995. *Mod. Phys. Lett. B*, **9**, 817-822.

[21] Takano, K., K. Kubo and H. Sakamoto. 1996. *J. Phys.: Condens. Matter*, **8**, 6405-6411.

[22] Alcaraz, F. C. and A. L. Malvezzi. 1997. *J. Phys. A: Math. Gen.*, **30**, 767-778.

[23] Niggemann, H., G. Uimin and J. Zittartz. 1997. *J. Phys.: Condens. Matter*, **9**, 9031-9042.

[24] Raposo, E. P. and M. D. Coutinho-Filho. 1999. *Phys. Rev. B*, **59**, 14384-14405.

[25] Nishide, H. 1995. *Adv. Mater.*, **7**, 937-941.

[26] Lieb, E. H. and D. Mattis. 1962. *J. Math. Phys.*, **3**, 749-751.

[27] Fukui, T. and N. Kawakami. 1997. *Phys. Rev. B*, **55**, R14709-R14712.

[28] Ivanov, N. B. and J. Richter. 1997. *Phys. Lett. A*, **232**, 308-312.

[29] Richter, J., N. B. Ivanov and J. Schulenburg. 1998. *J. Phys.: Condens. Matter*, **55**, 3635-3649.

[30] Drillon, M., J. C. Gianduzzo and R. Georges. 1983. *Phys. Lett. A*, **96**, 413-416.

[31] Seiden, J. 1983. *J. Phys. Lett. (France)*, **44**, L947.

[32] Georges, R., J. Curely, J. C. Gianduzzo, X. Qiang, O. Kahn and Y. Pei. 1988. *Physica B*, **153**, 77-84.

[33] de Vega, H. J. and F. Woynarovich. 1992. *J. Phys. A: Math. Gen.*, **25**, 4499-4516.

[34] Martins, M. J. 1993. *J. Phys. A: Math. Gen.*, **26**, 7301-7309.

[35] Nielsen, H. B. and S. Chandra. 1976. *Nucl. Phys. B*, **105**, 445.

[36] Halperin, B. I. and P. C. Hohenberg. 1969. *Phys. Rev.*, **188**, 898-918.

[37] Keffer, F. in: *Handbuch der Physik*, Vol. XVIII/2, Springer-Verlag, New York, 1.

[38] Wolf, W. P. 1961. *Rep. Prog. Phys.*, **24**, 212.

[39] Hohenberg, P. C. and J. Swift. 1974. *J. Phys. C*, **7**, 4009.

[40] Halperin, B. I. and W. M. Saslow. 1977. *Phys. Rev. B*, **16**, 2154-2162.

[41] Nakamura, T. and M. Bloch 1963. *Phys. Rev.* , **132**, 2528-2539.

[42] Harris, A. B., D. Kumar, B. I. Halperin and P. C. Hohenberg. 1971. *Phys. Rev. B*, **3**, 961-1024.

[43] Pati, S. K., S. Ramasesha and D. Sen. 1997. *Phys. Rev. B*, **55**, 8894-8904.

[44] Pati, S. K., S. Ramasesha and D. Sen. 1997. *J. Phys.: Condens. Matter*, **9**, 8707-8726.

[45] Brehmer, S, H.-J. Mikeska and S. Yamamoto. 1997. *J. Phys.: Condens. Matter*, **9**, 3921-3930.

[46] Oguchi, T. 1960. *Phys. Rev.*, **117**, 117-123.

[47] Ivanov, N. B., J. Richter and U. Schollwöck. 1998. *Phys. Rev. B*, **58**, 14456-14461.

[48] Yamamoto, S., S. Brehmer and H.-J. Mikeska. 1998. *Phys. Rev. B*, **57**, 13610-13616.

[49] Ivanov, N. B. 1998. *Phys. Rev. B*, **57**, R14024-R14027.

[50] Ivanov, N. B. 2000. *Phys. Rev. B*, in press.

[51] Kolezhuk, A. K., H.-J. Mikeska, K. Maisinger and U. Schollwöck. 1999. *Phys. Rev B*, **59**, 13565-13568.

[52] Ono, T., T. Nishimura, M. Katsumura, T. Morita and M. Sugimoto. 1997. *J. Phys. Soc. Jpn.*, **66**, 2576-2579.

[53] Bursill, R., G. A. Gehring, D. J. J. Farnell, J. B. Parkinson, T. Xiang, and C. Zeng. 1995. *J. Phys.: Condens. Matter*, **7**, 8605-8618.

[54] Chitra, R., S. Pati, H. R. Krishnamurthy, D. Sen, and S. Ramasesha. 1995. *Phys. Rev. B*, **52**, 6581-6587.

[55] Schollwöck, U., Th. Jolicoeur and T. Garel. 1996. *Phys. Rev. B*, **53**, 3304-3311.

[56] Kolezhuk, A., R. Roth and U. Schollwöck. 1996. *Phys. Rev. Lett.*, **77**, 5142-5145.

[57] White, S. R. and I. Affleck. 1996. *Phys. Rev. B*, **54**, 9862-9869.

[58] Sachdev, S. 1994. *Z. Phys. B*, **94**, 469-479.

[59] Adams, S. R. and M. J. Martins. 1993. *J. Phys. A: Math. Gen.* **26**, L529.

[60] Fujii, M., S. Fujimoto and N. Kawakami. 1996. *J. Phys. Soc. Jpn.*, **65**, 2381-2384.

[61] Richter, J., N. B. Ivanov, and K. Retzlaff. 1994. *Europhys. Lett.*, **25**, 545-550.

[62] Richter, J., U. Schollwöck and N. B. Ivanov. 1999. *Physica B*, in press.

[63] Azaria, P., B. Delamotte and T. Jolicœur. 1990. *Phys. Rev. Lett.* **64**, 3175-3178.

[64] Stephenson, J. 1970. *J. Math. Phys.*. **12**, 420.

[65] Lieb, E. H., T. Schultz and D. J. Mattis. 1961. *Ann. Phys. (N.Y.)*, **16**, 407.

[66] Chubukov, A. V. 1991. *Phys. Rev. B*, **44**, 4693-4696.

[67] Krivnov, V. Ya. and A. A. Ovchinnikov. 1996. *Phys. Rev. B*, **53**, 6435-6441.

[68] Kahn O., Y. Pei and Y. Journaux. 1995. in: *Inorganic Materials*, Wiley, New York.

[69] Hagiwara, M. K. Minami, Y. Narumi, K. Tatani and K. Kindo. 1999. *J. Phys. Soc. Jpn.*, **68**, 2214-2217.

[70] Bonner, J. C. and M. E. Fisher. 1964. *Phys. Rev.* , **135**, A640-A658.

[71] Yamamoto, S. 1997. *Int. J. Mod. Phys. C*, **8**, 609.

[72] Takahashi, M. 1986. *Prog. Theor. Phys. Supp.*, **87**, 233-245.

[73] Takahashi, M. 1987. *Phys. Rev. Lett.*, **58**, 168-170.

[74] Takahashi, M. 1989. *Phys. Rev. B*, **40**, 2494-2501.

[75] Hirsch, J. E. and S. Tang. 1989. *Phys. Rev. B*, **40**, 4769-4772.

[76] Tang, S., M. E. Lazzouni and J. E. Hirsch. 1989. *Phys. Rev. B*, **40**, 5000-5006.

[77] Rezende, S. M. 1990-1991. *Phys. Rev. B*, **42**, 2589.

[78] Hida, K. 1996. *J. Phys. Soc. Jpn.*, **65**, 594-600.

[79] Yamamoto, S. and T. Fukui. 1998. *Phys. Rev. B* **57**, R14008-R14011.

[80] Yamamoto, S., T. Fukui, K. Maisinger and U. Schollwöck. 1998. *J. Phys.: Condens. Matter*, **10**, 11033-11048.

[81] Wu, C., B. Chen, X. Dai, Y. Yu and Z. B. Su. 1999. *Phys. Rev. B* **60**, 1057-1063.

[82] Takahashi, M. and M. Yamada. 1985. *J. Phys. Soc. Jpn.*, **54**, 2808-2811.

[83] Yamada, M. and M. Takahashi. 1986. *J. Phys. Soc. Jpn.*, **55**, 2024-2036.

[84] Read, N. and S. Sachdev. 1995. *Phys. Rev. Lett.*, **75**, 3509-3512.

[85] Takahashi, M., H. Nakamura and S. Sachdev. 1996. *Phys. Rev. B*, **54**, R744-R747.

[86] Suzuki, M. 1976. *Prog. Theor. Phys.*, **56**, 1454-1469.

[87] Hirsch, J. E., R. L. Sugar, D. J. Scalapino and R. Blankenbecler. 1982. *Phys. Rev. B* **26**, 5033-5055.

[88] Yamamoto, S. 1995. *J. Phys. Soc. Jpn.*, **64**, 4051-4052.

[89] White, S. R. 1992. *Phys. Rev. Lett.*, **69**, 2863-2866.

[90] White, S. R. 1993. *Phys. Rev. B*, **48**, 10345-10356.

[91] Nishino, T. 1995. *J. Phys. Soc. Jpn.*, **64**, 3598.

[92] Bursill, R. J., T. Xiang and G. A. Gehring. 1996. *J. Phys.: Condens. Matter*, **8**, L583-L590.

[93] Shibata, N. 1997. *J. Phys. Soc. Jpn.*, **66**, 2221-2223.

[94] Wang, X. and T. Xiang. 1997. *Phys. Rev. B*, **56**, 5061-5064.

[95] Maisinger, K. and U. Schollwöck. 1998. *Phys. Rev. Lett.*, **81**, 445-448.

[96] Maisinger, K., U. Schollwöck, S. Brehmer, H.-J. Mikeska and S. Yamamoto. 1998. *Phys. Rev. B*, **58**, R5908-R5911.

[97] Affleck, I. 1990. *Phys. Rev. B*, **41**, 6697-6702.

[98] Tsvelik, A. 1990. *Phys. Rev. B*, **42**, 10499-10504.

[99] Takahashi, M. and T. Sakai. 1991. *J. Phys. Soc. Jpn.*, **60**, 760-763.

[100] Takahashi, M. and T. Sakai. 1991. *J. Phys. Soc. Jpn.*, **60**, 3615-3519.

[101] Takahashi, M. and T. Sakai. 1991. *Phys. Rev. B*, **43**, 13383-13393.

[102] Mila, F. 1998. *Eur. Phys. J. B*, **6**, 201.

[103] Chitra, R. and T. Giamarchi. 1997. *Phys. Rev. B*, **55**, 5816-5826.

[104] Kolezhuk, A. K., H.-J. Mikeska and S. Yamamoto. 1997. *Phys. Rev. B*, **55**, R3336-R3339.

[105] Fannes, M., B. Nachtergaele and R. F. Werner. 1989. *Europhys. Lett.* **10**, 633-637.

[106] Fannes, M., B. Nachtergaele and R. F. Werner. 1992. *Commun. Math. Phys.*, **144**, 443-490.

[107] Klümper, A., A. Schadschneider and J. Zittartz. 1991. *J. Phys. A*, **24**, L955-L959.

[108] Klümper, A., A. Schadschneider and J. Zittartz. 1992. *Z. Phys. B*, **87**, 281-287.

[109] Klümper, A., A. Schadschneider and J. Zittartz. 1993. *Europhys. Lett.*, **24**, 293-297.

[110] Korepin, V. E., N. M. Bogoliubov and A. G. Izergin. 1993. *Quantum inverse scattering method and correlation functions.*, Cambridge University Press.

[111] Klümper, A. 1993. *Z. Phys. B*, **91**, 507-519.

[112] Klümper, A. 1998. *Eur. Phys. J. B*, **5**, 677-685.

[113] Chaboussant, G., M.-H. Julien, Y.F.-Revurat, M. Hansen, L. P. Lévy and C. Berthier. 1998. *Eur. Phys. J. B*, **6**, 167-181.

[114] Hagiwara, M., H.A. Katori, U. Schollwöck and H.J. Mikeska. 1999. *Physica B*, in press.

HARMONIC APPROXIMATION
UNDER EXTERNAL RADIATION

Finite Temperature Properties of Electron-Phonon Systems in Semiconducting Heterostructures

Miguel Eduardo Mora-Ramos *

1. Introduction

If the motion of ions in a crystal lattice implies small deviations from their equilibrium positions, it is possible to describe it within the harmonic approximation and the subsequent quantization of the field associated to the ion displacements -considered as a system of uncoupled quantum harmonic oscillators- leads us to the concept of phonons as elementary excitations in the solid.

Electronic, optical, transport properties in the solid are more or less significatively affected by the scattering effect upon charge carriers due to phonon fields corresponding to one or more kinds of oscillation modes. To illustrate, let us just mention two particular phenomena: This interaction is responsible to a great extent of the electrical resistance but also determines the existence of superconducting states in certain materials at low temperatures.

A great variety of situations may appear depending on: the nature of the electronic subsystem (a single electron -or hole-, as is common to deal with in semiconductors and insulators, or many electrons as in the case of metals); the type of phonons under consideration (acoustic, optical, piezoelectric); the very strength of the carrier-phonon interaction; the dimensionality of the system as well as the presence of additional interactions like external radiation, applied electric and magnetic fields, electron-hole attraction, Coulombic forces between carriers or between the carriers and impurity centers, and so on. Actually, many important problems of the physics of solid state nowadays exhibit significant manifestations of phonon-related scattering processes of charge carriers.

Electron-phonon interaction in bulk polar materials

The dynamical problem of the motion of the ions in the crystal lattice can be microscopically solved by means of numerical calculations. Nevertheless, in the limit of large wavelengths (small wavevectors) oscillations it is possible to derive a theoretical treatment based on a macroscopic continuum model [1].

In all crystals the traslational invariance of the lattice gives rise to the existence of three oscillation branches whose frequencies tend to zero as the wavelength approaches infinity. These are known as acoustic vibration modes. However, additional branches are present when the unit cell contains two or more atoms. Such modes with frequencies that remain finite even for zero wavevector are known as optical ones. In the case where ion cores have charges of opposite signs, electrical neutrality is achieved only at the elementary cell level. The long range electrical interactions between the oppositely charged ions practically does not affect the long wave acoustic oscillation for in it ions displace approximately in phase without altering the electrical neutrality of the cell. On the contrary, the electrical interaction is decisive for the optical vibrations. The motion of charged ions about their

*Universidad Autónoma del Estado de Morelos Cuernavaca. Morelos México

equilibrium positions creates a polarization effect which causes the appearance of an electrical field interacting with the ions, thus affecting the oscillation process itself. The optical phonons associated to the resulting oscillation branches are usually known as polar optical phonons.

The macroscopic model of the polar optical oscillation modes in isotropic ionic or partially ionic (iono-covalent) crystals was developed a few decades ago [1], and the theory which describes the interaction of charge carriers (e.g. conduction electrons) with the electrostatic potential field associated to those phonon modes was established in the fifties and is known as Fröhlich model [2].

In this continuum phenomenological approach the -mechanical- field of ion displacements $\mathbf{u}(\mathbf{r})$ can be separated into two independent fields;

$$\mathbf{u}(\mathbf{r}) = \mathbf{u}_T(\mathbf{r}) + \mathbf{u}_L(\mathbf{r}), \tag{1}$$

in such a way that $\nabla \cdot \mathbf{u}_T(\mathbf{r}) = 0$, and $\nabla \times \mathbf{u}_L(\mathbf{r}) = 0$. The eigenfrequency of the solenoidal -transversal- vibrations is determined only by the elastic forces in the crystal and is typically labelled as ω_{TO}. The characteristic frequency of the irrotational -longitudinal- mode, ω_{LO}, is higher, and is related with ω_{TO} through the Lyddane-Sachs-Teller law

$$\omega_{LO}^2 = \omega_{TO}^2 \frac{\epsilon_0}{\epsilon_\infty}, \tag{2}$$

where ϵ_0 (ϵ_∞) is the static (high frequency) dielectric constant of the material. So far this macroscopic phenomenological approach excludes dispersion and constitutes the standard form of the Born-Huang model [1]. The only contribution of the mechanical oscillations to the macroscopic polarization field comes from the longitudinal part $\mathbf{u}_L(\mathbf{r})$ [2, 3], which means that the electrostatic potential originated from the long wave polar optical vibrations is merely associated to their longitudinal part. Hence the electrical scattering of charge carriers will be associated precisely to the longitudinal optical (LO) polar phonons.

Writing the general expression for the purely longitudinal vibration field as a normal mode expansion;

$$\mathbf{u}_L(\mathbf{r}) = \sum_{\mathbf{q}} \mathbf{q} \left(A_{\mathbf{q}} e^{i\mathbf{q}\cdot\mathbf{r}} + A_{\mathbf{q}}^* e^{-i\mathbf{q}\cdot\mathbf{r}} \right) \tag{3}$$

(with $A_{\mathbf{q}}^* = A_{-\mathbf{q}}$, ensuring that $\mathbf{u}_L(\mathbf{r})$ be real), the electrical polarization operator in the second quantized form has the form:

$$\mathbf{P}_L(\mathbf{r}) = \sum_{\mathbf{q}} \left(\frac{\hbar\omega_{LO}}{8\pi\epsilon^* V} \right)^{1/2} \frac{\mathbf{q}}{|\mathbf{q}|} \left(e^{i\mathbf{q}\cdot\mathbf{r}} \hat{b}_{\mathbf{q}} + \text{H.C.} \right). \tag{4}$$

In the former expression $\hat{b}_{\mathbf{q}}$ is the annihilation operator of a phonon with wavevector $\mathbf{q}$. V is the volume of the crystal, and $1/\epsilon^* = 1/\epsilon_\infty - 1/\epsilon_0$.

The electron-LO-phonon interaction in polar crystals can be described with the use of the following Hamiltonian, named after Fröhlich [2]:

$$\hat{H}_F(\mathbf{r}) = \sum_{\mathbf{q}} \left(C_{\mathbf{q}} e^{i\mathbf{q}\cdot\mathbf{r}} \hat{b}_{\mathbf{q}} + \text{H.C.} \right); \tag{5}$$

$$C_{\mathbf{q}} = -ie \left(\frac{2\pi\hbar\omega_{LO}}{\epsilon^* V} \right)^{1/2} \frac{1}{|\mathbf{q}|}. \tag{6}$$

Material	α
InSb	0.015
GaAs	0.068
InP	0.080
AlAs	0.17
CdTe	0.315
CdSe	0.460
CdS	0.527
AgCl	1.90
KI	2.60
KBr	3.22
RbI	3.35
KCl	3.69
RbCl	4.09

Table 1: The values of the Fröhlich constant for some partially ionic (top) and fully ionic crystals (bottom).

To provide a characterization of the strength of the coupling between an electron with effective mass m and the long wave LO phonons, Fröhlich introduced the dimensionless parameter α, wich is defined through

$$\alpha = \frac{e^2}{\epsilon^* \hbar} \left(\frac{m}{2\hbar\omega_{LO}} \right)^{1/2}; \tag{7}$$

from which, expression (6) is transformed to

$$C_{\mathbf{q}} = -i\hbar\omega_{LO} \left(\frac{4\pi\alpha\hbar}{2m\omega_{LO}} \right)^{1/2} \frac{1}{|\mathbf{q}|}.$$

The values of the Fröhlich constant for some ionic and partiallly ionic materials are shown in table 1.

Electron-phonon interaction in semiconducting heterostructures

As can be seen from the upper part of table 1, III-V and II-IV semiconductors are weakly polar materials in the sense that the magnitude of the coupling between electrons and long wave longitudinal optical phonons is small. Nonetheless, electron-LO-phonon scattering effects have been shown to be of much importance in the study of electronic and optical properties in these compounds [4].

In the case of low dimensional semiconducting structures, the spatial confinement is responsible for many new physical effects that concern to practically all properties in such systems [5]. In a particular manner, when these nanostructures are fabricated from polar materials, substantial differences are found in the optical phonon spectrum if compared with the bulk case, due to the presence of interfaces (the reader is referred to the concise but very complete review by Stroscio et al. [6]). As a consequence, electron-phonon interaction exhibits important modifications as well (see, for instance, [7] and references therein).

Among the most commonly used schemes for the description of single-electron states in a semiconducting structure bearing interfaces, it is the so-called *envelope function approximation* [5]. Usually, the procedure involves the solution of certain effective mass eigenproblem;

$$\hat{\mathcal{H}}\Psi_\nu(\mathbf{r}) = E_\nu \Psi_\nu(\mathbf{r}). \tag{8}$$

The information contained in the effective mass Hamiltonian $\hat{\mathcal{H}}$ normally includes the accomodation and change of profile of the energy bands pertaining to the materials separated by the interfaces -effect which usually gives rise to a confining potential-, plus some additional effects like the presence of external applied fields, interaction between electrons themselves and with certain scattering centers, as well as coupling between different energy bands. The Greek letter "ν" represents the set of quantum numbers (including spin) that completely describes the state with energy E_ν in the system.

Going over a formulation based on the second quantization, the quantum fields representing the electron subsystem can be written as expansions over the complete set of single-electron vector states $\Psi_\nu(\mathbf{r}) = \langle \mathbf{r}|\nu\rangle$:

$$\begin{aligned}\hat{\Psi}(\mathbf{r}) &= \sum_\nu \langle \mathbf{r}|\nu\rangle \hat{a}_\nu;\\ \hat{\Psi}^\dagger(\mathbf{r}) &= \sum_\nu \langle \mathbf{r}|\nu\rangle \hat{a}_\nu^\dagger.\end{aligned} \tag{9}$$

$\hat{a}_\nu$ ($\hat{a}_\nu^\dagger$) will be the annihilation (creation) operator of one electron in the state with quantum numbers ν.

Then, the non-interacting many-electron subsystem will be characterized by a second-quantized Hamiltonian $\hat{H}_e$ given by:

$$\hat{H}_e = \sum_\nu E_\nu \hat{a}_\nu^\dagger \hat{a}_\nu. \tag{10}$$

At the same time, the generic form of the Hamiltonian characterizing the non-interacting subsystem of optical phonons in a polar quantum heterostructures will be

$$\hat{H}_{ph} = \sum_s \hbar\omega_s \left(\hat{b}_s^\dagger \hat{b}_s + \frac{1}{2}\right). \tag{11}$$

The index "s" in (11) labels the set of quantum numbers associated to the phonon oscillation mode with characteristic dispersion law ω_s. This set -as happens for the ν-numbers as well- strongly depends on the geometry and design of the heterostructure.

In this context, a general expression for the static Fröhlich-like electron-polar-optical-phonon interaction Hamiltonian would be of the form

$$\hat{H}_F = \sum_s \left[\varphi_s(\mathbf{r})\hat{b}_s + \text{H.C.}\right]. \tag{12}$$

To mention some particular cases of the Hamiltonian (12), let us consider the specific situation in which the active region in the heterostructure is made -for instance- of $GaAs$, and is embedded in a material with considerably different dielectroelastic properties like $AlAs$. Within this framework, we could find:

¬ *Quasi-twodimensional layered double heterostructures* [7, 8]

$$\hat{H}_F = \sum_{\mathbf{q},m} 2e\left(\frac{\pi}{\hbar\epsilon^* S d}\right)^{1/2} \hbar\omega_{LO}\left[F_{\mathbf{q},m}(z)e^{i\mathbf{q}\cdot\mathbf{R}}\hat{b}_{\mathbf{q},m} + \text{H.C.}\right]. \tag{13}$$

In this expression, $s = \{\mathbf{q}, m\}$; $\mathbf{q} = (q_x, q_y)$, and m is a discrete index numbering the characteristic oscillation modes in the structure. $\mathbf{R} = (x, y)$; S is the transversal area of the *GaAs* layer and d is the quantum well width, assuming "z" to be the growth direction. $1/\epsilon^*$, and ω_{LO} correspond to the bulk *GaAs*.

¬ *Quasi-onedimensional cylindrical quantum wires*

In this structure, the set $s = \{q_z, n, m\}$. The interaction Hamiltonian for the oscillations perpendicular to the wire axis (z-axis), in the specific situation where $q_z = 0$, is [7]:

$$\hat{H}_F = \sum_{n,m} \pi e r_0 \left(\frac{2\rho}{\hbar\epsilon^* V}\right)^{1/2} \hbar\omega_{LO} \left[F_{n,m}(r) e^{in\theta} \hat{b}_{nm} + \text{H.C.}\right]; \tag{14}$$

where r_0 is the radius of the cylinder, θ the polar angular coordinate, V is the volume of the wire, and ρ is the *GaAs* mass density.

¬ *Quasi-zerodimensional spherical quantum dots*

The quantum numbers $s = \{n, l, m\}$, in this case, are all discrete. The corresponding interaction Hamiltonian is [7]:

$$\hat{H}_F = \sum_{n,l,m} e \left(\frac{2\pi}{\hbar\epsilon^* R}\right)^{1/2} \hbar\omega_{LO} F_{n,l}(r) Y_{lm}(\theta, \phi) \left(\hat{b}_{nm} + \hat{b}^\dagger_{nm}\right); \tag{15}$$

with R representing the sphere radius, and $Y_{lm}(\theta, \phi)$ being the spherical harmonics.

When passing to the complete second-quantized description for both electrons and phonons, the Hamiltonian (12) can be written in a general form, disregarding any particular case, which is;

$$\hat{H}_F = \sum_s \sum_{\mu,\nu} \left[P^s_{\mu\nu} \hat{a}^\dagger_\mu \hat{a}_\nu \hat{b}_s + \text{H.C.}\right]; \tag{16}$$

$$P^s_{\mu\nu} = \langle \mu | \varphi_s(\mathbf{r}) | \nu \rangle. \tag{17}$$

This will be the model Hamiltonian to be used in the next section to study the many-body electron-phonon system.

2. Many-body electron-phonon system in low dimensional heterostructures

In this part the analysis will focus on the many-body properties of the interacting electron-phonon system in a nanostructure made of ionic materials. The treatment is based upon the Matsubara formalism for finite temperatures [9]. By writing the imaginary-time-dependent electronic annihilation and creation operators through the Heisenberg representation,

$$\begin{aligned} \hat{a}^\dagger_\nu(\tau) &= e^{\hat{K}\tau} \hat{a}^\dagger_\nu e^{-\hat{K}\tau} \\ \hat{a}_\nu(\tau) &= e^{\hat{K}\tau} \hat{a}_\nu e^{-\hat{K}\tau}, \end{aligned} \tag{18}$$

the one-particle electron Green's function is defined by:

$$G_\nu(\tau - \tau') = -\langle \hat{T} \left[\hat{a}_\nu(\tau) \hat{a}^\dagger_\nu(\tau')\right] \rangle. \tag{19}$$

In this expression $\langle \ldots \rangle$ means a statistical average -at thermal equilibrium-, and $\hat{T}$ is the operator of chronological ordering. This is equivalent to write

$$G_\nu(\tau-\tau') = -Tr\left[e^{-\beta(\hat{K}-\Omega)}\hat{T}e^{\hat{K}\tau}\hat{a}_\nu e^{-\hat{K}(\tau-\tau')}\hat{a}_\nu^\dagger e^{-\hat{K}\tau'}\right]$$
$$= -\theta(\tau-\tau')Tr\left[e^{-\beta(\hat{K}-\Omega)}\hat{T}e^{\hat{K}(\tau-\tau')}\hat{a}_\nu e^{-\hat{K}(\tau-\tau')}\hat{a}_\nu^\dagger\right]$$
$$+\theta(\tau'-\tau)Tr\left[e^{-\beta(\hat{K}-\Omega)}\hat{T}e^{\hat{K}(\tau'-\tau)}\hat{a}_\nu^\dagger e^{-\hat{K}(\tau'-\tau)}\hat{a}_\nu\right]. \tag{20}$$

Besides, the new operator $\hat{K}$ is $\hat{K} = \hat{H} - \varepsilon\hat{N}$, where $\hat{H} = \hat{H}_e + \hat{H}_{ph} + \hat{H}_F$ is the total electron-phonon Hamiltonian, $\hat{N}$ is the operator of the total number of electrons in the system. and ε is the chemical potential. The Matsubara time τ is defined in the interval $-\beta \leq \tau \leq \beta$, where $\beta = (k_B T)^{-1}$ (k_B being the Bolztmann constant and T the absolute temperature).The symbol $Tr[...] = \sum_n \langle n|...|n\rangle$ means the trace over a complete set of the combined electron-phonon system.

The thermodynamical potential of the system, Ω, is given through

$$e^{-\beta\Omega} = Tr\left[e^{-\beta\hat{K}}\right] \tag{21}$$

As it is seen from (20), the right hand side of the equation (19) is a function of the difference $\tau - \tau'$ only. This fact allows to eliminate one of the imaginary time variables and write

$$G_\nu(\tau) = -\langle\hat{T}\left[\hat{a}_\nu(\tau)\hat{a}_\nu^\dagger(0)\right]\rangle = -Tr\left[e^{-\beta(\hat{K}-\Omega)}\hat{T}\left(e^{\hat{K}\tau}\hat{a}_\nu e^{-\hat{K}\tau}\hat{a}_\nu'\right)\right] \tag{22}$$

With the statistical sum defined as

$$\Xi = Tr\left[e^{-\beta\hat{H}}\right], \tag{23}$$

the statistical operator of the interacting system will be:

$$\hat{\text{w}} = \frac{e^{-\beta\hat{H}}}{\Xi} = \frac{e^{-\beta(\hat{H}_e+\hat{H}_{ph})}}{\Xi}\hat{T}\exp\left[-\int_0^\beta d\tau\hat{H}_F(\tau)\right]. \tag{24}$$

In the previous expression we have turned to the interaction representation to better deal with the electron-phonon part of the Hamiltonian. In the interaction picture, the corresponding τ-dependent second quantization operators have different defining expressions:

$$\hat{a}_\nu(\tau) = e^{\hat{H}_e\tau}\hat{a}_\nu e^{-\hat{H}_e\tau}$$
$$\hat{b}_s(\tau) = e^{\hat{H}_{ph}\tau}\hat{b}_s e^{-\hat{H}_{ph}\tau} \tag{25}$$

The value of the trace does not depend on the chosen representation. Consequantly, a suitable base for the combined electron-phonon states would be $|n\rangle = |ph\rangle|e\rangle$, from which we shall have

$$Tr\left[e^{-\beta(\hat{H}_e+\hat{H}_{ph}}\ldots\right] = Tr\left[e^{-\beta\hat{H}_e}\ldots\right]Tr\left[e^{-\beta\hat{H}_{ph}}\ldots\right] = Tr^{(e)}Tr^{(ph)}\ldots, \tag{26}$$

$Tr^{(e)}$ being taken over $|e\rangle$ states, while $Tr^{(ph)}$ is the trace over the phonon $|ph\rangle$ states.

At this stage, the idea is to carry out the statistical averaging with the use of a reduced statistical operator $\hat{\rho}$ that originates from performing the exact trace over phonon variables in $\hat{\text{w}}$:

$$\hat{\rho} = Tr^{(ph)}\hat{\mathrm{w}} = \frac{e^{-\beta\hat{H}_e}}{\Xi} Tr^{(ph)} \left\{ e^{-\beta\hat{H}_{ph}} \hat{T}_\tau \exp\left[-\int_0^\beta d\tau \hat{H}_F(\tau) \right] \right\}, \tag{27}$$

Therefore, $\hat{\rho}$ will correspond to the electronic subsystem only.

In order to evaluate the trace over the phonon states, $\hat{H}_F$ from equation (16) is substituted in the term between the braces in equation (27). Then it will written in the form

$$e^{-\beta\hat{H}_{ph}} \hat{T} \exp\left[-\int_0^\beta d\tau \hat{H}_F(\tau) \right] = \prod_s \exp\left(-\beta\hbar\omega_s \hat{b}_s^\dagger \hat{b}_s \right)$$
$$\hat{T} \exp\left\{ -\int_0^\beta d\tau \sum_{\mu,\nu} P^s_{\mu\nu} \hat{a}^\dagger_\mu(\tau) \hat{a}_\nu(\tau) \hat{b}_s(\tau) \right\} \exp\left\{ -\int_0^\beta d\tau \sum_{\mu,\nu} Q^s_{\mu\nu} \hat{a}^\dagger_\mu(\tau) \hat{a}_\nu(\tau) \hat{b}^\dagger_s(\tau) \right\}. \tag{28}$$

The argument of the integral in the last exponential function of the right hand side of equation (28) is no other than the Hermitian conjugated part indicated in the expression (16) for the interaction Hamiltonian $\hat{H}_F$.

The next step involves the introduction of the new operators

$$\hat{B}_s(\tau) = \sum_{\mu,\nu} Q^s_{\mu\nu} \hat{a}^\dagger_\mu(\tau) \hat{a}_\nu(\tau);$$
$$\hat{C}_s(\tau) = \sum_{\mu,\nu} P^s_{\mu\nu} \hat{a}^\dagger_\mu(\tau) \hat{a}_\nu(\tau);$$
$$\hat{F}_s(\tau) = \hat{A}_s(\tau) \hat{b}^\dagger_s(\tau) \hat{b}_s(\tau) + \hat{B}_s(\tau) \hat{b}^\dagger_s(\tau) + \hat{C}_s(\tau) \hat{b}_s(\tau), \tag{29}$$

together with the use of the following identity for the chronological ordering [10]:

$$\hat{T} \exp\left\{ -\int_0^\beta d\tau \hat{F}_s(\tau) \right\} = \exp\left\{ -\int_0^\beta d\tau \hat{A}_s(\tau) \hat{b}^\dagger_s \hat{b}_s \right\}$$
$$\exp\left\{ -\int_0^\beta d\tau \hat{B}_s(\tau) \exp\left[\int_0^\tau d\eta \hat{D}_s(\eta) \right] \hat{b}^\dagger_s \right\}$$
$$\exp\left\{ -\int_0^\beta d\tau \hat{C}_s(\tau) \exp\left[-\int_0^\tau d\eta \hat{D}_s(\eta) \right] \hat{b}s \right\}$$
$$\exp\left\{ \int_0^\beta d\tau_1 \int_0^{\tau_1} d\tau_2 \hat{B}_s(\tau_2) \hat{C}_s(\tau_1) \exp\left[\int_{\tau_1}^{\tau_2} d\eta \hat{D}_s(\eta) \right] \right\}, \tag{30}$$

in which, the new operator $\hat{D}_s(\tau)$ is

$$\hat{D}_s(\tau) = \hbar\omega_s + \hat{A}_s(\tau), \tag{31}$$

which in the case of our interest has $\hat{A}_s(\tau) \equiv 0$.

The use of (30) to perform the time ordering at the right of (28) yields

$$e^{-\beta\hat{H}_{ph}} \hat{T} \exp\left[-\int_0^\beta d\tau \hat{H}_F(\tau) \right] = \hat{T} \prod_s \exp\left(-\beta\hbar\omega_s \hat{b}^\dagger_s \hat{b}_s \right)$$

$$\exp\left\{-\int_0^\beta d\tau \hat{B}_s(\tau)e^{\hbar\omega_s\tau}\hat{b}_s^\dagger\right\}\exp\left\{-\int_0^\beta d\tau \hat{C}_s(\tau)e^{-\hbar\omega_s\tau}\hat{b}_s\right\}$$
$$\exp\left\{\int_0^\beta d\tau_1\int_0^{\tau_1} d\tau_2 \hat{B}_s(\tau_2)\hat{C}_s(\tau_1)e^{\hbar\omega_s(\tau_2-\tau_1)}\right\} \tag{32}$$

Now, the chronological operator $\hat{T}$ orders just the Fermi operators in $\hat{B}_s(\tau)$ and $\hat{C}_s(\tau)$ provided the Bose operators $\hat{b}$ in (32) are in the Schrödinger -τ-independent- representation.

By naming

$$\hat{\mathcal{B}}_s = \int_0^\beta d\tau e^{\hbar\omega_s\tau}\hat{B}_s(\tau)$$
$$\hat{\mathcal{C}}_s = \int_0^\beta d\tau e^{-\hbar\omega_s\tau}\hat{C}_s(\tau), \tag{33}$$

one finds the statistical operator of the electronic subsystem to be:

$$\hat{\rho} = \frac{e^{-\beta\hat{H}_e}}{\Xi}\prod_s Tr^{(ph)}\exp\left\{-(\beta\hbar\omega_s\hat{b}_s^\dagger\hat{b}_s + \hat{\mathcal{B}}_s\hat{b}_s^\dagger + \hat{\mathcal{C}}_s\hat{b}_s)\right\}$$
$$\hat{T}\exp\left\{\int_0^\beta d\tau_1\int_0^\beta d\tau_2 \hat{B}_s(\tau_2)\hat{C}_s(\tau_1)e^{\hbar\omega_s(\tau_2-\tau_1)}\right\}. \tag{34}$$

The trace over phonon variables is further evaluated to give:

$$Tr^{(ph)}e^{-(\beta\hbar\omega_s\hat{b}_s^\dagger\hat{b}_s+\hat{\mathcal{B}}_s\hat{b}_s^\dagger+\hat{\mathcal{C}}_s\hat{b}_s)} = \sum_{n=0}^{\infty} e^{-n\beta\hbar\omega_s}\langle n|e^{-\hat{\mathcal{B}}_s\hat{b}_s^\dagger}e^{-\hat{\mathcal{C}}_s\hat{b}_s}|n\rangle =$$
$$\sum_{n=0}^{\infty}\sum_{l=0}^{\infty}\frac{n!e^{-n\beta\hbar\omega_s}}{(n-l)!(l!)^2}(\hat{\mathcal{B}}_s\hat{\mathcal{C}}_s)^l. \tag{35}$$

Summing the double series yields,

$$\sum_{n=0}^{\infty}\sum_{l=0}^{\infty}\frac{n!}{(n-l)!(l!)^2}x^n y^l = \frac{1}{1-x}\exp\left(\frac{xy}{1-x}\right),$$

provided $|x| < 1$. Substituting $x = e^{-\beta\hbar\omega_s}$, and $y = \hat{\mathcal{B}}_s\hat{\mathcal{C}}_s$, the desired result for (35) is found to be

$$Tr^{(ph)}\exp\left\{-(\beta\hbar\omega_s\hat{b}_s^\dagger\hat{b}_s + \hat{\mathcal{B}}_s\hat{b}_s^\dagger + \hat{\mathcal{C}}_s\hat{b}_s)\right\} = \frac{1}{1-e^{-\beta\hbar\omega_s}}\exp\left[\frac{\hat{\mathcal{B}}_s\hat{\mathcal{C}}_s}{e^{\beta\hbar\omega_s}-1}\right] \tag{36}$$

With the use of all of these previous results, the reduced statistical operator becomes

$$\hat{\rho} = \frac{e^{-\beta\hat{H}_e}}{\Xi}\left(\prod_s\frac{1}{1-e^{-\beta\hbar\omega_s}}\right)\hat{T}\exp\left\{\sum_s\sum_{\mu,\nu}\sum_{\mu',\nu'}P^s_{\mu\nu}Q^s_{\mu'\nu'}\right.$$
$$\left[\int_0^\beta d\tau_1\int_0^{\tau_1}d\tau_2\left(e^{-\hbar\omega_s(\tau_1-\tau_2)}+\frac{e^{\hbar\omega_s(\tau_1-\tau_2)}}{e^{\beta\hbar\omega_s}-1}\right)\hat{a}^\dagger_\mu(\tau_1)\hat{a}_\nu(\tau_1)\hat{a}^\dagger_{\mu'}(\tau_2)\hat{a}_{\nu'}(\tau_2)\right.$$
$$\left.\left.+\int_0^\beta d\tau_1\int_{\tau_1}^\beta d\tau_2\frac{e^{\hbar\omega_s(\tau_1-\tau_2)}}{e^{\beta\hbar\omega_s}-1}\hat{a}^\dagger_\mu(\tau_1)\hat{a}_\nu(\tau_1)\hat{a}^\dagger_{\mu'}(\tau_2)\hat{a}_{\nu'}(\tau_2)\right]\right\} \tag{37}$$

In the obtention of this expression - and in the process that inmediately follows it-, the mathematical handling takes into account the commutation properties of Fermi-type operators and the appropriate making of such exchanges like $\tau_1 \leftrightarrow \tau_2$, $\mu \leftrightarrow \mu'$, or $\nu \leftrightarrow \nu'$ in the double integrals. The assumption that these exchanges do not alter the values of $P^s_{\mu\nu}$, $Q^s_{\mu\nu}$ or ω_s is made throughout. This is what seems to occur in most situations, mainly in the case of isotropic constituent media. Then, the further manipulation of the second double integral in (37), and the additional introduction of the function

$$\Gamma^{\mu'\nu'}_{\mu\nu}(s;\tau_1-\tau_2) = P^s_{\mu\nu}Q^s_{\mu'\nu'}e^{-\hbar\omega_s(\tau_1-\tau_2)}\left[\frac{\theta(\tau_1-\tau_2)}{1-e^{-\beta\hbar\omega_s}}+\frac{\theta(\tau_2-\tau_1)}{e^{\beta\hbar\omega_s}-1}\right]$$
$$+P^s_{\mu'\nu'}Q^s_{\mu\nu}e^{\hbar\omega_s(\tau_1-\tau_2)}\left[\frac{\theta(\tau_1-\tau_2)}{e^{\beta\hbar\omega_s}-1}+\frac{\theta(\tau_2-\tau_1)}{1-e^{-\beta\hbar\omega_s}}\right], \qquad (38)$$

lead to the final expression for the reduced statistical operator;

$$\hat{\rho} = e^{-\beta\hat{H}_e}\frac{\Xi_{ph}}{\Xi}\hat{T}\exp\left\{\frac{1}{2}\sum_s\sum_{\mu,\nu}\sum_{\mu',\nu'}\int_0^\beta d\tau_1\int_0^\beta d\tau_2\Gamma^{\mu'\nu'}_{\mu\nu}(s;\tau_1-\tau_2)\hat{a}^\dagger_\mu(\tau_1)\hat{a}_\nu(\tau_1)\hat{a}^\dagger_{\mu'}(\tau_2)\hat{a}_{\nu'}(\tau_2)\right\}. \qquad (39)$$

The product between brackets in (37) is no other than the phonon statistical sum Ξ_{ph}. That is

$$\Xi_{ph} = \prod_s \frac{1}{1-e^{-\beta\hbar\omega_s}}.$$

Indeed, the thermodynamical potential corresponding to the electronic subsystem is defined through

$$e^{\beta\Omega_e} = \frac{\Xi_{ph}}{\Xi},$$

allowing to write equation (39) in a more compact form:

$$\hat{\rho} = e^{\beta(\Omega_e-\hat{H}_e)}\hat{T}e^{\hat{\mathcal{V}}}. \qquad (40)$$

The operator $\hat{\mathcal{V}}$ here introduced is

$$\hat{\mathcal{V}} = \frac{1}{2}\sum_s\sum_{\mu,\nu}\sum_{\mu',\nu'}\int_0^\beta d\tau_1\int_0^\beta d\tau_2\Gamma^{\mu'\nu'}_{\mu\nu}(s;\tau_1-\tau_2)\hat{a}^\dagger_\mu(\tau_1)\hat{a}_\nu(\tau_1)\hat{a}^\dagger_{\mu'}(\tau_2)\hat{a}_{\nu'}(\tau_2). \qquad (41)$$

Taking the trace over a complete set of $|e\rangle$ sates on both sides of equation (40), the resulting expression for the thermodynamical potential of the electron subsystem interacting via $\hat{\mathcal{V}}$ is

$$\Omega_e = -\frac{1}{\beta}\ln Tr\left[e^{-\beta\hat{H}_e}\hat{T}e^{\hat{\mathcal{V}}}\right] \qquad (42)$$

At this point the question can be summarized as follows: The statistical average over the phonon states is performed in such a way that the electron-phonon interaction reduces -exactly- to a retarded effective interaction between electrons.

According to (41), the elementary act of interaction may be represented by the product which is the argument of the sums and integrals. It involves two scattering pocesses, each with the annihilation of an electron in certain state with quantum numbers ν, and the consequent creation of an electron in a state μ. Moreover, this electron-electron interaction through the crystal lattice is also applied to the electron's self-interaction; i.e., the excitation created at τ_2 (expressed in terms of phonon absorption and emission) propagates in the lattice and acts over the same electron at time τ_1.

The idea resumed in expressions (40)-(42) is, in fact, not new. By using his variational path-integral formalism, Feynman performed for the first time an exact reduction of the electron-phonon interaction into an effective electron-electron one in a work dedicated to the study of slow electrons in bulk polar crystals (the polarons) [11]. Posteriorly, Bogolubov (father and son) proposed an alternative -and mathematically rigorous- method for the elimination of the phonon variables in the bulk polaron problem based on second quantization [12].

The approach that has been put forward above constitutes an extension of Bogolubov's method to the many-electron system interacting with long wave polar optical phonons in semiconducting heterostructures. The purpose is to look after a possible generalization of Feynman's variational principle for the polaron free energy to the thermodynamical potential of the system of interacting electrons in these structures. This aspect shall be discussed later in this chapter. For the moment, let us go back to the one-electron Green's function and see how it is expressed along the lines of this effective statistics.

In the interaction representation, statistical averaging with the use of the reduced statistical operator $\hat{\rho}$ implies the temperature Green's function (22) to become [9, 13]

$$G_\nu(\tau) = -\frac{\langle \hat{T}\left[\hat{a}_\nu(\tau)\hat{a}^\dagger_\nu(0)e^{\hat{\mathcal{V}}}\right]\rangle_0}{\langle \hat{T}e^{\hat{\mathcal{V}}}\rangle_0}. \tag{43}$$

In this case, $\langle\ldots\rangle_0$ means that the average is carried out with the use of the statistical operator that describes the non-interacting electron subsystem at thermal equilibrium. Furthermore, the weak character of the interaction coupling supports a perturbative treatment by means of a diagrammatic expansion which results from the series

$$G_\nu(\tau) = \sum_{n=1}^{\infty} G^{(n)}_\nu(\tau).$$

As usual, the contribution from disconnected diagrams (those in which all of their parts are not topologically connected between each other [9, 13]) cancels de denominator in (43), resulting in

$$G^{(n)}_\nu(\tau) = -\frac{1}{n!}\langle \hat{T}\left[\hat{a}_\nu(\tau)\hat{a}^\dagger_\nu(0)\mathcal{V}^n\right]\rangle_0^{(con)} \tag{44}$$

It is not difficult to formulate the corresponding rules for constructing th n-th order diagrams, starting from the application of Wick's theorem for the statistical average of a chronological ordering of a product of creation and annihilation operators. They are completely analogous to the typical diagram rules that can be found for the electron-phonon Green's function in the standard texts on many-particle physics. Instead, let us concentrate in the thermodynamical potential.

Taking the thermodynamical potential of the non-interacting electron subsystem to be

$$\Omega_0 = -\frac{1}{\beta}\ln Tr^{(e)}e^{-\beta\hat{H}_e};$$

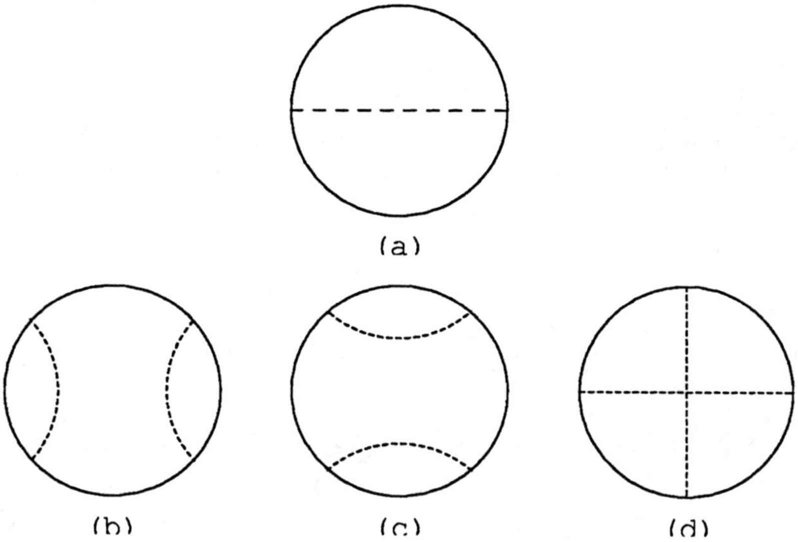

Figure 1: First and second order diagrams for the thermodynamical potential

the correction $\Delta\Omega = \Omega_e - \Omega_0$ to it, coming from the electron-phonon interaction is -according to the scheme developed by Abrikosov et al. [13]- given by;

$$\Delta\Omega = -\frac{1}{\beta}\langle Tr^{(e)} e^{\hat{\mathcal{V}}}\rangle_0 \tag{45}$$

and

$$\langle Tr^{(e)} e^{\hat{\mathcal{V}}}\rangle_0 = \exp\left[\langle Tr^{(e)} e^{\hat{\mathcal{V}}}\rangle_0^{(con)} - 1\right]. \tag{46}$$

This allows to express $\Delta\Omega$ in terms of a connected-diagram expansion:

$$\Delta\Omega = -\frac{1}{\beta}\left[\langle Tr^{(e)} e^{\hat{\mathcal{V}}}\rangle_0^{(con)} - 1\right]. \tag{47}$$

One of the possible ways of writing expression (47) in a more practicable form exploits the expansion of $e^{\hat{\mathcal{V}}}$ as a series in the electron-phonon coupling constant α; taking into account the fact that the n-th order in the expansion of the Green's function is proportional to the n-th power of α. This feature allows to obtain

$$\Delta\Omega = \frac{1}{\beta}\int_0^\alpha \frac{d\alpha}{\alpha}\sum_\nu\left\{\frac{G_\nu(\tau)}{G_\nu^{(0)}(\tau)} - 1\right\}, \tag{48}$$

thus expressing the correction to the thermodynamical potential due to the interaction with the phonons through the one-electron Green's function. From (48) the series of Feynman-like diagrams for $\Delta\Omega$ can be derived. The figure 1 shows the first order (fig. 1(a)) and second order (figs. 1(b)-1(d)) diagrams. Solid lines represent unperturbed Green's functions

$$G_\nu^{(0)}(\tau) = [\theta(\tau) - f_\nu]e^{-E_\nu\tau}, \tag{49}$$

where

$$f_\nu = \frac{1}{\exp(\beta E_\nu) + 1} \tag{50}$$

is the Fermi occupation number ($\xi_\nu = E_\nu - \varepsilon$).

The dashed lines represent the effective interaction field. Each of them contains a Γ-function which can be viewed as the amplitude of the elementary act of interaction described through (41). As an example, the first correction term -in the τ-representation- in the series $\Delta\Omega = \sum_n \Delta\Omega^{(n)}$ is

$$\Delta\Omega^{(1)} = \frac{1}{\beta}\sum_s \sum_{\nu,\mu} \sum_{\nu',\mu'} \int_0^\beta d\tau_1 \int_0^\beta d\tau_2 \Gamma_{\mu\nu}^{\mu'\nu'}(s;\tau_1-\tau_2)\theta_{\nu\mu'}\theta_{\nu'\mu} G_\nu^{(0)}(\tau_1-\tau_2) G_{\nu'}^{(0)}(\tau_2-\tau_1) \tag{51}$$

The factors $\theta_{\nu\mu}$ appearing in (51) are related to certain restrictions on the quantum numbers ussually imposed by the conservation laws of the magnitudes involved. Their effect will be that of eliminating some of the summations as the reader shall see in the next section.

The next step is the derivation of a set of rules for writing the n-th order contribution to $\Delta\Omega$. To do this it is better to go over the Fourier representation. The Fourier expansion in the interval $-\beta \leq \tau \leq \beta$ for the Green's function is [9]:

$$G_\nu(\tau) = \frac{1}{\beta}\sum_n G_\nu(i\omega_n) e^{-i\omega_n \tau} \tag{52}$$

$$G_\nu(i\omega_n) = \int_0^\beta d\tau G_\nu(\tau) e^{i\omega_n \tau}; \qquad \omega_n = \frac{(2n+1)\pi}{\beta}. \tag{53}$$

In the particular case of the non-perturbed Green's function $G^{(0)}$, the use of (52) together with (49) leads to the very simple expression

$$G_\nu^{(0)}(i\omega_n) = \frac{1}{i\omega_n - \xi_\nu}. \tag{54}$$

On the other hand, the Fourier transform of the factor $\Gamma_{\mu\nu}^{\mu'\nu'}(s;\tau)$;

$$\Gamma_{\mu\nu}^{\mu'\nu'}(s;i\omega_l) = \frac{1}{2}\int_{-\beta}^{\beta} d\tau \Gamma_{\mu\nu}^{\mu'\nu'}(s;\tau) e^{i\omega_l \tau} = \frac{P^s_{\mu'\nu'} Q^s_{\mu\nu}}{i\omega_l + \hbar\omega_s} - \frac{P^s_{\mu\nu} Q^s_{\mu'\nu'}}{i\omega_l - \hbar\omega_s} \tag{55}$$

reveals that the allowed values for $i\omega_l$ are $2l\pi/\beta$. indicating that the Γ-factor is, at least, proportional to the unperturbed phonon propagator usually found in the conventional electron-phonon Green's function theory.

The diagram rules for the n-th order electron-phonon contribution to the thermodynamical potential are:

- Construct the $(2n-1)!!$ circular diagrams with $2n$ vertices regularly distributed on. In each vertex should meet two solid lines and one dashed line.
- Associate one set of s-type quantum numbers to each dashed line in the diagram, and two sets of (ν,μ)-quantum numbers to each solid line, in such a way that a $\theta_{\nu\mu}$ factor should appear for each vertex ensuring the accomplishment of the appropriate conservation laws if there should be some.

- Associate a frequency to each diagram line. Only $n+1$ frequencies are independent, and the conservation of the frequencies must be guaranteed at each vertex.

- Include a factor $G_{\nu}^{(0)}(i\omega_n)$ given through (54) for each solid line of the diagram, and a factor $\Gamma_{\mu\nu}^{\mu'\nu'}(s;i\omega_l)$ (from eqn. (55)) for each dashed line.

- Sum over all independent frequencies and over all quantum numbers of each kind (some summations may be eliminated by the presence of the θ-factors!).

- Multiply the resulting expression by $1/n\beta^{n+1}$.

Let us write now the expressions for the diagrams of the figure 1 as result from these rules. In the case of the first order diagram (fig. 1(a)), one shall have:

$$\Delta\Omega^{(1)} = \frac{1}{\beta^2}\sum_{s}\sum_{\mu,\nu}\sum_{\mu',\nu'}\sum_{i\omega_l}\sum_{i\omega_n}\Gamma_{\mu\nu}^{\mu'\nu'}(s;i\omega_l)\theta_{\nu\mu'}\theta_{\nu'\mu}G_{\nu}^{(0)}(i\omega_n)G_{\nu'}^{(0)}(i\omega_n - i\omega_l), \tag{56}$$

while for the second order diagrams the corresponding expressions are:

$$\Delta\Omega^{(2)} = \frac{1}{2\beta^3}\sum_{s,s'}\sum_{\mu,\nu}\sum_{\mu',\nu'}\sum_{\eta,\zeta}\sum_{\eta',\zeta'}\sum_{i\omega_{l_1},i\omega_{l_2}}\sum_{i\omega_n}\Gamma_{\mu\nu}^{\mu'\nu'}(s;i\omega_{l_1})\Gamma_{\eta\zeta}^{\eta'\zeta'}(s';i\omega_{l_2})\theta_{\nu\eta}\theta_{\zeta\nu'}\theta_{\zeta'\mu'}\theta_{\nu'\mu}$$
$$G_{\nu}^{(0)}(i\omega_n)G_{\zeta}^{(0)}(i\omega_n - i\omega_{l_2})G_{\zeta'}^{(0)}(i\omega_n)G_{\nu'}^{(0)}(i\omega_n - i\omega_{l_1}), \tag{57}$$

(fig. 1(b));

$$\Delta\Omega^{(2)} = \frac{1}{2\beta^3}\sum_{s,s'}\sum_{\mu,\nu}\sum_{\mu',\nu'}\sum_{\eta,\zeta}\sum_{\eta',\zeta'}\sum_{i\omega_{l_1},i\omega_{l_2}}\sum_{i\omega_n}\Gamma_{\mu\nu}^{\mu'\nu'}(s;i\omega_{l_1})\Gamma_{\eta\zeta}^{\eta'\zeta'}(s';i\omega_{l_2})\theta_{\nu\mu'}\theta_{\nu'\eta}\theta_{\zeta'\eta'}\theta_{\zeta'\mu}$$
$$G_{\nu}^{(0)}(i\omega_n - i\omega_{l_1})G_{\nu'}^{(0)}(i\omega_n)G_{\zeta}^{(0)}(i\omega_n - i\omega_{1_2})G_{\zeta'}^{(0)}(i\omega_n), \tag{58}$$

(fig. 1(c));

$$\Delta\Omega^{(2)} = \frac{1}{2\beta^3}\sum_{s,s'}\sum_{\mu,\nu}\sum_{\mu',\nu'}\sum_{\eta,\zeta}\sum_{\eta',\zeta'}\sum_{i\omega_{l_1},i\omega_{l_2}}\sum_{i\omega_n}\Gamma_{\mu\nu}^{\mu'\nu'}(s;i\omega_{l_1})\Gamma_{\eta\zeta}^{\eta'\zeta'}(s';i\omega_{l_2})\theta_{\nu\eta}\theta_{\zeta\mu'}\theta_{\nu'\eta'}\theta_{\zeta'\mu}$$
$$G_{\nu}^{(0)}(i\omega_n)G_{\zeta}^{(0)}(i\omega_n - i\omega_{l_2})G_{\nu'}^{(0)}(i\omega_n - i\omega_{l_1} - i\omega_{1_2})G_{\zeta'}^{(0)}(i\omega_n - i\omega_{l_1}), \tag{59}$$

(fig. 1(d)).

In many cases, the study of the electron-phonon correction to the thermodynamical potential restricts to the first order contributions. This is the situation one finds when dealing with such systems made of III-V semiconducting materials (with, perhaps, the sole exceptions of GaN and AlN). By summing over the frequencies ω_l and ω_n in (56) [14], the following formula is obtained;

$$\Delta\Omega^{(1)} = \sum_{s}\sum_{\mu,\nu}\sum_{\mu',\nu'}(f_{\nu'} - f_{\nu})\left[\frac{g_{\nu'\nu} - n_s}{E_{\nu'} - E_{\nu} + \hbar\omega_s}P_{\mu'\nu'}^{s}Q_{\mu\nu}^{s} + \frac{n_s + 1 - g_{\nu'\nu}}{E_{\nu'} - E_{\nu} - \hbar\omega_s}P_{\mu\nu}^{s}Q_{\mu'\nu'}^{s}\right]\theta_{\nu\mu'}\theta_{\nu'\mu}. \tag{60}$$

Expression (60) constitutes the extension to low-dimensional semiconducting structures of the so-called Buckingham-Schafroth formula [15]. In it, n_s represents the Bose occupation number

$$n_s = \frac{1}{\exp(\beta\hbar\omega_s) - 1},$$

and

$$g_{\nu'\nu} = \frac{1}{\exp[\beta(E_{\nu'} - E_\nu)] - 1}.$$

¿From the equation (60), first order thermodynamical properties of the interacting electron-phonon system in nanostructures are readily derived. For instance, the entropy and heat capacity of the system would come given by:

$$S = -\left(\frac{\partial\Delta\Omega^{(1)}}{\partial T}\right)_{V,\varepsilon}; \qquad C_V = T\left(\frac{\partial S}{\partial T}\right)_{V,\varepsilon}. \tag{61}$$

To finish this section, let us remark that if we had derived the diagram rules for the one-electron Green'n function $G_\nu(i\omega_n)$ instead of those corresponding to $\Delta\Omega$, the latter could be easily obtained by eliminating the first external particle line in each diagram of $G_\nu^{(n)}(i\omega_n)$, and closing back the second external particle line over the first vertex at the far left. Additionally, the sum over the external particle line indexes (ν, μ) must be performed and the result be multiplied by $1/n\beta$. This fact allows to realize why in the set of diagrams for the second order corrections there are two which are topologically equivalent. Indeed, they come from topologically different second order diagrams of the one-particle Green's function, and their contributions have necessarily to be taken into account [13].

Further, the Fourier-transformed analog of equation (48) makes possible to write another general expression for $\Delta\Omega$ in the form

$$\Delta\Omega = \frac{1}{\beta}\int_0^\alpha \frac{d\alpha}{\alpha} \sum_{\nu, i\omega_n} \frac{\Sigma_\nu(i\omega_n)}{i\omega_n - \xi_\nu - \Sigma_\nu(i\omega_n)}, \tag{62}$$

where $\Sigma_\nu(i\omega_n)$ is the electron-phonon self-energy. This means that, if by some way you are able to calculate the self-energy of the system, then it can be used to obtain the electron-phonon corrections to its thermodynamical properties.

3. The polaron free energy in a double semiconducting heterostructure

As it is commonly known, the quantum numbers designating the electronic states in the case of a double heterostructure in the envelope function approximation are $\nu = \{\mathbf{k}, r\}$. The two-dimensional wavevector $\mathbf{k} = (k_x, k_y)$ characterizes the unconfined motion in the xy plane, with energy $E(\mathbf{k}) = \hbar^2\mathbf{k}^2/2m^*$. The index r represents confined states along the growth direction of the structure, z; having discrete or continuum values depending on the potential model chosen to describe these states in the quantum well as well as on the interval for the energy of the motion.

As it was already seen in section 1, phonon modes in this kind of structure are distinguished between them by their characteristics dispersion laws $\omega_m(\mathbf{q})$. The vector $\mathbf{q} = (q_x, q_y)$ is the two-dimensional phonon wavevector, while m is an index numbering the different oscillation modes. In equation (13), this number runs over a discrete set of values due to the mechanically confined character of the modes in a $GaAs/AlAs$ double heterostructure. But if a full matching of the mechanical field is carried out, there will appear half-space modes propagating along the semiinfinite barrier regions. In this situation, m will become a continuum variable.

One of the effects of the electron-phonon interaction on the single condcution electron states in polar materials is the decrease of the ground energy. The magnitude of this decrement is what it is called *polaron binding energy.* In this section, the model Hamiltonian (13) is going to be used to calculate the polaron binding energy as a function of the temperature in a $GaAs/AlAs$ quantum well. This quantity is known as the *polaron free energy.*

The evaluation of the polaron free energy is one of the direct results of the path-integral formalism in bulk systems [11, 16 – 18]. However, such a quantity does not seem to have been so far widely calculated in low-dimensional heterostructures perhaps due to explicit difficulties for the application of a Feynman integrals procedure in systems bearing interfaces. A different choice is the use of the perturbative method based on the Green's function approach to the electron-phonon thermodynamical potential.

The electronic state in the system we are dealing with come from the solution of a Schrödinger-like effective mass differential equation, which includes a finite barrier rectangular potential function along the z-axis. The conduction band in the system is considered to be parabolic., and the one-dimensional states corresponding to the motion along this direction can be chosen to have a definite parity. we shall have

¬ Even (symmetric) states.

- Discrete spectrum ($E_z = E_n < V_0$);

$$\langle z|n\rangle = \begin{cases} A_n^e \cos(k_n z), & |z| \le d/2; \\ A_n^e \cos(\frac{k_n d}{2}) e^{-\kappa_n(|z|-d/2)}, & |z| > d/2. \end{cases} \tag{63}$$

In this expression, $k_n = \sqrt{2m_w E_n}/\hbar$, and $\kappa_n = \sqrt{2m_b(V_0 - E_n)}/\hbar$. V_0 is the height of the barrier; d is the quantum well width; m_w is the effective mass in the well region ($GaAs$), while m_b is the electron effective mass in the barriers ($AlAs$). The discrete energy levels $E_n = 1, 3, 5, \ldots$, are obtained by finding the roots of the transcendental equation

$$x_n \sin x_n - \sqrt{c(b^2 - x_n^2)} \cos x_n = 0, \tag{64}$$

in the interval $x_n < b$. Here, $x_n = k_n d/2$; $c = m_w/m_b$, and $b^2 = m_w V_0 d^2/2\hbar^2$. The normalization constant A_n^e is given by

$$A_n^e = \left(\frac{2k_n}{k_n d + \sin(k_n d) + 2\frac{k_n}{\kappa_n}\cos^2(k_n d/2)} \right)^{1/2}. \tag{65}$$

- Continuum spectrum ($E_z = \mathcal{E} > V_0$);

$$\langle z|\mathcal{E}\rangle = \frac{A'_e}{\sqrt{L_z}} \begin{cases} \cos(k_w z), & |z| \le d/2; \\ C_e \cos(k_b z) + D_e \sin(k_b |z|), & |z| > d/2, \end{cases} \tag{66}$$

where $k_w = \sqrt{2m_w\mathcal{E}}/\hbar$; $k_b = \sqrt{2m_b(\mathcal{E}-V_0)}/\hbar$. L_z is the total sample length along the z-direction, and is assumed that $L_z \to \infty$. The constants A'_e, C_e, D_e are given, respectively by:

$$A'_e = \left(\frac{2}{\cos^2(k_w d/2) + \left(\frac{m_b k_w}{m_w k_b}\right)^2 \sin^2(k_w d/2)} \right); \tag{67}$$

$$C_e = \cos\left(\frac{k_w d}{2}\right)\cos\left(\frac{k_b d}{2}\right) + \frac{m_b k_w}{m_w k_b}\sin\left(\frac{k_w d}{2}\right)\sin\left(\frac{k_b d}{2}\right); \tag{68}$$

$$D_e = \cos\left(\frac{k_w d}{2}\right)\sin\left(\frac{k_b d}{2}\right) - \frac{m_b k_w}{m_w k_b}\sin\left(\frac{k_w d}{2}\right)\cos\left(\frac{k_b d}{2}\right). \tag{69}$$

As previously stated, the eigenenergy has a continuum of values above the barrier V_0.

¬ Odd (antisymmetric) states.

- Discrete spectrum;

$$\langle z|n\rangle = \begin{cases} B_n^o \sin(k_n z), & |z| \le d/2; \\ \frac{z}{|z|} B_n^o \sin(\frac{k_n d}{2}) e^{-\kappa_n(|z|-d/2)}, & |z| > d/2, \end{cases} \tag{70}$$

with

$$B_n^o = \left(\frac{2k_n}{k_n d - \sin(k_n d) + 2\frac{k_n}{\kappa_n}\sin^2(k_n d/2)} \right)^{1/2}. \tag{71}$$

The odd discrete energy levels E_n; $n = 2, 4, 6, \ldots$ are determined by solving the transcendental equation

$$x_n \cos x_n + \sqrt{c(b^2 - x_n^2)}\sin x_n = 0, \qquad (x_n < b). \tag{72}$$

- Continuum spectrum;

$$\langle z|\mathcal{E}\rangle = \frac{B'_o}{\sqrt{L_z}} \begin{cases} \sin(k_w z), & |z| \le d/2; \\ \frac{z}{|z|}\left[C_o \cos(k_b z) + D_o \sin(k_b |z|)\right], & |z| > d/2; \end{cases} \tag{73}$$

$$B'_o = \left(\frac{2}{\sin^2(k_w d/2) + \left(\frac{m_b k_w}{m_w k_b}\right)^2 \cos^2(k_w d/2)} \right); \tag{74}$$

$$C_o = \sin\left(\frac{k_w d}{2}\right)\cos\left(\frac{k_b d}{2}\right) - \frac{m_b k_w}{m_w k_b}\cos\left(\frac{k_w d}{2}\right)\sin\left(\frac{k_b d}{2}\right); \tag{75}$$

$$D_e = \sin\left(\frac{k_w d}{2}\right)\sin\left(\frac{k_b d}{2}\right) - \frac{m_b k_w}{m_w k_b}\cos\left(\frac{k_w d}{2}\right)\cos\left(\frac{k_b d}{2}\right). \tag{76}$$

Then, the wave function

$$\langle \mathbf{r}|\mathbf{k},\mathcal{E}\rangle = \frac{1}{\sqrt{S}} e^{i\mathbf{k}\cdot\mathbf{R}} \begin{cases} \langle z|n\rangle, & E_z < V_0; \\ \langle z|\mathcal{E}\rangle, & E_z > V_0 \end{cases} \tag{77}$$

completely describes the electronic state with energiy

$$E = E_\perp + E_z. \tag{78}$$

The energy of the motion in the xy plane, $E_\perp$ is given by:

$$E_\perp = \frac{\hbar^2 \mathbf{k}^2}{2m_\perp}, \tag{79}$$

with the effective mass in the xy plane, $m_\perp$ being defined as:

$$m_\perp = \frac{m_w m_b}{P_w m_b + P_b m_w}. \tag{80}$$

In this expression, P_w (P_b) being the probability of finding the electron inside (outside) the quantum well [19].

Regarding the phonon states, the dispersive electroelastic continuum model from which the interaction Hamiltonian (13) is derived separates them in both even and odd potential states as well, according to the symmetry of the electrostatic field with respect to the $z = 0$ plane [7].

The Even Potential States (EPS) of the *GaAs* oscillation modes are such that the electrostatic Fröhlich amplitude is given by [20]:

$$F^{(e)}_{qm}(z) = \mathcal{A}_e \begin{cases} \cos(k_L z) - \frac{C_L}{q} e^{-qd/2}\cosh(qz), & |z| \le d/2; \\ C_0 e^{-q|z|}, & |z| > d/2. \end{cases} \tag{81}$$

The normalization constant $\mathcal{A}_e$ is obtained from;

$$|\mathcal{A}_e|^2 \int_0^{d/2} \left[|f_e(z)|^2 + |g_e(z)|^2 \right] dz = \frac{d}{4}, \tag{82}$$

where the functions $f_e(z)$, $g_e(z)$ are the components of the mechanical oscillation field in the xy plane. These functions are taken to be zero outside the *GaAs* quantum well, due to the assumption of complete mechanical confinement made in the derivation [7, 20]. Meanwhile, in the quantum well region;

$$f_e(z) = q\cos(k_L z) - k_T \frac{C_L}{S_T}\left(\frac{\omega_{LO}^2 - \omega^2}{\omega_{TO}^2 - \omega^2}\right)\cos(k_T z) + C_L\left(\frac{\omega_{LO}^2 - \omega_{TO}^2}{\omega_{TO}^2 - \omega^2}\right) e^{-qd/2}\cosh(qz); \tag{83}$$

$$g_e(z) = -\left[k_L \sin(k_L z) + q\frac{C_L}{S_T}\left(\frac{\omega_{LO}^2 - \omega^2}{\omega_{TO}^2 - \omega^2}\right)\sin(k_T z) - C_L\left(\frac{\omega_{LO}^2 - \omega_{TO}^2}{\omega_{TO}^2 - \omega^2}\right) e^{-qd/2}\sinh(qz)\right]. \tag{84}$$

Additionally,

$$S_{L(T)} = \sqrt{q^2 + k_{L(T)}^2} \sin\left[\frac{k_{L(T)}d}{2} + \vartheta_{L(T)}\right]; \tag{85}$$

$$C_{L(T)} = \sqrt{q^2 + k_{L(T)}^2} \cos\left[\frac{k_{L(T)}d}{2} + \vartheta_{L(T)}\right]; \tag{86}$$

$$\tan\vartheta_{L(T)} = \frac{k_{L(T)}}{q}; \tag{87}$$

$$k_L = \sqrt{\frac{\omega_{LO}^2 - \omega^2}{\beta_L^2} - q^2}; \tag{88}$$

$$k_T = \sqrt{\frac{\omega_{TO}^2 - \omega^2}{\beta_T^2} - q^2}. \tag{89}$$

β_L and β_T are parameters estimated from the LO and TO bulk dispersion relations, respectively, of the bulk constituent material either by fitting *ab initio* calculations or neutron scattering data around the Brillouin zone center. The coefficient C_0 is,

$$C_0 = \cos\left(\frac{k_L d}{2}\right)\sinh\left(\frac{qd}{2}\right) + \frac{k_L}{q}\sin\left(\frac{k_L d}{2}\right)\cosh\left(\frac{qd}{2}\right), \tag{90}$$

and the eigenfrequencies of these EPS are the solutions of the transcendental equation

$$(\omega_{LO}^2-\omega^2)\left[q\cos\left(\frac{k_L d}{2}\right)-k_L\sin\left(\frac{k_L d}{2}\right)\right]\left[k_T\cos\left(\frac{k_T d}{2}\right)\sinh\left(\frac{qd}{2}\right)-q\sin\left(\frac{k_T d}{2}\right)\cosh\left(\frac{qd}{2}\right)\right]$$

$$=(\omega_{TO}^2-\omega^2)\left[q\sin\left(\frac{k_T d}{2}\right)+k_T\cos\left(\frac{k_T d}{2}\right)\right]\left[k_L\sin\left(\frac{k_L d}{2}\right)\cosh\left(\frac{qd}{2}\right)+q\cos\left(\frac{k_L d}{2}\right)\sinh\left(\frac{qd}{2}\right)\right]. \tag{91}$$

On the other hand, the electrostatic amplitude for the *GaAs* Odd Potential States (OPS) will be [20]:

$$F_{qm}^{(o)}(z) = \mathcal{A}_o\begin{cases}\sin(k_L z) - \frac{S_L}{q}e^{-qd/2}\sinh(qz), & |z| \le d/2;\\ S_0\frac{z}{|z|}e^{-q|z|}, & |z| > d/2;\end{cases} \tag{92}$$

$$S_0 = \sin\left(\frac{k_L d}{2}\right)\cosh\left(\frac{qd}{2}\right) - \frac{k_L}{q}\cos\left(\frac{k_L d}{2}\right)\sinh\left(\frac{qd}{2}\right); \tag{93}$$

$$|\mathcal{A}_o|^2\int_0^{d/2}\left[|f_o(z)|^2 + |g_o(z)|^2\right]dz = \frac{d}{4}, \tag{94}$$

where

$$f_o(z) = q\sin(k_L z) + k_T\frac{S_L}{C_T}\left(\frac{\omega_{LO}^2-\omega^2}{\omega_{TO}^2-\omega^2}\right)\sin(k_T z) + S_L\left(\frac{\omega_{LO}^2-\omega_{TO}^2}{\omega_{TO}^2-\omega^2}\right)e^{-qd/2}\sinh(qz); \tag{95}$$

$$g_o(z) = k_L\cos(k_L z) - q\frac{S_L}{C_T}\left(\frac{\omega_{LO}^2-\omega^2}{\omega_{TO}^2-\omega^2}\right)\cos(k_T z) + S_L\left(\frac{\omega_{LO}^2-\omega_{TO}^2}{\omega_{TO}^2-\omega^2}\right)e^{-qd/2}\cosh(qz), \tag{96}$$

in the region $|z| \leq d/2$, and zero outside.

The OPS characteristic frequencies come out when solving the equation

$$(\omega_{LO}^2-\omega^2)\left[k_L\cos\left(\frac{k_L d}{2}\right)+q\sin\left(\frac{k_L d}{2}\right)\right]\left[k_T\sin\left(\frac{k_T d}{2}\right)\cosh\left(\frac{qd}{2}\right)+q\cos\left(\frac{k_T d}{2}\right)\sinh\left(\frac{qd}{2}\right)\right]$$

$$=(\omega_{TO}^2-\omega^2)\left[q\cos\left(\frac{k_T d}{2}\right)-k_T\sin\left(\frac{k_T d}{2}\right)\right]\left[k_L\cos\left(\frac{k_L d}{2}\right)\sinh\left(\frac{qd}{2}\right)-q\sin\left(\frac{k_L d}{2}\right)\cosh\left(\frac{qd}{2}\right)\right]. \quad (97)$$

With all this information, the polaron free energy is evaluated from the first order correction to the thermodynamical potential. But first, let us examine which are the quantum indexes (ν, μ) in this particular problem.

Starting with equation (13), and using the electronic eigenstates (77), the construction of the interaction Hamiltonian in the complete second-quantized form (16) shows that the P and Q factors in this case are:

$$P^{\mathbf{q},m}_{\{\mathbf{k}+\mathbf{q},n'\},\{\mathbf{k},n\}} = \langle \mathbf{k}+\mathbf{q}, n'|F_{\mathbf{q}m}(z)|n, \mathbf{k}\rangle \quad (98)$$

$(\nu = \mathbf{k}, n;\ \mu = \mathbf{k}+\mathbf{q}, n')$;

$$Q^{\mathbf{q},m}_{\{\mathbf{k}-\mathbf{q},n'\},\{\mathbf{k},n\}} = \langle \mathbf{k}-\mathbf{q}, n'|F^*_{\mathbf{q}m}(z)|n, \mathbf{k}\rangle \quad (99)$$

$(\nu = \mathbf{k}, n;\ \mu = \mathbf{k}-\mathbf{q}, n')$.

Once the indexes have been identified, the process for the evaluation of the polaron free energy begins with the equation (51). It is the proper time to show how the θ-factors work. By naming $\nu = \mathbf{k}, n$; $\nu' = \mathbf{k}', l$, it can be shown that:

$$\theta_{\nu\mu'} = \delta_{n,l'}\delta_{\mathbf{k},\mathbf{k}'-\mathbf{q}}$$
$$\theta_{\nu'\mu} = \delta_{l,n'}\delta_{\mathbf{k}',\mathbf{k}+\mathbf{q}}. \quad (100)$$

In these expressions, the δ-factors are Kronecker symbols. The second ones are the consequence of the conservation of the electron quasi-momentum $\mathbf{k}$ in both vertices.

The presence of $\theta_{\nu\mu'}\theta_{\nu'\mu}$ eliminates some summations in (51), and transforms the Γ-coefficient in

$$\Gamma^{n,l}_{\mathbf{k},\mathbf{k}+\mathbf{q}}(\mathbf{q}, m; \tau_1-\tau_2) = I^m_{nl}(\mathbf{q})D_{\mathbf{q}m}(\tau_1-\tau_2), \quad (101)$$

where $I^m_{nl}(\mathbf{q})$ is the square modulus of the electron-phonon matrix elements

$P^{\mathbf{q},m}_{\{\mathbf{k}+\mathbf{q},l\},\{\mathbf{k},n\}}$, and $D_{\mathbf{q}m}(\tau_1-\tau_2)$ results to be the unperturbed phonon propagator in the τ-representation:

$$D_{\mathbf{q}m}(\tau_1-\tau_2) = \frac{e^{-\hbar\omega_m(q)|\tau_1-\tau_2|}}{1-e^{-\beta\hbar\omega_m(q)}} + \frac{e^{\hbar\omega_m(q)|\tau_1-\tau_2|}}{e^{\beta\hbar\omega_m(q)}-1}. \quad (102)$$

Finally, the formula for the first order contribution to $\Delta\Omega$ in a *GaAs*/*AlAs* double heterostructure is written as

$$\Delta\Omega^{(1)} = \frac{1}{\beta}\sum_{\mathbf{q},m}\sum_{\mathbf{k},\sigma}\sum_{n,l}\int_0^{\beta} d\tau_1 \int_0^{\beta} d\tau_2 I_{nl}^{m}(\mathbf{q}) D_{\mathbf{q}m}(\tau_1-\tau_2) G_n^{(0)}(\mathbf{k},\sigma;\tau_1-\tau_2) G_l^{(0)}(\mathbf{k}+\mathbf{q},\sigma;\tau_2-\tau_1). \tag{103}$$

Until this moment, the spin quantum number σ had not been explicitly considered. However, it should be clear that this index is common to all electron states (corresponding θ-factors would equalize all spin indexes of the different states to a single σ), and summing over it simply results in the multiplication of the whole expression by the quantity $(2S+1) = 2$, the total number of spin projections for the electron.

In order to simplify the evaluation we will restrict ourselves to calculate the finite temperature electron-phonon corrections to the ground level in the quantum well; i.e., only the term with $n = 1$ in the series of equation (102) will be considered. On the other hand, the calculation will take place in the high temperature limit ($\beta \longrightarrow 0$), which justifies approximating the Fermi function by an expansion in powers of the *virial* $e^{\beta\varepsilon}$ in which just the lowest order term is kept [21]:

$$f_\nu = e^{\beta(\varepsilon-E_\nu)}\left[1+e^{\beta(\varepsilon-E_\nu)}+\ldots\right] \simeq e^{\beta\varepsilon}e^{-E_\nu} \tag{104}$$

A key aspect in this process is to determine the value of $e^{\beta\varepsilon}$. To do that, another important assumption is made in combination with (104). That is, our polaron belongs to a quasi-twodimensional electron gas whose total number of particles will be given by

$$N_e = e^{\beta\varepsilon}\sum_n e^{-\beta E_n}\sum_{\mathbf{k}} e^{-\beta E_\perp}, \tag{105}$$

which means that the motion of the whole gas along the z-axis is confined, having discrete values for its energies. Now, taking the one-particle (polaron) limit $N_e = 1$ (which retains only one of the spin projections in the summation over σ in (103)) it is obtained;

$$e^{\beta\varepsilon} = \frac{\pi}{S\gamma_d}\frac{\hbar^2\beta}{m_\perp}, \tag{106}$$

where

$$\gamma_d(t) = \sum_n e^{-tE_n/\hbar\omega_{LO}} \qquad (t = \beta\hbar\omega_{LO}) \tag{107}$$

is -obviously- a function of the $GaAs$ layer width d as well.

With these approximations in mind, $D_{\mathbf{q}m}(\tau_1-\tau_2)$ from (102), and of $G_\nu^{(0)}(\tau)$ given by (49), are substituted in equation (103). Some straightforward manipulations, which include multiplying by the inverse of the first order polaron free energy in $GaAs$ bulk material;

$$1/\Delta\Omega_B^{(1)} = -\frac{\sqrt{\pi}}{\alpha\hbar\omega_{LO}}\left[\int_0^{t/2}\left(\frac{e^x}{e^t-1}+\frac{e^{-x}}{1-e^{-t}}\right)\frac{dx}{\sqrt{x(1-x/t)}}\right]^{-1} = -\frac{\sqrt{\pi}}{\alpha\hbar\omega_{LO}}\frac{1}{\Theta_B}, \tag{108}$$

yield the following expression for the relative polaron free energy $\mathcal{F}_1 = \Delta\Omega_1^{(1)}/\Delta\Omega_B^{(1)}$

$$\mathcal{F}_1(t) = \frac{4\sqrt{\pi}e^{-tE_1/\hbar\omega_{LO}}}{\alpha\gamma_d(t)\Theta_B d}\sum_m\left(\sum_l f(t,E_l) + \int_{V_0}^{\infty}\rho(\mathcal{E})f(t,\mathcal{E})d\mathcal{E}\right). \tag{109}$$

The function

$$\rho(\mathcal{E}) = \frac{2\sqrt{2}L_z}{\pi\hbar}\frac{\sqrt{m_b}}{\sqrt{\mathcal{E}-V_0}}, \tag{110}$$

is the density of states in the region $V_0 \leq \mathcal{E} \leq \infty$. The argument function $f(t, E)$ is given by:

$$f(t,E) = \frac{R_p(E)}{\alpha_\perp(E)}\int_0^{t/2} e^{-\Delta_1(E)x}dx$$
$$\int_0^{Q_{max}} dq \frac{qI^m_{1,E}(q)}{\omega_{qm}}\left(\frac{e^{\omega_{qm}x}}{e^{\omega_{qm}t}-1} + \frac{e^{\omega_{qm}x}}{1-e^{-\omega_{qm}t}}\right) e^{-q^2R_p^2x(1-x/t)}. \tag{111}$$

The dimensionles quantity in the exponent of the first integral $\Delta_1(E)$ is defined as $\Delta_1(E) = (E - E_1)/\hbar\omega_{LO}$, and

$R_p(E) = (\hbar/2m_\perp\omega_{LO})^{1/2}$ is the two-dimensional polaron radius. $\alpha_\perp(E)$ is a Fröhlich-like constant (7) with the *GaAs* parameters except for the effective mass which is substituted by the xy plane mass $m_\perp$. Besides, ω_{qm} is a dimensionless quantity defined by $\omega_{qm} = \omega_m(\mathbf{q})/\omega_{LO}$.

The upper limit Q_{max} in the second integral of equation (111) restricts the integration to the range of values of the wavevector $\mathbf{q}$ where the long-wave continuum model is conceptually valid. Following the analysis presented in references [7] and [20], the value $Q_{max} = 10^7$ cm^{-1} is adopted in this case.

The sum over intermediate states indicated by $\sum_l$ in (103) must necessarily include the contribution of the continuum states above the barrier. This is represented by the integral between brackets in (109). However, in the numerical results presented in figures 2 and 3 this sum restricts to the contributions coming from the discrete states; that is, only the summation over the index l is included. This obeys to two main reasons: To be consequent with the approximation made for $e^{\beta\varepsilon}$, and because the exponential factor $e^{-x(E-E_1)/\hbar\omega_{LO}}$ in the integration of $f(t, E)$ together with the values of the corresponding electron-phonon matrix elements make the contributions from higher energies to be very small. In fact, the results shown contain nothing but the $l = 1, 2$ terms.

The values of the different parameters used in the evaluation are the following: $\omega_{LO} =$ 291.9 cm^{-1}; $\omega_{TO} = 273.8$ cm^{-1}; $\beta_L = 3.2 \times 10^5$ cm/s; $\beta_T = 3.3 \times 10^5$ cm/s; $\epsilon_0 = 12.53$; $\epsilon_\infty = 10.89$; $m_w = 0.0665m_0$ (m_0 being the bare electron mass); $m_b = 0.15m_0$. It is worth noting that the value of m_b has been adopted assuming that the electron is in the Γ-conduction band of the *AlAs*. However, this material has, in fact, an indirect gap, and the minimun of its conduction band is located at the X point. Avoiding this additional complication, the calculation restricts to consider the polaron dynamics in the vecinity of the Brillouin zone center, and for that reason the value for the barrier height to be used is $V_0 = 915$ meV [22].

Results for the relative polaron free energy are resumed in figures 2 and 3. In figure 2, the quantity $\mathcal{F}_1(t)$ is shown as a function of the dimensionless variable t for different quantum well widths. The lower value of this dimensionless quantity in the graphic corresponds to the Debye temperature of *GaAs* ($T_D = 613$ K). $t = 1$ is that of the LO-phonon temperature in the material; $T = 420$ K, and the range in the horizontal axis is extended to a value where one can expect that the non-degenerate electron gas approximation (104) still works good. It is seen that this quantity exhibits the same decreasing behavior as it has for the

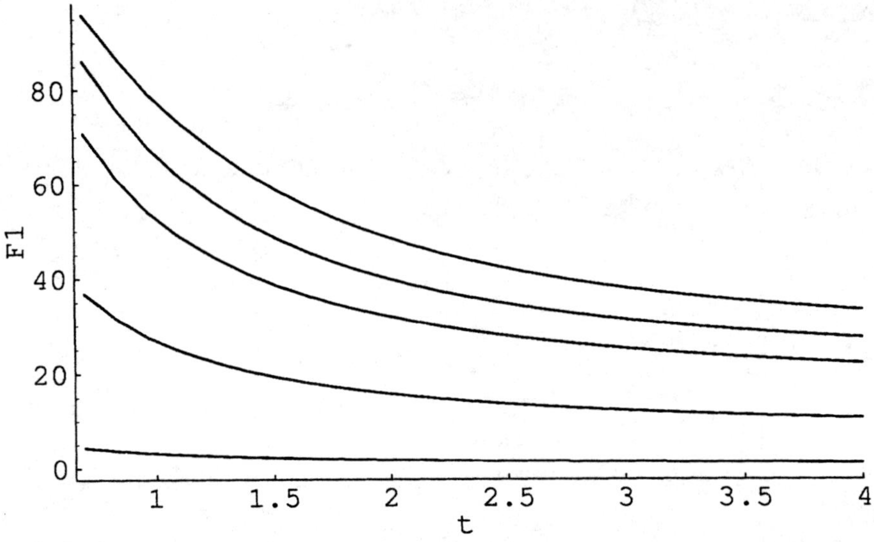

Figure 2: Relative first-order polaron free energy correction to the first subband in a *GaAs/AlAs* double heterostructure, associated to the interaction with *GaAs*-like modes. Curves from top to bottom correspond to $d = 10$ nm, 8 nm, 6 nm, 4 nm, and 2 nm.

case of bulk materiasl in first order of α. However, the magnitude of the correction is well above the bulk value in the whole range of temperatures considered, indicating that, for certain values of the well width, second-order free energy corrections corrections would no be neglectable.

On the other hand, figure 3 shows the variation of $\mathcal{F}_1(t)$ with respect to the well width d for different temperatures. It includes the curves corresponding to the two above mentioned values plus that with $T = 300$ K. For the sake of comparison, a lowest curve, corresponding to the $T = 0$ K polaron relative binding energy in the leading term approximation has been included. This curve is obtained from the expression resulting when taking the limit $t \to \infty$ of equation (109). The result of this limit is precisely the $T = 0$ Rayleigh-Schrödinger-perturbation-theory equation previously presented in reference [8]. As expected, the polaron free energy raises with increasing quantum well width. The rate of increment varies for different temperatures indicating that the way by which the *GaAs*-like phonon modes contribute depends on the temperature as well. Nevertheless, it can be observed a tendency to change the monotony, which reverses towards a limiting bulk-value for d sufficiently large. This decreasing behavior is stronger at high temperatures -as it is shown in the curve corresponding to T_D-, while is much more subtle for $T = 0$ K. The range of well width values selected represents the region of greater interest for the effect here studied.

4. The variational principle for the thermodynamical potential

In this last section let us briefly explore the possibility of generalizing the Feynman's variational principle for the polaron free energy to the thermodynamical potential of an interacting electron-phonon system.

We consider two real operators $\hat{V}$ and $\hat{V}_0$, and construct

$$\hat{V}_\lambda = \lambda\hat{V} + (1-\lambda)\hat{V}_0; \tag{112}$$

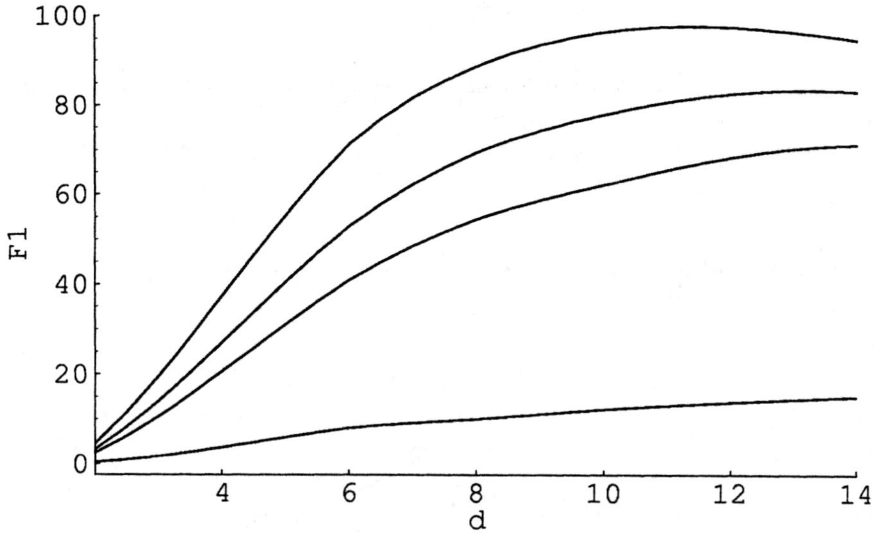

Figure 3: *GaAs*-like first-order polaron free energy correction to the first subband in a *GaAs/AlAs* double heterostructure. Curves from top to bottom correspond to $T = 613$ K, $T = 420$ K, $T = 300$ K, and $T = 0$ K.

with which, a λ-dependent thermodynamical potential is defined through

$$\Delta\Omega(\lambda) = -\frac{1}{\beta}\left\{\langle\hat{T}\exp(\hat{V}_\lambda)\rangle_0^{(con)} - 1\right\}. \tag{113}$$

It is easy to see that

$$\begin{aligned}\frac{d\Delta\Omega(\lambda)}{d\lambda} &= -\frac{1}{\beta}\langle\hat{T}\left(\exp(\hat{V}_\lambda)(\hat{V} - \hat{V}_0)\right)\rangle_0^{(con)} \\ \frac{d^2\Delta\Omega(\lambda)}{d\lambda^2} &= -\frac{1}{\beta}\langle\hat{T}\left(\exp(\hat{V}_\lambda)(\hat{V} - \hat{V}_0)^2\right)\rangle_0^{(con)} \leq 0,\end{aligned} \tag{114}$$

from which it is inmediate that

$$\Delta\Omega(1) \leq \Delta\Omega(0) - \frac{1}{\beta}\langle\hat{T}\left(\exp(\hat{V}_0)(\hat{V} - \hat{V}_0)\right)\rangle_0^{(con)}; \tag{115}$$

In particular, desiganations are changed to:

$$\begin{aligned}\Delta\Omega(1) &= \Delta\Omega \\ \Delta\Omega(0) &= \Delta\Omega_0 = -\frac{1}{\beta}\left\{\langle\hat{T}\exp(\hat{V}_0)\rangle_0^{(con)} - 1\right\},\end{aligned} \tag{116}$$

as a result of what one finally gets:

$$\Delta\Omega \leq \Delta\Omega(0) - \frac{1}{\beta}\langle\hat{T}\left(\exp(\hat{V}_0)(\hat{V} - \hat{V}_0)\right)\rangle_0^{(con)}. \tag{117}$$

The above inequality is, in in the case of a single electron, equivalent to that obtained by Feynman for the polaron free energy [11], and constitutes an extension of his variational principle to a many-body electron-phonon system. As it is pointed out in [23, 24], this inequality is stronger than the so called Bogolubov's variational principle [25]. However, it differs from that of Bogolubov in the fact it is only valid when $\hat{V}$ and $\hat{V}_0$ are, necessarily, real operators.

To better understand this equivalence let us move out from the low-dimensional systems to the traditional bulk electron-phonon theory. As the interaction Hamiltonian is exactly of the same form in the second quantization operators, the method outlined in section 2 applies in this case without major modifications. The effective interaction potential operator will be [24]:

$$\hat{\mathcal{V}} = \frac{1}{2}\sum_{\mathbf{q}}\sum_{\mathbf{k}_1}\sum_{\mathbf{k}_2}|\gamma_{\mathbf{q}}|^2\int_0^\beta d\tau_1\int_0^\beta d\tau_2 D_{\mathbf{q}}(|\tau_1-\tau_2)\hat{a}^\dagger_{\mathbf{k}_1+\mathbf{q}}(\tau_1)\hat{a}_{\mathbf{k}_1}(\tau_1)\hat{a}^\dagger_{\mathbf{k}_2-\mathbf{q}}(\tau_2)\hat{a}_{\mathbf{k}_2}(\tau_2). \quad (118)$$

here, $\mathbf{k}$ and $\mathbf{q}$ are three dimensional wavevectors, and $D_{\mathbf{q}}$ is the bulk phonon propagator. This operator is the second quantized representation of

$$\hat{V} = \frac{1}{2}\sum_{\mathbf{q}}|\gamma_{\mathbf{q}}|^2\int_0^\beta d\tau_1\int_0^\beta d\tau_2 D_{\mathbf{q}}(|\tau_1-\tau_2|)\sum_{i,j=1}^{N}\exp\left(i\mathbf{q}\cdot\hat{\mathbf{r}}_i(\tau_1)\right)\exp\left(i\mathbf{q}\cdot\hat{\mathbf{r}}_j(\tau_2)\right), \quad (119)$$

where N is the total number of electrons.

Written in this form, it clearly appears the analogy with the Feynman's path integral method, in which the statistical operator of the polaron ($N = 1$), is given by [11]:

$$\langle\mathbf{r}'|\hat{\rho}|\mathbf{r}\rangle = \frac{1}{Z}\int_{\mathbf{x}(0)=\mathbf{r}}^{\mathbf{x}(\beta)=\mathbf{r}'} D\mathbf{x}\exp\left\{-\frac{1}{2}\int_0^\beta d\tau\dot{\mathbf{x}}^2(\tau)+V[x]\right\}, \quad (120)$$

and

$$V[x] = \frac{1}{2}\sum_{\mathbf{q}}|\gamma_{\mathbf{q}}|^2\int_0^\beta d\tau_1\int_0^\beta d\tau_2 D_{\mathbf{q}}(|\tau_1-\tau_2)\exp\left[i\mathbf{q}\cdot(\mathbf{x}(\tau_1)-\mathbf{x}(\tau_2))\right]. \quad (121)$$

It is precisely this analogy that which suggests the generalization of the Feynman's polaron model to the many-electron system. In Feynman's theory, an exactly solvable model plays the role of a trial functional in the variational calculation of the polaron free energy. This specific model has the potential energy given by the functional

$$V_0[x] = -\frac{1}{2}\int_0^\beta d\tau_1\int_0^\beta d\tau_2\mathcal{D}_{\mathbf{q}}(|\tau_1-\tau_2|)\dot{\mathbf{x}}(\tau_1)\dot{\mathbf{x}}(\tau_2), \quad (122)$$

where $\mathcal{D}_{\mathbf{q}}(\tau) = \mathcal{D}_{\mathbf{q}}(\beta-\tau)$. The extension of this potential to the many- electron system will be:

$$\hat{V}_0 = -\frac{1}{2}\sum_{\mathbf{k}_1}\sum_{\mathbf{k}_2}\mathbf{k}_1\cdot\mathbf{k}_2\int_0^\beta d\tau_1\int_0^\beta d\tau_2\mathcal{D}_{\mathbf{q}}(|\tau_1-\tau_2|)\hat{a}^\dagger_{\mathbf{k}_1}(\tau_1)\hat{a}_{\mathbf{k}_1}(\tau_1)\hat{a}^\dagger_{\mathbf{k}_2}(\tau_2)\hat{a}_{\mathbf{k}_2}(\tau_2). \quad (123)$$

Equations (119) and (123) provide the potential operators to be used in a variational calculation for $\Delta\Omega$. In this case, the effective phonon propagator $\mathcal{D}(\tau)$ will be determined within the variational procedure in order to give the best possible result. The reader can realize that the extension to the case of low-dimensional microstructures is nothing but formal after the theory presented in section 2. This might constitute a new subject of research within the problem of the electron-phonon interaction in semiconducting heterostructures.

References

[1] M. Born and K. Huang, *Dynamical Theory of Crystal Lattices*, Clarendon Press, Oxford (1988).

[2] H. Fröhlich, Adv. Phys **3**, 325 (1954).

[3] A. S. Davydov, *Teoriia Tverdogo Tela*, Naúka, Moscow (1976).

[4] A. I. Anselm, *Introduction to Semiconductor Theory*, Mir, Moscow (1981).

[5] C. Weisbuch and B. Vinter, *Quantum Semiconductor Structures*, Academic Press, London (1991).

[6] M. A. Stroscio, G. J. Iafrate, H. O. Everit, K. W. Kim, Y. Sirenko, S. Yu, M. A. Littlejohn, and M. Dutta, *Confined and interface optical and acoustic phonons in quantum wells, superlattices and quantum wires* in *Properties of III-V Quantum Wells and Superlattices*, EMIS Datareviews Series No. **15**, Ed. P. Bhattacharya, INSPEC, London (1996).

[7] C. Trallero-Giner, R. Pérez-Alvarez, and F. García-Moliner, *Long wave polar modes in semiconductor heterostructures*, Pergamon/Elsevier Science, London (1998).

[8] M. E. Mora-Ramos and D. A. Contreras-Solorio, Physica B **253** 325 (1998).

[9] G. D. Mahan, *Many-Particle Physics*, 2nd. Ed., Plenum, New York (1990).

[10] V. K. Strukov and V. K. Fedyanin, Preprint JINR R17-11954, Dubna (1978).

[11] R. P. Feynman, Phys. Rev. **97**, 660 (1955).

[12] N. N. Bogolubov and N. N. Bogolubov Jr., Sov. Phys. Part. Nucl. **11**, 93 (1980).

[13] A. A. Abrikosov, L. P. Gorkov, and I. E. Dzyalozhinskii, *Methods of Quantum Field Theory in Statistical Physics*, Prentice-Hall, New Tork (1963).

[14] T. Matsubara, Prog. Theor. Phys. **14**, 351 (1955).

[15] M. J. Buckingham and M. R. Schafroth, Proc. Roy. Soc. (London) **A67**, 828 (1954).

[16] K. Arisawa and M. Saitoh, phys. stat. sol. (b) **120**, 361 (1983).

[17] C. Rodríguez and V. K. Fedyanin, Sov. Phys. Part. Nucl. **15**, 4 (1984).

[18] D. P. L. Kastrigiano, N. Kokiantonis, and H. Stierstorfer, Phys. Lett. **104A**, 364 (1984).

[19] G. Q. Hai, F. M. Peeters, and J. T. Devreese, Phys. Rev. B **48**, 4666 (1993).

[20] F. Comas, R. Pérez-Alvarez, C. Trallero-Giner, and M. Cardona, Superlatt. and Microstruct. **14**, 95 (1993).

[21] H. Haug and S. W. Koch, *Quantum Theory of the Optical and Electronic Properties of Semiconductors*, World Scientific, Singapore (1994).

[22] G. Hai, F. M. Peeters and J. T. Devreese, Phys. Rev. B **42** (1990), 11063.

[23] S. N. Gorshkov, A. V. Zabrodin, C. Rodríguez, and V. K. Fedyanin, Teor. Mat. Fiz. **62**, 304 (1985).

[24] C. Rodríguez and M. E. Mora-Ramos, Rev. Cubana de Física (unpublished).

[25] S. V. Tyablikov, *Methods of the Quantum Theory of Magnetism*, Plenum Press, New York (1967).

AS COLLEAGUE AND FRIEND, I HAVE TO WARN YOU AGAIN ABOUT HOW DANGEROUS YOUR NMR TIME-REVERSAL EXPERIMENTS COULD BE FOR A MAN IN YOUR POSITION...
DEAN
POKÉMON

Quantum Chaos: An Answer to the Boltzmann-Loschmidt Controversy? An Experimental Approach

H. M. Pastawski, G. Usaj and P. R. Levstein *

Abstract

We revive the long standing Boltzmann-Loschmidt polemic which confronts irreversible Thermodynamics with reversible Mechanics. While this is solved within classical physics by the theory of Chaos, until now, there is no quantum definition of dynamical chaos. A nice possibility offered by NMR is to invert the direction of time (quantum Loschmidt daemon) in a many-body spin system. The efficiency of this reversion $M_{PE}(t_R)$ decays with the elapsed time t_R before reversal, with a characteristic irreversibility time τ_ϕ. What controls this time? By changing the time scales of the system in several ways, we observe cases where τ_ϕ is determined by the reversible spin-spin interactions. A natural definition of quantum chaos in our experimental context is the instability of dynamical quantum correlations with respect to infinitesimal perturbations in the Hamiltonian. Therefore, our whole set of experiments indicates that *there is quantum chaos* intrinsic to the many-body spin dynamics in the thermodynamic limit. It manifests as an instability towards perturbations (decoherence) with a τ_ϕ independent on how small the interaction with the environment is. We also argue that $-\ln[M_{PE}(t_R)]$ represents a measure of the increasing entropy. This new evidence in support to Boltzmann's physical insight also serves as a landmark to direct the theoretical and experimental work in the new field of quantum chaos.

1. The Boltzmann-Loschmidt Controversy

1.1. *Importance of basic questions.*

As a new century starts, we still hear the echoes of two fundamental polemics that deeply affect our vision of the physical world and shape our expectancies from science and technology for the centuries ahead.

• From the XIX century, we inherit the unsolved problem of the origin of time irreversibility, whose central aspects can be represented in the arguments between Ludwig Boltzmann and Joseph Loschmidt. The former stated that an arrow of time appears[1] when mechanical laws are applied to macroscopic systems with appropriate approximations. The latter, doubted[2] of that demonstration on the basis that successive applications of time reversible equations of motion could not lead to macroscopic irreversibility.

• From the XX century we receive the paradigmatic polemic between Albert Einstein and Niels Bohr over the interpretation and range of applicability of Quantum Mechanics (QM) in apparent conflict with the realistic view of Classical Mechanics. The latter advocated a cut

*LaNAIS deRMS, FaMAF, Universidad Nacional de Córdoba Ciudad Universitaria, 5000 Córdoba, Argentina between classical (macroscopic) and quantum (microscopic) observables[3,4], while the former claimed unsatisfied about an intrinsically probabilistic description of natural phenomena and the possibility of macroscopically entangled states[5].

A common thread in both conflicts can be found in the fact that both confronted laws developed to describe the experimental evidence from their respective macroscopic and microscopic domains against the very solid foundations of Newton's Classical Mechanics. Indeed, transformations that made possible the modern world were fueled by a Laplacean dream of scientists and engineers inspired in the classical laws. The conflicts are summarized in Fig. 1.

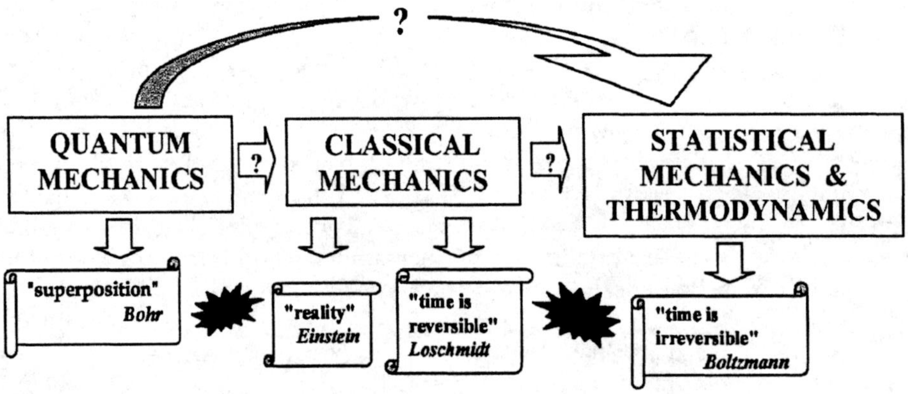

Fig. 1. Disciplines in conflict as involved in the Boltzmann-Loschmidt and Bohr-Einstein controversies.

While most physicists, and all the experimental facts, take the side of Boltzmann and Bohr, the polemic is not over yet. Those apparently philosophical questions take new relevance as we enter into a new century and people unleash their imagination to create the new goals for science and technology. There are licit questions such as: Is it physically possible to travel in time? Will teleportation, as shown in science fiction movies, ever be possible? Is it possible to create computers with the qualities of human mind such as consciousness? Will quantum computers ever be capable of performing computations of great complexity? When we read about them in the leading journals, it is not just fashion. In fact, the deep understanding, required even to get partial answers, contains the seed of technological development. Moreover, since these questions are intrinsically mingled with the Boltzmann-Loschmidt polemic as extended to the quantum world to include the Bohr-Einstein conflict, it is not clear that their answer should be a straight **no**.

Our point in this chapter is to show that within Solid State Physics, we have tools to contribute substantially to the evolution of these concepts. We present material that can be used to guide the reader through the details of our more technical articles.

1.2. *Colliding views*

For scientist adhering to the atomistic view of matter, the conflict started when the macroscopic laws of Thermodynamics were established. The intrinsic irreversibility of the second law determined an arrow of time which appeared contradictory with the Newtonian laws of motion. Boltzmann attempted to reconcile both views in his formulation of the H theorem. By noting that a complex system prepared out of equilibrium and left to its own would evolve in a diffusive way, Boltzmann hinted that complexity played a fundamental role in the onset of irreversibility[6]. While he considered a gas, a situation easier to visualize is a petri dish with clear water where a drop of ink would end coloring all the water. Key to his proof was his hypothesis of "molecular chaos" or *stosszahlansatz.* He assumed that after each molecular collision (characterized by a typical time) no memory of the previous state is retained. Loschmidt raised the objection that for every possible trajectory that leads to equilibrium there will be another trajectory, equally possible, that would lead to the initial low entropy state. Therefore, it would be possible to revert the velocities of every particle to get again a low entropy state. As an answer, Boltzmann improved his theory transforming it into a probabilistic description. His analysis introduced a precise relation between Ω, the phase space volume of the macrostate, and its probability[7]. He emphasized on the extreme practical difficulty of achieving the time reversed evolution proposed by Loschmidt by answering "it is *you* who would invert the velocities". In fact, nobody has ever prepared inked water ending as a single drop in the surface... much less jumping up to the pipette! The fact that by watching a movie for a while we could easily say whether it was played forward or backward clarifies intuitively that *there is* an arrow of time. Therefore, such unrealistic experiments remained a task of a hypothetical *Loschmidt's daemon,* a close relative to Maxwell's creature. The appearance of a state that violates the law of entropy growth $S(t) = k_B \ln \Omega(t)$ is highly improbable except for very small fluctuations. This maintains an objectively definite description in which detailed information is merely unknown prior to observation. Boltzmann probabilistic arguments laid the grounds for the development of non-equilibrium Statistical Mechanics whose conditions have began to be clarified[8] by the modern theory of Chaos.

Modern mathematical views of non-linear equations and chaotic dynamics[9] take us closer to a formal understanding of Boltzmann's views. The Kolmogorov-Arnold-Moser theorem says that in a general mechanical system, outside small stable islands, there are large parametric regions where chaos sets up. In the chaotic regime, trajectories in the phase space starting arbitrarily close from each other will separate at an exponential rate $\exp[\lambda t]$ where the typical inverse time λ is a Lyapunov characteristic exponent. In fact, it is the intrinsic chaos of complex systems which provides for the mixing and growth of Kolmogorov-Sinai entropy that, though with some limitations, justifies[8] the ergodic hypothesis of Statistical Mechanics. An illustrative example which shows the chaotic instability consists in an idealized american billiard game: after a single hit, the sixteen balls move dissipationless and with perfectly elastic collisions with the walls and with each other. However, a sequence of collisions will start to differ from another considering the gravitational interaction with an electron at the further edge of the galaxy, in just one minute[10]! As a consequence of local instability, a similar situation would appear when the initial positions of the balls are varied infinitesimally. Thus, even a simplified model for the dynamics of a drop of ink in water[11], if solved in a computer with great but finite precision would show an increasing difficulty[12] to return to the initial state.

Therefore, the conflict between Boltzmann and Loschmidt seems to be solved in the realm of classical physics. However, Boltzmann could not have foreseen what his academic descendant Erwin Schrödinger contributed to create: A new vision of the world based on microscopic laws governed by Quantum Mechanics. Every experimental evidence is consistent with Quantum Mechanics, even beyond[13] the realm of microscopic physics (as the superconducting magnet recalls us every day). Thus, the unsolved polemic between Bohr and Einstein, over the application of classical kinematic concepts to Quantum Mechanics, re-opens the problem of time asymmetry and forces us to review another conflicting[14] field.

Quantum Mechanics implies a probabilistic description of the world objectively indefinite, i.e. not associated with noise or hidden variables. As a consequence, a state of a system is built as an arbitrary superposition of basic states whose entanglement can become very important. While the correspondence principle suggests that Classical Mechanics is in some way contained in Quantum Mechanics as a special case, this argument leaves unexplained why macroscopic objects come only in narrow wave packets, even when the superposition principle would allow for far more "non-classical" states. This problem is best represented in the Schrödinger's cat paradox. A decaying nucleus could trigger a superposition between two states: one with an alive creature and the other with a dead one. This defies our common sense whether we are allowed to observe it or not!.

Reactions to this difficulty have been very diverse. On one extreme it is formalized in the *many-worlds* interpretation of Quantum Mechanics developed by Hugh Everett III and rescued[15] from oblivion by Bryce DeWitt. This view, accepting that all the universe is described by Quantum Mechanics, considers its state as continuously splitting in a superposition of alternative histories which constitute a basis for this state. In this way, the many-worlds interpretation precludes the need for a quantum/classical border. Yet our perception seems to detect a single alternative of these possible histories with superpositions existing only in a very limited range of space and time in the laboratory. Clearly, this option would not satisfy Einstein at all.

Alternatively, other authors affirm that the Schrödinger equation is only applicable to closed systems (this is true at least in a practical sense). But any real system within the universe has at least a very weak interaction with the external world. As a result, those systems should suffer some leak of information towards the environment. This would introduce some decoherence in the quantum description of this "almost closed" system producing the collapse into pointer states[16] which have some localization in space and momentum compatible with a classical behavior. This superselection is justified because the weak interactions with the "environment" are local in space and they closely conserve momentum. Hence, the Wigner coordinates are the natural basis to separate macroscopic and microscopic variables that describe correlations between initial and final states. In cases where the splitting of energy spectrum is greater than the energy uncertainty produced by the environment, the pointer states would be eigen-energy states compatible with quantum behavior. This is essentially the point of view advocated by Zurek[16] to which we personally subscribe[17]. However, the price is to renounce to the exact description of the Schrödinger equation in favor of some less fundamental quantum master equation or Keldysh equation where an irreversible "ad-hoc" description of the "environment" is injected[18]. Are decoherence and irreversibility contained in a complete quantum description? While Zurek proposal seems to solve the problem of reality, it does not clarify the origin of irreversibility.

The other problem introduced by Quantum Mechanics seems to be even worse: in spite of the probabilistic character of quantum dynamics, quantum effects seem to suppress the chaotic properties of classical dynamics.

2. Possible Solution

2.1. *Does Quantum Chaos exist?*

In a system whose classical description is chaotic, when one prepares two wave packets with very close initial conditions, one would see that there is no such a thing as the positive Lyapunov exponent[19] in their quantum evolution. The distance of any two initial states remains the same as long as its evolution is described by the Schrödinger equation.

Without entering in the quantum regime, let us recall the amazing series of experiments by M. Fink[20]. A sound pulse traveling in a random (chaotic) media is re-emitted at receptor sites in a time reversed sequence producing a pulse back in the emitter. This experiment, and their enormous fields of applications, would not be possible if sound were described by particles instead of waves. It should be noted however, that this experiment is not a time reversal in the sense of a Loschmidt daemon but rather a constructive interference of signals.

Numerical simulations of the Quantum Loschmidt daemon were performed in one-body systems by a number of authors[21]. It is found that quantum versions of systems which are chaotic in the classical limit present accurate reversibility in consistency with unitarity of Quantum Mechanics and the absence of local instabilities in the forward evolution. These properties raise the question about if there is such a thing as Quantum Chaos, originating various reactions.

On a conservative side, one could focus on the characterization of the quantum properties of systems whose classical versions are chaotic. On a radical side, invalidity of Quantum Mechanics is invoked.

Among the first group, there are nuclear physicists who had to deal with the lack of a solvable Hamiltonian describing scattering properties of complex nuclei. In this context, it was natural to assume that those Hamiltonians would be described by the Random Matrix Theory[22] (RMT). It was later realized[23] that they share common characteristics, such as level repulsions, with classically chaotic systems. In particular, for real symmetric matrices (orthogonal ensembles) the normalized spacing between eigen-energies $s = (\varepsilon_{i+1} - \varepsilon_i)/\Delta_E$, with Δ_E the mean level spacing, has a probability given by the Wigner-Dyson distribution $p(s) = (\pi s/2)\ \exp(-\pi s^2/4)$. This spectral rigidity is pretty well satisfied once the spectrum is unfolded to get rid of the degeneracies induced by remaining symmetries. In bounded systems, the spectral rigidity should produce a break-down of the diffusive law and ergodicity at long times. Hence, it was predicted the revival of a density excitation[24] at the Heisenberg time, $\hbar/\Delta_E$, which is called a mesoscopic echo.

Another related problem is how eigen-energies of $\mathcal{H}+\Sigma$ depend on an external parameter α on which $\Sigma[\alpha]$ depends monotonically. A correlation function C_α of the energies or velocities is then studied. Numerical and analytical calculations[25] of finite systems show that after certain critical value $\Sigma_c \equiv \Sigma[\alpha_c]$ one could not know what the original energies or velocities were. The larger the system, the smaller Δ_E and Σ_c.

In complementary studies[26], single particle eigenfunctions in the parameter range that would correspond to classical chaos show a sort of homogeneous distribution that indicates that a particle will visit almost any point in the phase space. There are also some less random wave functions or "scars" that are signatures of isolated periodic orbits.

The consequences of these properties to dynamical chaos, the central issue at stake, are not obvious. In view of this, other authors emphasize the need to obtain some dynamical manifestations of chaos, apparently not present in the linear Schrödinger Equation. A possibility is the generalization of the Schrödinger equation into a non-linear one[27]. This is suggested by the fact that approximations, such as the Hartree mean field, introduce Chaos. The need for a completely different description[28] of quantum phenomena is also invoked.

2.2. *Our theoretical proposal, a possible way out.*

Our point of view is the following. On one side, one should not assume that a microscopically reversible Quantum Mechanics, when applied all the way up to build up macroscopical many-body objects will have time reversible behavior. Therefore, time asymmetry is not the attribute of any individual interaction, but should result as an emergent property in the composition of large number of intrinsically time-symmetric individual couplings. We think this property must be searched in the instability of the dynamical description toward perturbations. In fact, we know that regardless of how big and perfect a system could be, there will be always a small perturbation not considered in the description (recall the 16 balls billiard). This is, of course, implicit in the great tradition[29] of classical physics associated with the names of Boltzmann, Kelvin, Maxwell, Smoluchowski, Ehrenfersts, Einstein and Szilard and in the Landauer's description of quantum transport[30]. On another side, we know that by studying the sensitivity of dynamics to changes in the initial conditions, one is asking the wrong question. Which is the signature of this property in quantum systems? Could we have better luck with sensitivity to perturbations?

The way to measure if a state $|0\rangle$ that evolves under the action of $\mathcal{H}$ differs from the same original state evolving in the presence of a perturbation $(\mathcal{H}+\Sigma)$, is to evaluate the overlap:

$$M(t) = |\,\langle 0|\exp[\mathrm{i}(\mathcal{H}+\Sigma)t/\hbar]\exp[-\mathrm{i}\mathcal{H}t/\hbar]\,|0\rangle\,|^2, \qquad (1)$$

Consistently, more than 10 years ago A. Peres [31] proposed that dynamical signatures of quantum chaos should be searched on the sensitivity of dynamics to perturbations in the Hamiltonian. For a one-body classically chaotic system, he found a decay within a typical time $\tau_\phi(\Sigma)$ not observed in integrable systems. Clearly, $\tau_\phi(\Sigma)$ is a decoherence time. It is the typical scale at which the phase of the wave function evolved with $\mathcal{H}$ becomes meaningless because it is totally modified by effect of a perturbation Σ. The same idea can be formulated in terms of the trace of the product of evolved density matrices

$$M(t) = \mathrm{Tr}[\rho_{\mathcal{H}+\Sigma}(t)\rho_{\mathcal{H}}(t)]/\mathrm{Tr}[\rho(0)\rho(0)]. \qquad (2)$$

These results contrast with the absence of sensitivity to initial conditions

$$|\,\langle 0+\delta|\exp[\mathrm{i}\mathcal{H}t/\hbar]\exp[-\mathrm{i}\mathcal{H}t/\hbar]\,|0\rangle\,|^2 \equiv \mathrm{cte}.$$

Actually, for a classically chaotic system, a perturbation in the initial condition is equivalent to a perturbation in the Hamiltonian.

It is remarkable that the left evolution operator can also be seen as an "*imperfect time reversal*" of the wave function! i.e. where $t \to -t$ but the Hamiltonian changed a bit $\mathcal{H} \to (\mathcal{H}+\Sigma)$. It is also equivalent to the return probability after subsequent forward evolution with an imperfectly inverted Hamiltonian $\mathcal{H} \to -(\mathcal{H}+\Sigma)$.

Since the simplest real gas has 10^{23} particles, a problem on the sequential order of the limits will appear as the number of particles goes to infinity and the perturbation strength to zero. A similar situation occurs if the volume explored by a *single* particle becomes unbounded. The point is that if in the chaotic regime the two possible sequences do not give the same result (i.e. *the limits are not uniform*), we would be in the presence of an emergent property as occurs in phase transitions. Hence, it is possible to obtain a macroscopically irreversible dynamics from microscopic reversible individual laws. This is the point we are going to clarify in both the analytical and experimental studies described in this chapter.

There is not much hope to calculate Eq. (1) for an actual many-body system, but we did a quantum calculation[32] for a particle in an unbounded space with a concentration of impurities well below the localization threshold. There, the collisions determine, as in the classical version[33], a typical elastic collision time τ_o and a single classical Lyapunov exponent $\lambda \propto 1/\tau_o$. In the range of high energies we can build a wave packet that has minimal uncertainty for its position and velocity. A calculation of its evolution was done[32] in terms of the Green's function in the semiclassical approximation[34] where the phases are determined by the action of classical trajectories. The divergence of classical trajectories, under the effect of the perturbation, transforms in growing phase differences. This leads to an exponential decay of the overlap.

$$\lim_{\Sigma \to 0} \lim_{t \to \infty} -\frac{1}{t} \ln M(t) = \frac{1}{\tau_\phi} \tag{3}$$

One striking point is that while the decay at short times depends on the perturbation, the asymptotic decay *depends only on the Hamiltonian.* The other, is that the *inverse decay time* is *just* the *classical Lyapunov exponent*:

$$\frac{1}{\tau_\phi} = \lambda. \tag{4}$$

This establishes a new connection between classical and quantum chaos. Implicit here is the order of the limits mentioned before. The continuous spectrum used to expand the wave packet implies an infinite dimension of the available Hilbert space before the perturbation is made arbitrarily small. The result of Eq. (3) is confirmed by numerical calculations[35]. They also clarify the connection between the thermodynamic limit and the decay of parametric correlations and Eq. (3): In order to obtain Eqs.(3) and (4) in a practical finite size calculation, the perturbation must be above the critical value[25] corresponding to its size and spectral region. This means that the time scale associated with the perturbation, τ_Σ, is much smaller than the Heisenberg time. Therefore, dynamical quantum chaos is definitely connected with steady state properties. On the basis of the physical meaning of Eq.1 as a measure of the inefficiency of a quantum Loschmidt daemon, these results state that *chaos provides a path to both decoherence and irreversibility.*

A fundamental aspect of this approach is that it connects chaos with irreversibility without getting out from the unitary description of Quantum Mechanics. In fact $\mathcal{H}+\Sigma$ can be a perfectly unitary operator. In particular, it provides a justification for an irreversible description of the decoherent bath proposed by Zurek. The non unitary case, $\Sigma = \Delta + \mathrm{i}\Gamma$, not included in our demonstration, would involve replacing $\langle 0|\exp[\mathrm{i}(\mathcal{H} + \Sigma)t/\hbar]$ by the solution of the corresponding Keldysh equation[18].

So far, we have dealt with a single particle in a chaotic potential. What about a many-body system in the thermodynamic limit? A classical reasoning shows that the dynamical instability of a billiard where only *one* ball is mobile should be quantitatively *different* from one where all *fifteen* are free to move around. A qualitative difference (phase transition) occurs only in the thermodynamic limit. What would be the quantum behavior of a many-body system? The dimension of the available Hilbert space is now elevated to a power given

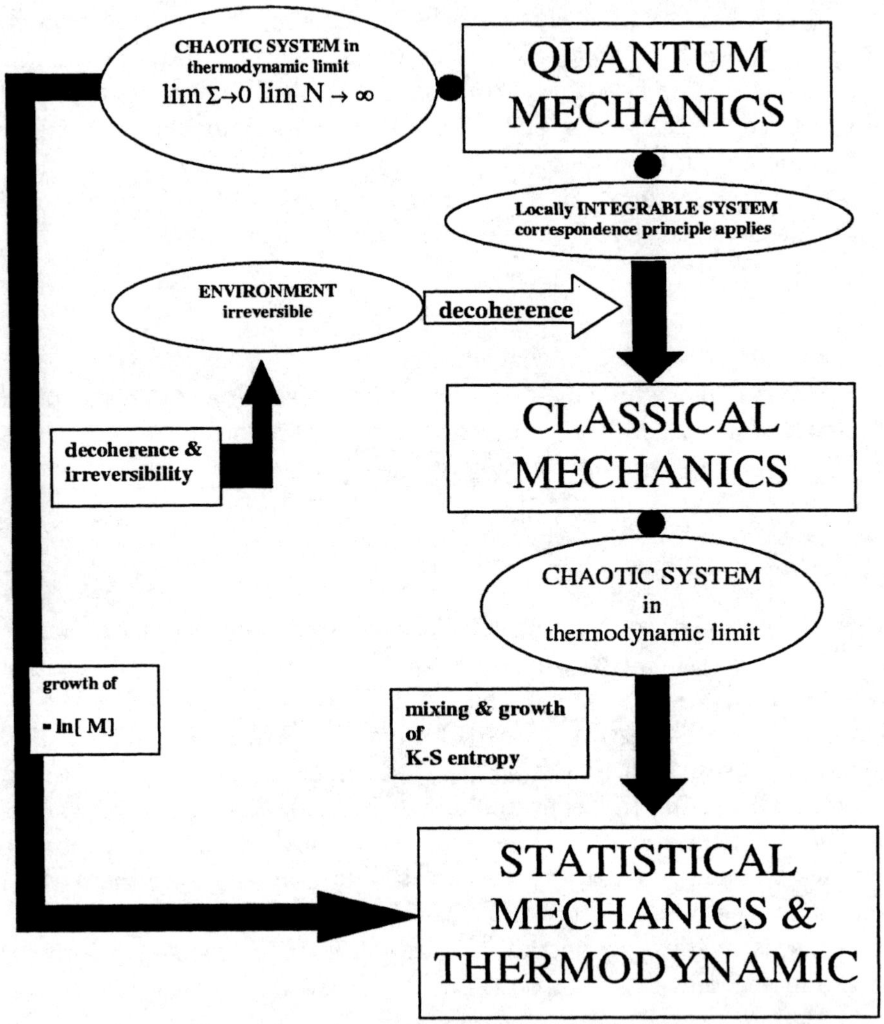

Fig. 2. Conceptual map showing the connections between disciplines as they result from the arguments in the text. The introduction of Chaos in its classical and quantum versions, which produce irreversibility and decoherence, solves the conflicts schematized in Fig. 1. Notice the feedback loop with Quantum Chaos: A quantum chaotically system satisfies Statistical Mechanics and may serve as a bath for simpler quantum systems.

by the number of particles. Immediately, our analytical and numerical tools become very limited to follow even the first quantitative changes. The only option at the moment is to resort to some ingenious experiments that verify and extend the hints obtained from the semiclassical calculation. This will be the aim of the next section. Let us first review the conceptual tree we have obtained in Fig 2.

3. The Experimental Approach

3.1. *High temperature Spin Dynamics as a Quantum Zeno Effect.*

Presently, there is no way to revert the velocities and interactions of all the particles in a box containing a gas or a liquid. However, there is a system that we can control quite well in a standard laboratory: the nuclear spins in a solid. Interactions between spins are so weak that the interaction times are much longer than typical molecular motions. Thus, the coupling with environmental degrees of freedom tends to average to zero. Nuclear spins are hence quite isolated systems. On the other side, the interaction energy is always much smaller than the available temperatures. From a quantum sense these systems are in an infinite temperature regime, far away from the ground state. Therefore, in the absence of an external magnetic field the net polarization is zero. In particular, if one imagines the crystal divided in a number of identical pieces and we assign the label 0th to a particular nucleus in the center of each piece and we sum the magnetic polarization in such sites is also zero. In the presence of an external magnetic field B_0, the fraction between up and down spins is $\exp[\hbar\gamma B_0/k_B T]$, where γ is the nuclear gyromagnetic ratio.

Now, let us imagine an ideal situation in which there is a set of some perfect local probes that allow us to excite the system orienting those selected nuclei at certain time. This is equivalent to have an ensemble of water dishes and dropping a single molecule of ink at the center of each one. One expects the concentration of ink to diffuse from the center. The magnetic polarization centered just at the selected sites of the ensemble would also spread away but now governed by the Schrödinger equation. About the initial state, we only know that spin 0th is "up". The Hamiltonian will typically be given by the general form:

$$\mathcal{H}_{II} = \sum_{k,j>k} J_{jk} \left[\alpha 2 I_j^z I_k^z - \frac{1}{2}\left(I_j^+ I_k^- + I_j^- I_k^+\right)\right]. \tag{5}$$

where I are the usual 1/2 spin operators with subscripts indicating spin sites j and k in a network. For $\alpha = -\frac{1}{2}$, it describes the Heisenberg model (isotropic exchange), $\alpha = 0$ defines the XY model, while $\alpha = 1$ is a truncated dipolar Hamiltonian. In this last case, standard NMR literature adopts d_{jk} ($\equiv J_{jk}$) to denote the coupling parameter.

In this subsection we consider a d-dimensional (hyper)-cubic lattice of spins where $J_{jk} = J_\mu$ are nearest neighbors interaction parameters depending only upon the direction $\widehat{\mu}$ along sites j and k at distance $a_\mu \equiv a = 1$.

From a macroscopic scope, i.e. for long times and large wave-lengths, one expects that excitations should evolve irreversibly. The resulting hydrodynamic equations[36] describe the "spin diffusion". In this regime, a local excitation decays as $(Dt)^{-d/2}$. From the microscopic point of view, however, it is clear that the dynamics is governed by reversible Quantum Mechanics as long as quantum coherence is maintained.

We are going to show that under certain approximations this dynamics can be described by a diffusive process. This is just one of the goals of the many-body techniques, i.e. to allow the match between the quantum and hydrodynamic regimes providing the machinery for the calculation of the diffusion constant and other macroscopic observables. This is still an active field of research[37]. However, it involves implicit approximations which sometimes obscure the nature of irreversibility.

The calculation of the dynamics of a local spin polarization in the high temperature regime can be mapped to a system of particles on a lattice. Up spins are identified with particles and down spins with holes. A particle is localized at the excited site with the rest of the sites being occupied with probability one half. The initial state is an incoherent superposition of all the possible initial states a few of which are shown in Fig. 3. The particle initially at 0th site starts to evolve having a finite probability amplitude to jump into empty neighboring sites. Meanwhile, other particles also can move in and out of neighboring sites creating a fluctuating effective potential. If this is approximated by a stochastic potential, one obtains an irreversible equation satisfying the hydrodynamic limits. The essential point is that the correlation times of the potential are the same that characterize the dynamics of a particle in its fluctuating environment. This leads to a *self-consistent* equation which is the core of classic many-body calculations[38] of spin "diffusion" for symmetric lattices. Let us do a simplified calculation, valid for both symmetric and asymmetric systems, which will clarify the many-body effect and the dimensional crossover in the dynamics.

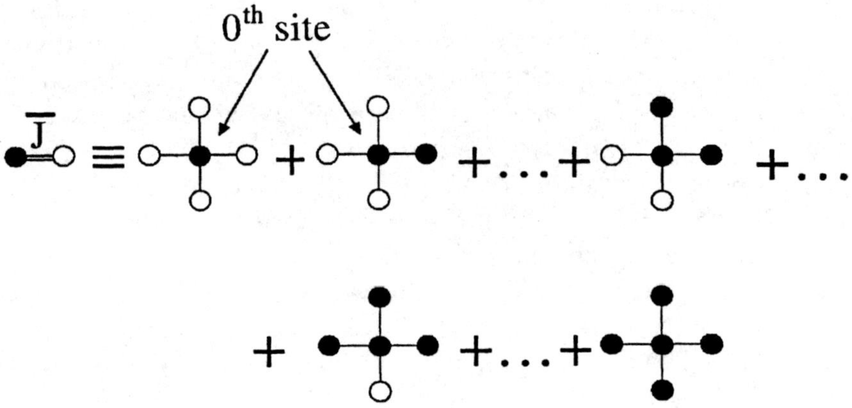

Fig. 3. Up spins are identified with particles (solid circles) and down spins with holes (open circles). The initial state is an incoherent superposition of all the possible initial states. Only 5 of the 16 configurations corresponding to a local excitation in the high temperature limit are sketched. The particle initially at 0th site starts to evolve having a finite probability amplitude to jump into empty neighboring sites. On the left, an effective two sites system is represented. This system should include either a non-linear or a decoherent dynamics.

Let us first notice that one spin up, in a lattice with all spins down, propagates with a typical "ballistic" velocity v proportional to the exchange constant J. At high temperature, there are as many up spins as down ones. Hence, this propagation in the sublattice of spins down is interrupted by the modification of the lattice. This produces a "collision" rate $(1/\tau_{\text{mb}})$ also proportional to J. If we neglect quantum interferences (*stosszahlansatz*) produced by multiple collisions, a diffusion constant $D \propto v^2\tau_{\text{mb}} \propto J$ is obtained. Again, the point is that a diffusive behavior is a good approximation for the dynamics. It will be stabilized by later irreversible interactions. Keeping this in mind, we may *approximate* each "*collision*" *as a measurement* process.

Quantum dynamics is strongly modified by recurrent measurements. This phenomenon is known as the *Quantum Zeno Effect* (QZE)[39]. In simple words, it affirms that *if the evolution* of a quantum system *is observed too frequently, there will be no evolution* to be seen! While this might sound paradoxical from a classical point of view, within the quantum logic it does not: a measurement involves a collapse of the wave function and the start of a new quantum evolution. For short times, the probability of staying in the initial state is $P_{i,i}(t) = 1 - (\overline{J}t/2\hbar)^2 + \ldots$, with $\overline{J}$ being an average exchange coupling. The lack of a linear term on this expansion has very important consequences. If the evolution in a time t_s is interrupted by N observations, the final probability of stay is $\widetilde{P}_{i,i}(t = t_s)$ $=[P_{i,i}(t_s/N)]^N \rightarrow 1$ when $N \rightarrow \infty$. Here, $t_s/N = \tau_\phi$ defines a decoherence time. Hence, one sees why a frequent collapse (i.e. $\omega_\phi = 1/\tau_\phi \gg 1$) hinders evolution[40]. The result of the successively interrupted evolution is shown schematically in Fig. 4.

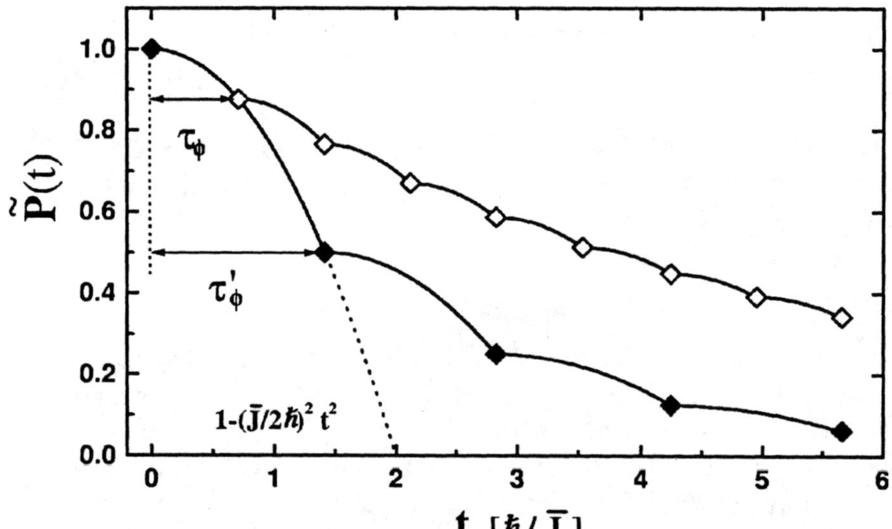

Fig. 4. Probability of stay for a particle whose quantum evolution is interrupted by a measurement process with a period τ_ϕ and $\tau'_\phi = 2\tau_\phi$. The coherent evolution is approximated up to the quadratic term and the return probability is neglected. The increase of the probability of stay is a manifestation of the Quantum Zeno Effect. The evolution shown is equivalent to the average evolution of the ensemble in Fig. 3.

For each initial state, $\mid i\rangle$, with 0-th site polarized (i.e. one of the states shown in Fig. 3), the probability of finding the same site polarized in the state $\langle f \mid$ after a time t, is

$$W_{f,i}(t) = \left| \langle f \mid \exp[-\frac{\mathrm{i}}{\hbar}\mathcal{H}_{II}t] \mid i\rangle \right|^2 . \tag{6}$$

A total ensemble averaged probability that a spin initially up at position 0-th is still up at time t can be calculated summing over all the N_i and N_f possible initial and final states which, for a macroscopic sample is arround 10^{20}:

$$\langle P(t)\rangle \;=\; \sum_f^{N_f}\sum_i^{N_i}\frac{1}{N_i}W_{f,i}(t)$$

$$= 1 - \frac{1}{4}\sum_{\mu}^{d} \langle Z_\mu \rangle \, J_\mu^2 t^2/\hbar^2 + O(t^4) + \ldots . \qquad (7)$$

$\langle Z_\mu \rangle = 1$ is the average number of neighbors along direction μ with spin down. In the second order term only the flip–flop terms produce the exchange of the originally polarized 0th spin with its neighbors. This corresponds to the *one-body* dynamics of one up spin in a lattice in which the other up spins remain frozen. Higher order terms contain the dynamics of those other spins and the *many-body* interactions.

A normalized magnetization can be calculated from the spin selfcorrelation function as

$$P_M(t) = \langle I_0^z(t) I_0^z \rangle \, / \, \langle I_0^z I_0^z \rangle = 2(\langle P(t) \rangle - \frac{1}{2}). \qquad (8)$$

This magnitude is experimentally accessible.

For the case α=0, (XY-model) an exact mapping[41] to a *non-interacting* fermions system, allows us to write the magnetization as the square modulus of an effective *single particle* wave function: $P_M(t) = \left| \langle 0 \mid \exp[-\frac{i}{\hbar}\mathcal{H}_{II}t] \mid 0 \rangle \right|^2$ where the $\mid 0 \rangle$ state has spin I_0 "up" and all the others "down". In contrast, dipolar ($\alpha = 1$) and Heisenberg ($\alpha = -1/2$) cases mapping to *interacting* particles maintain their many-body nature.

For short times, the mean square displacement of the magnetization is written in terms of the nearest neighbor spins correlation functions $\left\langle r_\mu^2 \right\rangle \propto a_\mu^2 \left\langle I_{0\pm a_\mu}^z(t) I_0^z \right\rangle$. The truncated quantum dynamics given by Eq. (7) gives:

$$\left\langle r_\mu^2 \right\rangle_{\text{quant.}} = \frac{1}{2} a_\mu^2 J_\mu^2 t^2/\hbar^2. \qquad (9)$$

However, the perturbative expansion of Eq.(7) is not a practical way to obtain the long time dynamics of complex many-body systems. It is used only for a few simple one dimensional systems[37]. Therefore, we need an entirely different approach. While evolution with (5) is not an irreversible process, it is close enough to a diffusive evolution so that a self-consistent condition is already achieved for intermediate times.

When $\langle P(t) \rangle$ has decayed substantially, let us say when the second term in Eq.(7) is one half, the environment has changed completely and essentially no phase coherence with the initial *one-body* state is retained. This defines the spin collision time

$$\tau_{\text{mb}} = \hbar 2 / \sqrt{2 \sum_{\mu}^{d} J_\mu^2} = 1/\omega_{\text{mb}} \qquad (10)$$

$$= \hbar 2 / \sqrt{\sum_k J_{0,k}^2}, \qquad (11)$$

which also can be interpreted as a one-body dephasing time. The second line, valid for a general network, is included for future use. It is important to note that this functional dependence on J_μ does not depend on the amount of decay chosen to determine τ_{mb}. The change in the environment, following itself a quantum dynamics according to Eq. (7), is slower than linear at early times, becoming important only at about τ_{mb}. Then, it can be described by a discrete time Markovian process leading to a classical random walk. At the dephasing time the mean square displacement of the magnetization is

$$\left\langle r_\mu^2 \right\rangle_{class.} = 2 D_\mu \tau_{\text{mb}}. \qquad (12)$$

At $t = \tau_{\rm mb}$, both quantum and classical diffusive regimes must coincide. This statement was rigorously proved in Ref.17, where we used the Keldysh's formalism[18] to achieve a *non-perturbative* description of the cross-over from the quantum to the diffusive regime for a particle interacting with a dephasing field. In our model, this field acts at a typical time with an interaction probability $p(t)\mathrm{d}t = \delta(t - \tau_{\rm mb})\mathrm{d}t$. Then, the coherence with the initial state has a survival probability $\theta(\tau_{\rm mb} - t)$. This step function is more appropriate to describe the quantum dynamics of the dephasing field than the usual $\exp[-t/\tau_{\rm mb}]$. According to Ref.17, within this approximation the *self-consistent* propagation of density excitations (satisfying the integral Keldysh equation) requires that both quantum and Markovian descriptions[40] give the same probability distribution at $t = \tau_{\rm mb}$. This condition is equivalent to equate (9) and (12), from which we obtain the diffusion constant for each direction:

$$D_\mu = \frac{a_\mu}{2\hbar} \times J_\mu^2 / \sqrt{2\sum_\mu^{\rm d} J_\mu^2} = \frac{a_\mu^2}{4\hbar^2} J_\mu^2/\omega_{\rm mb} \tag{13}$$

While our procedure has been mainly qualitative, we believe it catches the fundamental phenomena, and therefore the correct functional dependence. For the symmetric three dimensional lattice this gives

$$D = Ja^2/(\hbar 2\sqrt{6}),$$

in fair agreement with the values calculated by a number of previous authors[38,42] and consistent with simulations[38] in a system of classical spins. The diffusion constant is proportional to the magnetic coupling parameter.

A paradoxical aspect of the dimensional dependence in the dynamics of magnetic excitations can be seen[43] at the light of the QZE. This is evident for a cubic lattice where the interaction along direction y grows ($J_y/J_{x(z)} \to \infty$). The diffusion coefficient in that direction will also grow: $D_y \propto J_y$. However, the spreading rate in the other directions will decrease $D_{x,(z)} \propto 1/J_y$. The essence of the argument is that spins spreading in each xz-plane register the evolution in the other parallel planes through the many-body coupling J_y (i.e. they "observe" each other with a frequency $\omega_{\rm mb} \to J_y/\hbar$), producing the reduction of the dynamics within the planes. This is a general behavior also valid for the non-perturbative limit of $J_y \approx J_{x(z)}$.

While the numerical solution of the quantum dynamics of systems with a large number of spins is a formidable task, we can attempt to see signatures of the discussed phenomena in small two dimensional systems. We consider a nine spins system with periodic boundary conditions and evaluate the eigenstates of all the spin configurations to study the dynamics according to Eqs. (5) and (6) with $\alpha = -1/2$. In Fig. 5 we show the local polarization in this lattice up to intermediate times. For longer times, weak mesoscopic beats[44] appear at the Heisenberg time of the one-body system. At even longer times, trains of beats indicate the non ergodic nature of the quantum evolution.

3.2. *Experiments on Spin "diffusion".*

It is not simple to test all the features of local quantum dynamics for spin systems. Traditional methods to obtain signatures of the *electron* spin dynamics are neutron scattering and electron paramagnetic resonance (EPR) experiments. The first measures directly the spin correlation function. In the last, the information is contained in the linewidth of the EPR spectrum. Different local environments in a system of non-interacting spins produce an inhomogeneously broadened absorption line, where each frequency corresponds to spins

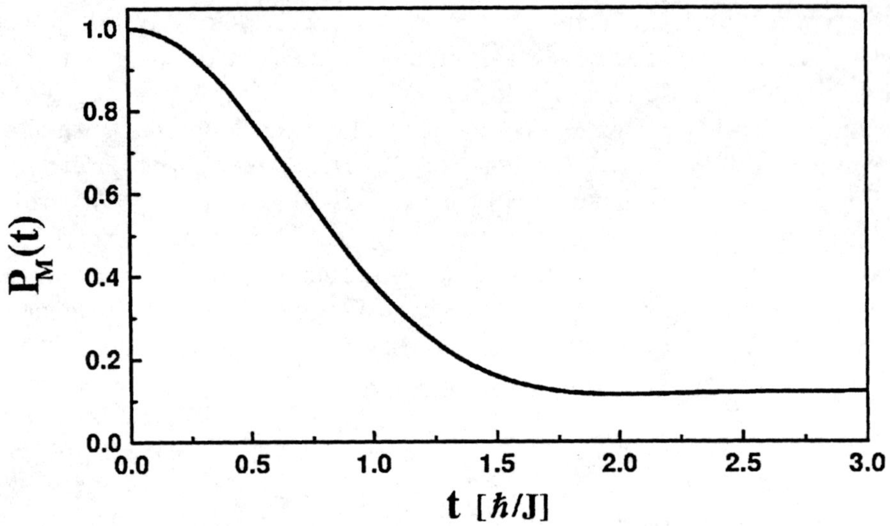

Fig. 5. Exact quantum evolution of a spin autocorrelation function in a finite symmetric (J_y=J_x =J) 2-d lattice with Heisenberg interaction. Interferences starting at the Heisenberg time lie out of the plotting range.

seeing a different local field. However, through the flip-flop mechanism, each spin excitation explores different lattice sites producing an averaging of the field. Thus, the narrowed linewidth [45] is: $\Delta\nu \propto \int_0^\infty \langle I_0^z(t) I_0^z \rangle \, dt$ (eventually the integral extends up to a cut-off time). Then, the slower the dynamics the wider the line. It is not usually possible to control the magnitude of the exchange interaction for electronic spins. However, this is just the situation observed[46] for a family of $Cu(aa)_2$ single crystals (aa stands for amino acid). Since in this family of compounds it is possible to pass gradually from symmetric to asymmetric interaction networks, the spin diffusion manifests the features of QZE discussed above.

A novel approach to test nuclear spin diffusion in homogeneous solids involves sophisticated application of pulsed rf with controlled spatial gradients [47]. The diffusion constant in *each direction* can then be measured independently. While this has enormous potentialities in measuring previously unavailable quantities of the many-body dynamics, it involves a novel non-standard apparatus. Fortunatly, an ingenious technique to measure the local polarization in a system of protons using standard equipment was introduced[48] by Zhang, Meier and Ernst (ZME). From this, the local diffusive properties are extracted, though without information on directionality. They proposed to use a rare ^{13}C nucleus (in its 1.1% natural abundance in the sample) as a *local probe* to inject and detect polarization in a system of ^1H spins interacting through the dipolar interaction. Candidates for this experiment are organic molecular crystals. Among them, Ferrocene, $(C_5H_5)_2Fe$, or Cobaltocene, $(C_5H_5)_2Co$, have optimal properties that allow one the measurement of oriented crystallites in the powder sample. Both have a metallic atom surrounded by cyclopentadienil rings (C_5H_5) as shown in figure 6. Since the rings rotate very fast[49] ($\tau_R \approx 10^{-12}$s) all sites in the molecule become magnetically equivalent yielding a single time-averaged[50] intramolecular $d_{j,k}$. As a consequence, the magnitude of the interaction depends only on the nuclear distances and the angle between the molecular axis and the magnetic field. This contrasts with the general case where each pair of nuclei can have a different interaction parameter depending on the angle between the internuclear vector and the magnetic field.

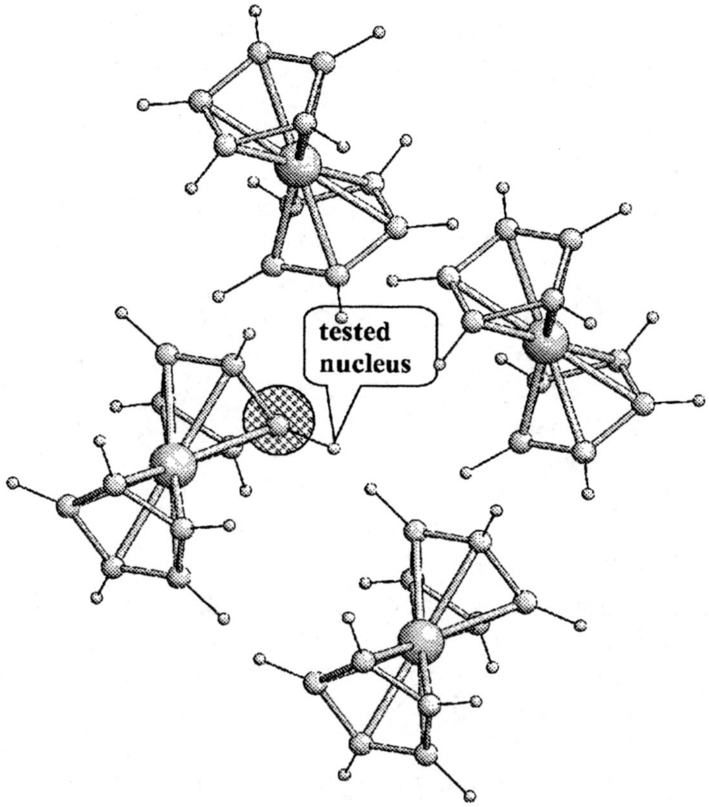

Fig. 6. Crystalline structure of ferrocene, $Fe(C_5H_5)_2$ in its (room temperature) monoclinic form with space group $P2_1/a$ as described in Ref. 49. Two unit cells are included to give a better idea of the ^{1}H network. A rare ^{13}C nucleus (spin S), identified with a circle, serves to probe the polarization of the directly bonded ^{1}H (spin I_0).

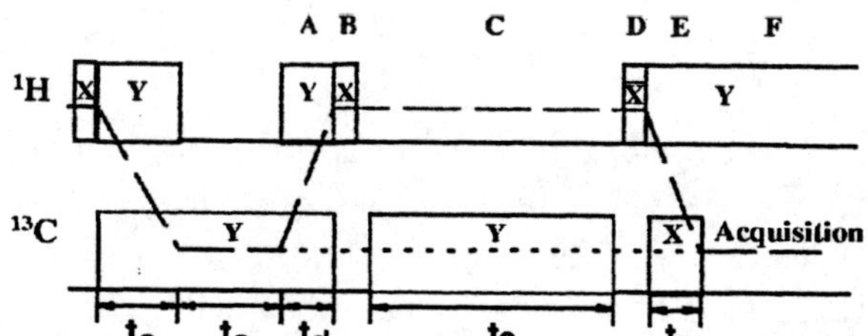

Fig. 7. Pulse sequence for proton spin 'diffusion' in the *laboratory frame*. A $(\pi/2)_x$ pulse on the abundant I spins creates polarization. This is transferred during t_C to a rare ^{13}C spin S. After decay of proton spin coherence during t_S: **A)** The S magnetization is transferred to the y-axis of I_0 through a CP pulse of duration t_d. **B)** A $(\pi/2)_x$ pulse tilts the polarization to the *laboratory frame* where **C)** it evolves freely during t_2. **D)** A $(\pi/2)_{-x}$ pulse tilts it back to y and **E)** a CP pulse of length $t_p = t_d$ transfers the polarization to the x-axis of S where **F)** it is recorded. Dashed and dotted lines show main and secondary probability amplitude pathways for the polarization.

The NMR radio-frequency pulse sequence developed by ZME, involves a two-way transfer of polarization[51] among the spin S of a rare ^{13}C and the I_0 spin of a directly bonded ^{1}H. Then, the ^{13}C can be used to inject magnetization in the proton and to capture it later on. The set of coupled protons constitutes the *system* where the spin dynamics can be monitored.

The complete pulse sequence is schematized in Fig. 7. A $(\pi/2)_x$ pulse on the abundant ^{1}H spins system creates a polarization in the plane perpendicular to the magnetic field. This polarization is transferred through a cross polarization (CP) procedure during t_C to the rare ^{13}C system when both ^{1}H and ^{13}C are irradiated at their respective resonant frequencies. The rf field strengths must fulfill the Hartmann-Hahn[51] condition $\gamma_I B_{1I}^y = \gamma_S B_{1S}^y = \omega_1$, i.e. the Zeeman splittings of ^{1}H and ^{13}C in the rf field are degenerate enabling the 'flip-flop' mechanism. After decay of the proton spin coherence during time $t_S \simeq 1ms$ we are left with a polarized ^{13}C and an unpolarized ^{1}H system. Then the most relevant part begins:

A) The magnetization from an initially y polarized S spin is transferred to the y-axis of the I_0 spin through a new CP pulse of duration t_d which is the shortest maximizing the polarization transfer[52,53] (coherent 'flip-flop' mechanism) for the selected orientation.

B) A $(\pi/2)_x$ pulse tilts the polarization to the laboratory frame.

C) The I-spins evolve freely during a time t_2. Thus, the relevant evolution of the spin-diffusion sequence occurs in the laboratory frame while S-irradiation prevents the $I - S$ (system-probe) coupling.

D) A $(\pi/2)_{-x}$ pulse tilts the polarization back to the rotating xy plane.

E) Another CP pulse of length $t_p = t_d$ is applied to transfer back the polarization to the x-axis of S.

F) The S polarization is detected while the I-system is kept irradiated (high-resolution condition) which means that the ^{13}C's do not interact with the protons during this period.

In Fig. 8 we present measurements of this spin dynamics in a single crystal of ferrocene. The curve has a number of very interesting features. For short times, we see that superimposed to a parabolic curve with a maximum around t_m there are high frequency oscillations. This is a *quantum phenomenon*[54] analogous to a time domain of a Aharonov-Bohm electrostatic interferometer[55]. Local polarization in ^{13}C splits in a superposition of two states: in one the polarization stays in the ^{13}C in the other it wanders in the ^{1}H system. When they rejoin they interfere manisfesting the difference of quantum phase $(\omega_1 t_2)$ acquired by the states in each environment. We are going to explain the origin of the maximum shifts in the next section.

Local polarization shows a global decreasing trend following the general cross-over from the cuadratic decay to the diffusive law discussed in the previous section. However, a *quantum beat* appears with a clearly developed maximum at $t \approx 0.6\text{h}/d$. A second maximum which, on the basis of numerical analysis, we identify with the *mesoscopic dynamical echo* corresponding to the Heisenberg time $\hbar/\Delta_E$ in the isolated ring[44], develops at $1.1\text{h}/d$. The last is overcome by an overall attenuation associated with diffusion to neighboring molecules. The solid line shows an ideal evolution, $P_M(t_2)$, with Hamiltonian (5) for an isolated molecule (10 spins in Fig. 6). An important conclusion to be drawn from the comparison of this experiment with the numerical calculations is that while changes in the short time evolution, $P_M(t_2) \simeq 1 - \frac{1}{2}(t_2/\tau_{\text{mb}})^2$, are not much apparent, the effect of those interactions in blurring out the long range interferences is decisive. This is evident in the decrease of the dynamical echo. The above arguments also lead to interpret the additional attenuation observed in the experimental curve as due to intermolecular interactions.

In summary, although polarization dynamics shows some "diffusive" trends already at intermediate times[48], a careful analysis[44] can recognize particular patterns due to quantum interferences. Its experimental observation[54,56] constitutes a fingerprint of the bounded regions where the short time dynamics occurs.

3.3. *An experimental Loschmidt daemon.*

There is an article[57] entitled "*Atomic Memory*" which shares some concepts with the present section. One of the authors is the "Wizard of Resonance", Erwin Hahn, who discovered and explained the spin echo, on which most of present NMR imaging and spectroscopy is based. Basically, the polarization decay in materials is originated in the different precession rates of spins in different environments. This produces a decrease of polarization equivalent to the spreading of runners along a field. However, if the velocities are reversed they will return together to the starting point. In QM the phase acquired by a given state is its energy multiplied by the elapsed time. One can see a Hahn's echo as a change in the sign of the Zeeman energy associated with each spin, i.e. a change of sign of the Hamiltonian linear in spin (spin-field interaction). On this, however, the *many-body* part of the spin Hamiltonian plays the role of an *uncontrolled* irreversible part. This produces an attenuation of the spin echo with a time scale called $T_2 \simeq \tau_{mb}$ which is determined from the many-body Hamiltonian given in Eq. (10). In summary, the Hahn's echo is a realization of the Loschmidt daemon for the very simple system of independent spins rotating in the presence of their local magnetic field. In a way, this is a model for an ideal gas: a box containing point-like particles starting at the same position. Having some distribution of velocities they would end filling all the box. Inverting all velocities simultaneously will produce a revival of concentration. Now, if the spins are going to serve as a model for a non ideal gas or liquid having collisions the problem becomes more complex.

We come back to an important feature observable in the spin diffusion experiment. As seen in Fig. 8 the curve does not decay immediately but needs a while to reach its maximum. It seems that this time is used to refocus some evolution occurring during cross polarization.

To understand why the sign of the many-body Hamiltonian may change, let us recall that the magnetic dipolar interaction is anisotropic. Any network of homologous spins can be described, in the quantization axis of the external field, by the Hamiltonian (5). If a $\pi/2$ pulse is applied and then irradiation is maintained in the direction of the polarization (spin-lock) with a reasonable rf power, the quantization axis should be taken along the rf field. During this time the spin dynamics is described, after truncations of the non-secular terms, by[58]

$$\mathcal{H}'_{II} = -[\frac{1}{2}]\mathcal{H}_{II}. \tag{14}$$

This is an entirely quantum effect. A feeling of its nature can be grasped from a classical analysis of two magnets constrained to a plane. If they are aligned along their internuclear vector they attract each other (negative energy goes to zero with distance). But if both are rotated in a $\pi/2$ angle they repel each other (positive energy).

Now, changing the sign of the Hamiltonian is equivalent to change the sign of time. In more precise terms, in the pulse sequence of Fig. 7, during t_2 the spins evolve with Hamiltonian (5), while during the CP periods t_d and t_p the proton spins evolve with a Hamiltonian (14). Thus, a part of the free evolution time t_2 (around $t_m = [\frac{1}{2}](t_d + t_p)/2$) is spent going backwards in time (the Lochsmidt's daemon[59] is working!) compensating the undesired evolution.

This observation[48] leads to the design of the next experiment. After a time t we apply an NMR pulse that changes again the sign of the many-body Hamiltonian[60]. A revival of

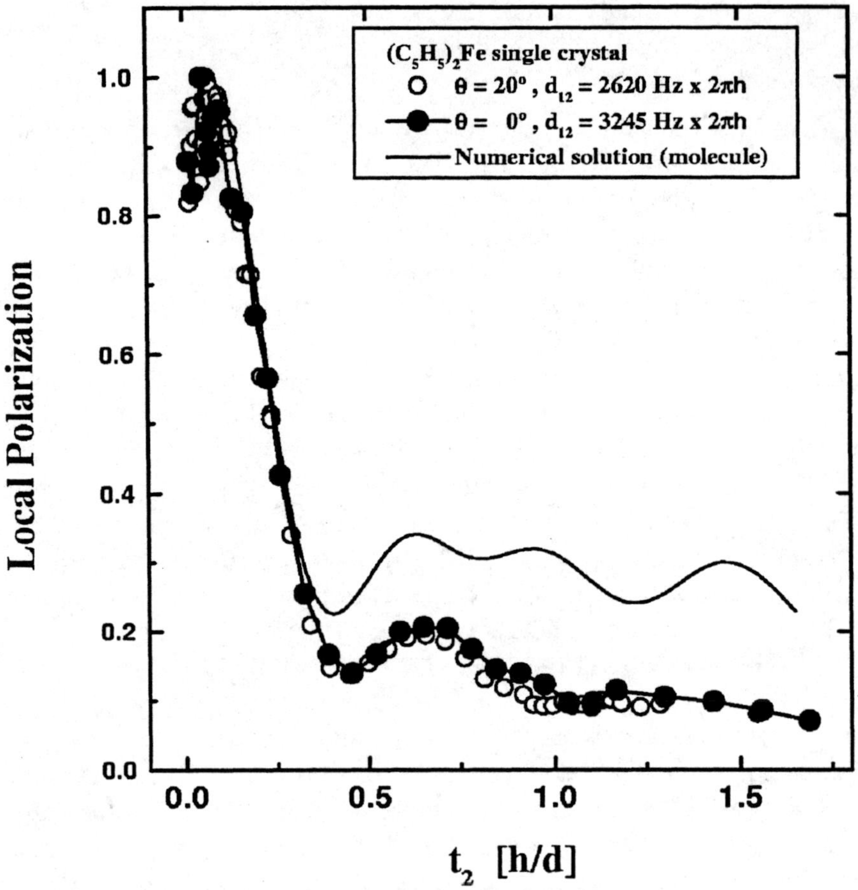

Fig. 8. Evolution of the laboratory frame local spin polarization in a single crystal of ferrocene. Experiments performed at two orientations, θ, of the molecular five-fold axis with respect to the external magnetic field are superimposed in a unified time scale. The solid line represents the calculated evolution of the local polarization in a complete molecule (see Fig. 6) where the rings rotate independently in a staggered configuration.

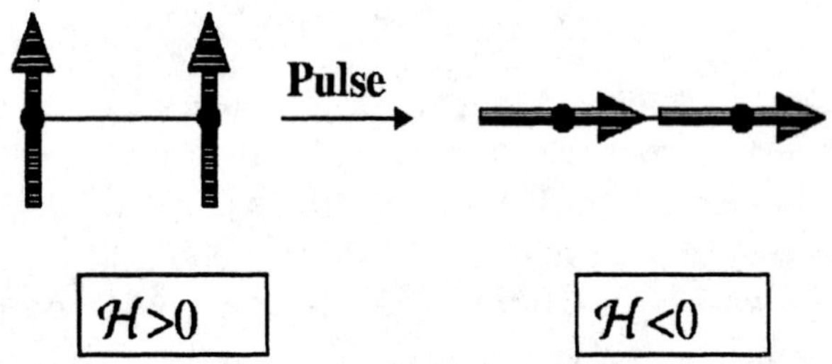

Fig. 9. Schematic representation of the effect of a $\pi/2$ pulse followed by spin lock irradiation in order to change the sign of the magnetic dipolar Hamiltonian. In this pictorial view two magnetic moments repel or attract each other depending on the angle between their orientation and the internuclear vector.

the polarization that retraces the forward evolution is then obtained with a maximum M_{PE} at $t + [2]t$ called *Polarization Echo* (PE). Notice that the reversed Hamiltonian induces a factor in the time scale which we always write within brackets.

Studying a local polarization defined as an excitation on the (high temperature) mixed state, the Polarization Echo results:

$$M_{PE}(t) = 2\mathrm{Tr}[\rho_{\mathcal{H}+\Sigma}(t)\rho_{\mathcal{H}}(t))]/\mathrm{Tr}[\rho(0)^2] - 1 \tag{15}$$

where Σ accounts for the non-inverted interactions. This is equivalent to the definition of Eq. (2) valid for a pure state, but for spins in the mixed state.

In Fig. 10 we present a set of experiments evidencing the build up of the Polarization Echoes. The spreading of the original polarization (forward evolution)during time t_1 is indicated by points joined by the lower thin line. During this time, evolution occurs in the rotating frame. The time reversal pulse is applied at each of these times. From each point, a polarization evolves in the laboratory frame (backwards in time) and it grows reaching a maximum , M_{PE}, when $t_2 \simeq [\frac{1}{2}]t_1$. The decay of the local polarization occurs with a law symmetrical to its build up, except for the small quantum interferences associated with the measurement process. This is a manifetation of time reversal symmetry at the microscopic scale. However, it turns out that M_{PE} is smaller the longer we wait to apply the reversal sequence. This attenuation represents the failure of quantum interferences in fully rebuilding the local excitation i.e. decoherence. In this case, there is no known interaction which could account for the observed time scale of decoherence.

There are specific situations[53] where the attenuation of $M_{PE}(t)$ can be attributed to a fast direct action of the environment such as interactions with quadrupolar nuclei. This is the case of the cymantrene molecule $(C_5H_5)Mn(CO)_3$. This crystal has a more diluted proton network than ferrocene. Besides, the quadrupolar manganese nuclei (spin 5/2), having a good coupling with both spins and lattice vibrations, provide an efficient pathway to the environment which essentially determines Σ. In this case $-\ln[M_{PE}(t)] \sim (t/\tau_\phi)$ and $\tau_\phi \approx \tau_\Sigma < \tau_{\mathrm{mb}}$.

In ferrocene and cobaltocene, where $\tau_{\mathrm{mb}} \ll \tau_\Sigma$, we observe $-\ln[M_{PE}(t)] \sim (t/\tau_\phi)^2$. This can be appreciated in Fig. 11 for ferrocene. Since ferrocene is a diamagnetic molecule and Fe has no quadrupolar moment, it is reasonable to expect very weak dissipative mechanisms. Here, the experimental spin-lattice relaxation time in the rotating frame is $\mathrm{T}_{1\rho} \geq 40ms$ from which $\tau_\Sigma \approx \mathrm{T}_{1\rho} \gg \tau_\phi \simeq 4\tau_{\mathrm{mb}}$. Besides, by changing the orientation of the sample we observe that $\tau_\phi \propto \hbar/d$ revealing a dipolar origin for the decoherence. Therefore, the relation between the time scales of the many-body dynamics and that of the perturbation seems to play a relevant role in fixing the decay law. In order to control this ratio, without modifying the system, we devised a new pulse sequence to progressively reduce the dipolar dynamics (increase of τ_{mb}).

The idea is to reduce the degree of complexity imposed by the many-body dynamics while the interaction time with the environment is kept constant. The resulting decay laws can then be compared. This is schematized in Fig. 12. There, the vertical axis is a mean space coordinate that represents the boundaries of the diffusion front. The horizontal axis is time in a unified scale. At $t = 0$ the initial local polarization starts to spread out (identified with straight lines) under the action of Hamiltonian $\mathcal{H}$ until time t_R. Then, a pulse starts the reversed evolution with Hamiltonian -$\mathcal{H}$. The shadowed region indicates a first step to reduce the polarization spreading. The spreading is reversed by a suitable pulse at time $t_R/2$, again at t_R and again at $3t_R/2$ i.e. in a time $2t_R$ we detect a PE. This

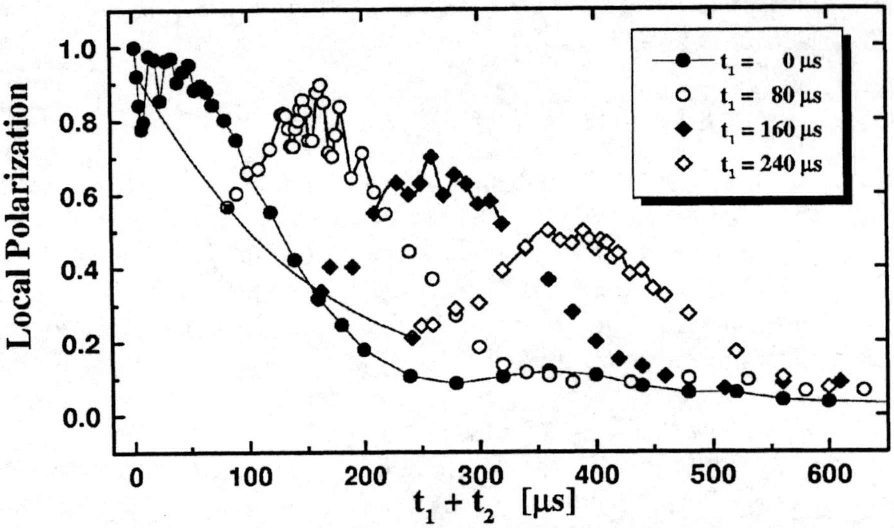

Fig. 10. Polarization echoes in a polycrystalline sample of ferrocene. The experimental data is plotted as a function of $t_1 + t_2$, where t_1 (interval not shown in the sequence of Fig. 7) is the time during which evolution occurs with the Hamiltonian $\mathcal{H}'_{II} = -\left[\frac{1}{2}\right] \mathcal{H}_{II}$, while during t_2, $\mathcal{H}_{II}$ is operative. In these experiments $\omega_{1I}/2\pi$=44.6kHz, $t_c = 2ms$, t_s=1ms, $t_d = t_p = 85\mu s$. Notice the interference phenomena at the top of the echoes.

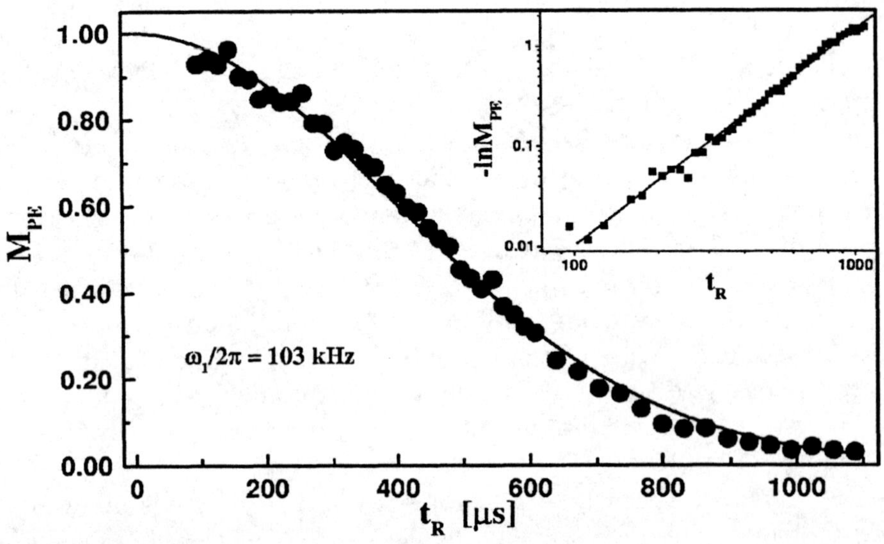

Fig. 11. Attenuation of the Polarization Echoes in a single crystal of ferrocene as a function of t_R. The data correspond to an orientation where the two molecules per unit cell are magnetically equivalent. The solid line corresponds to a Gaussian fitting yielding a characteristic time $\tau_\phi = (400 \pm 10)\mu s$ as the single free parameter. The inset shows $-\ln M_{PE}$ vs t_R in a log-log plot. The slope of the resulting line is 2.1 ± 0.08.

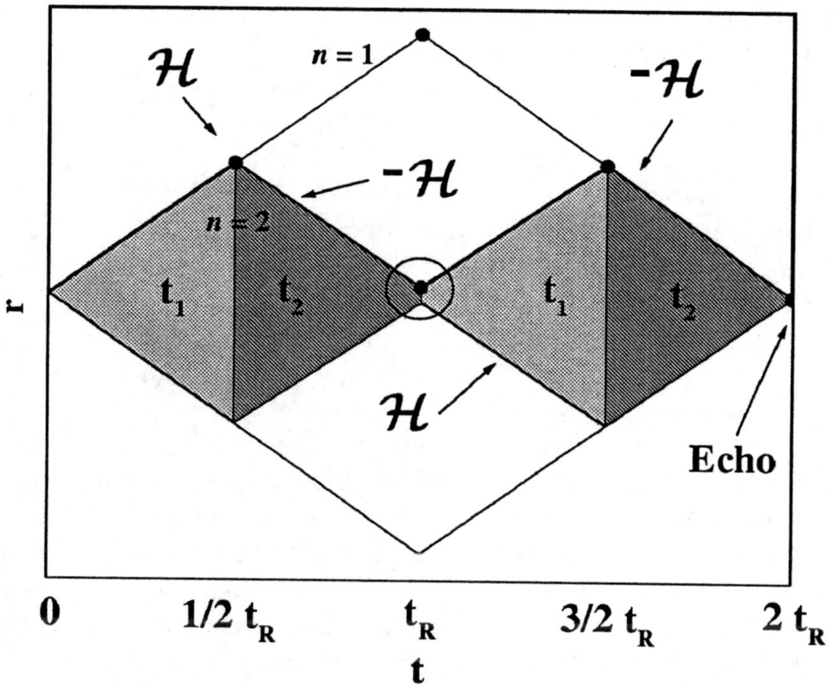

Fig. 12. Scheme to interpret the effect of the Reduced Evolution of the Polarization Echoes (REPE) pulse sequence. Horizontal axis is time in a unified scale. Vertical axis represents some mean square spreading of the polarization whose diffusion front is indicated by solid lines. Bullets represent reversal pulses.

evolution forward•backward•forward•backward can also be compared with the sequence forward•backward-backward•forward which, by saving one rf pulse, allows us to rule out experimental artifacts. Clearly, the number of spin sites involved in the shadowed region ($n = 2$) is much smaller than in the single forward•backward experiment ($n = 1$). This structure can be repeated recursively up to an experimentally available limit ($n \simeq 16$). Since the pulse sequence **R**educes the **E**volution of the **P**olarization **E**cho it is referred as REPE. For more technical details on the REPE sequence we refer to Ref.60.

As we mentioned above, we have observed two extreme behaviors for the Polarization Echo decay: exponential and Gaussian. The last is the more striking since it occurs in systems with a strong (reversible) spin dynamics such as ferrocene and cobaltocene. The application of the REPE sequence could reveal a possible underlying "irreversible" mechanism.

Figure 13 shows the results of the Polarization Echo atenuation in ferrocene in the REPE sequence. The striking fact is that when the many-body dipolar dynamics is reduced, we still observe Gaussian decays whose characteristic times $\tau_{\phi,n}$ are linear with n. The linear growth of $\tau_{\phi,n}$ on n leads to define the effective many-body time $\tau_{\text{mb}}^{n} = \kappa^{-1}\tau_{\phi,n}$ with $\kappa \simeq 4$.

Figure 14 shows the PE attenuation in the cobaltocene sample as a function of t_R for several factors (n) of dynamical reduction. It should be emphasized here that ferrocene and cobatocene have essentially the same proton dipolar networks. Differences in the time scales of Figs. 13 and 14 are explained in terms of the different orientations of the crystals which, determining different scales for the couplings, change $\tau_{\text{mb}} \approx \hbar/d$. There is a clear cross-over from the Gaussian decay at small n to a limiting exponential one. This characteristic time τ_ϕ can be associated to the paramagnetic and quadrupolar nature of the Co atom. Experimentally, it also manifests as $T_{1\rho} \geq \tau_\Sigma \simeq \tau_\phi$.

What controls the irreversibility/decoherence time τ_ϕ? The same basic question is buzzing in the NMR community since the development of the magic echoes[59], the precursor of the Polarization Echoes. There, a complex many-body evolution is also reverted by a complex pulse sequence but a strong irreversibility was observed which could not be explained by any available mechanism[61]. Our experiments, as described above, strongly suggest that $\tau_\phi = \min[\tau_{\text{mb}}, \tau_\Sigma]$. The non trivial answer, consistent with our whole set of experiments, seems to be that for $\tau_{\text{mb}} < \tau_\Sigma$ the decoherence time scale is determined by the many-body spin dynamics in the thermodynamic limit! In our context, the natural definition of *quantum chaos is the instability of dynamical quantum correlations with respect to infinitesimal perturbations in the Hamiltonian as seen in a time reversal experiment.* The existence of hidden mechanisms capable to produce PE attenuation would rule out the experimental support to such a strong statement. Hence, we tested all reasonable alternatives for uncontrolled processes: power droop, number of pulses, phase switching times, etc. The last important candidates which remained to be tested were the non-secular (non-inverted) parts of the homonuclear $^{1}H-^{1}H$ and heteronuclear $^{13}C-^{1}H$ dipolar interactions. These non-inverted terms have a dependence $\Sigma \sim d^2/\hbar\omega_1$, where $\hbar\omega_1$ is proportional to the rf field amplitude, hence they decrease with increasing rf power. Calculations in molecules ($N = 11$ spins) show that this perturbation produces a weak attenuation of M_{PE} in the presence of spin dynamics and none in its absence. However, no numerical calculation is able to hint what would be the result for a macroscopic ($N \to \infty$) number of spins. We resort to the quantum computer[62] we have at hand: the NMR spectrometer.

Thus, we progressively reduced the already small non-inverted interactions in order to see how much the attenuation of M_{PE} depends on them. This can be achieved by performing the ZME pulse sequence[48] at increasingly higher radio frequency powers. Experimental

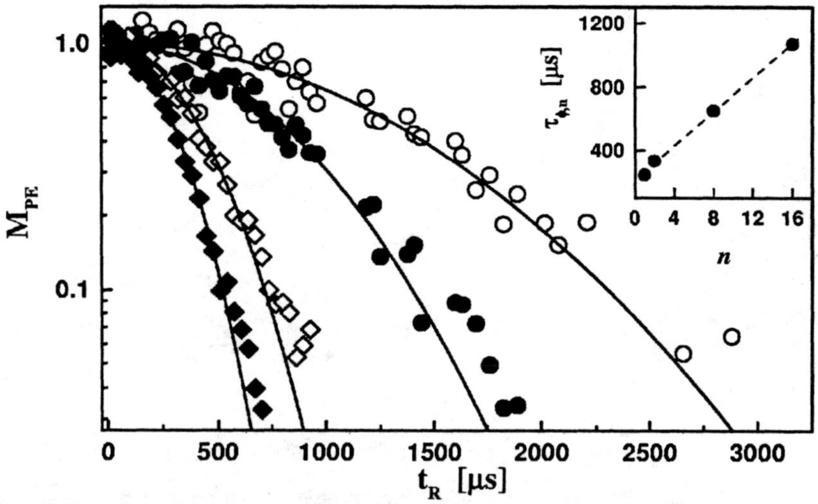

Fig. 13. Attenuation of the Polarization Echoes in a single crystal of ferrocene as a function of t_R, using the REPE sequence with $n = 1, 2, 8, 16$. The attenuation is slowed down as the complexity reached by the system is reduced. The lines represent Gaussian fittings $M_{PE}(t_R) = \exp[-\frac{1}{2}(t_R/\tau_{\phi,n})^2]$ with the characteristic time $\tau_{\phi,n}$ as the only free parameter. **Inset:** Characteristic time $\tau_{\phi,n}$ dependence on n. This linearity and angular dependence ($\tau_{\phi,n} \propto \hbar/d$) defines the effective many-body time $\tau^n_{\text{mb}} = \kappa^{-1}\tau_{\phi,n}$.

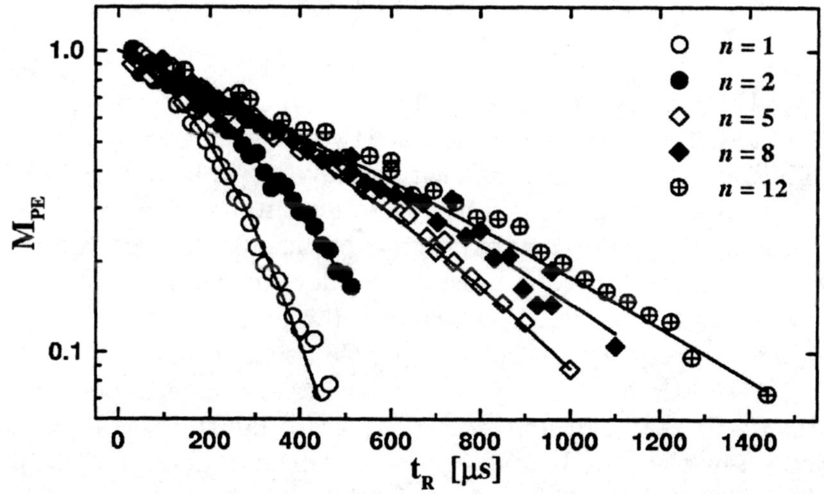

Fig. 14. Attenuation of the Polarization Echoes in a cobaltocene polycrystalline sample as a function of t_R, using the REPE sequence with $n = 1, 2, 5, 8, 12$. The $n = 16$ experiment cannot be distinguished from the $n = 12$ one. There is a clear crossover between a Gaussian and an exponential decay. This crossover results from a change of the relative importance between τ^n_{mb} and τ_Σ. The solid lines represent fittings of the whole set of data with only three free parameters to the equation $M_{PE}(t_R) = \exp[-t_R/\tau_\Sigma - \frac{1}{2}(t_R/\kappa\tau_{\text{mb},n})^2]$. Based on the results obtained for ferrocene we set $\tau_{\text{mb}.n} = an + b$.

details of the procedure can be found in Refs. 53 and 60.

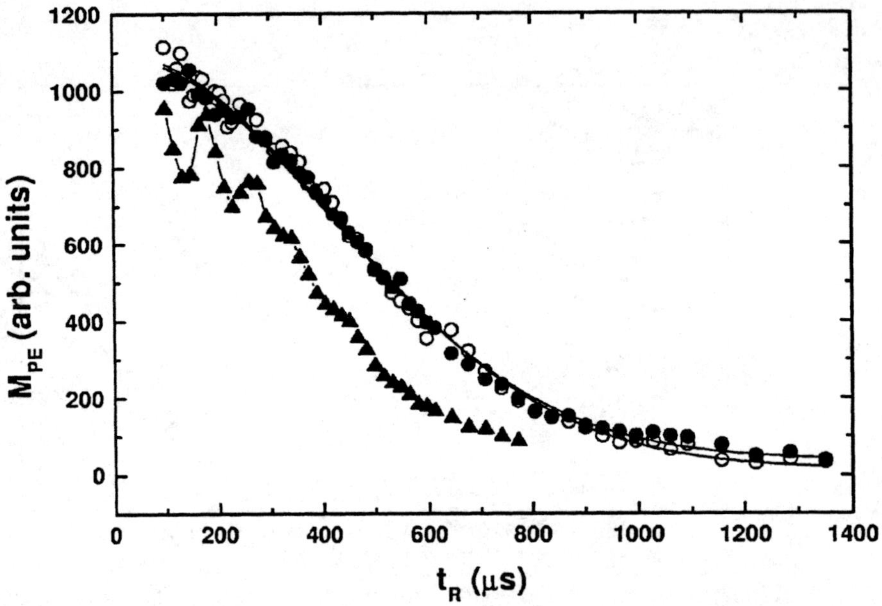

Fig. 15. Attenuation of the Polarization Echoes as a function of t_R in a single crystal of ferrocene, with $\omega_1/2\pi$ = 22 kHz (triangles), 54 kHz (open circles) and 103 kHz (solid circles). The solid lines represent least squares fitting to a Gaussian function $M_{PE} = A\exp[\frac{1}{2}(t_R/\tau_\phi)^2] + B$ giving $\tau_\phi = 420 \pm 7\mu s$ and $\tau_\phi = 413 \pm 6\mu s$ for $\omega_1/2\pi$ =54 kHz and 103 kHz respectively.

We performed the measurements in a single crystal of ferrocene in an orientation with respect to the external magnetic field where the two molecules per unit cell are magnetically equivalent. This doubles the signal intensity achieving the best signal to noise ratio. The values for $\omega_1/2\pi$, used during the relevant evolution times of the sequence were 22, 54 and 103 kHz as carefully determined by nutation experiments. In Fig. 15, we present the raw results keeping intensities and noise baselines as they come out from the peak amplitude of the Fourier transformed free induction decays (FID).

At all rf powers we observe[63], as before, clear Gaussian decays of M_{PE}. In the experiment at $\omega_1/2\pi$ = 22 kHz it is possible to observe oscillations superimposed to the decay with a frequency $\omega_1/2$. These quantum oscillations are originated in a polarization coherent tunneling induced by the ^{13}C-^{1}H couplings as explained in detail in Ref. 64. They are also observable at the other rf powers using appropriate time scales.

The remarkable result that can be observed in Fig. 15 is that attenuation saturates with increasing rf power (decreasing Σ) giving a characteristic time of $\tau_\phi = (420 \pm 7)$ μs which is about $4\hbar/d$. Measurements at several orientations of the single crystal and in powder are consistent with this dependence of τ_ϕ on the dipolar interaction d. We found that $M_{PE}(t) = \exp[-\frac{1}{2}((t/\tau_\phi)^2]$ with $\tau_\phi(\Sigma)$ depending very weakly on Σ and reaching a finite value $\tau_\phi \cong 4\tau_{\text{mb}}$ as $\Sigma \to 0$. Thus, decoherence time becomes independent on the only identified external perturbation and has a time scale intrinsic of $\mathcal{H}$. These experiments give further support to the relevant role of the *many-body* interactions in decoherence.

3.4. *Loschmidt daemon and Entropy*

As we have shown above, the attenuation of the Polarization Echoes results from the inefficiency to reverse the dynamics. As a measure of this inefficiency we can define the magnitude

$$S_L(t) = -\ln M_{PE}(t)$$

which increases with the time t elapsed until the dynamics is reversed. The following simple argument could convince one that this is a measure of the entropy. When the polarization is a conserved magnitude within the observed time scale, $M_{PE}(t)$ is a measure of the inverse of the volume Ω_t occupied by the polarization at time t, the Polarization Echo time. Hence $-\ln M_{PE}(t) \sim \ln \Omega_t$ has an entropic meaning. To further visualize this, one might assume that after a few pulses the shape of the distribution function for the refocused polarization remains with the same functional form (e.g. a Gaussian distribution of width $\Omega_t^{1/\mathrm{d}}$ in dimension d or a square distribution $P_M(x,t) = 1/\Omega_t$ for distances $|x| < \Omega_t/2$ in $\mathrm{d} = 1$). This is, the echo represents a scale transformation. In this case, the effectively available space for the polarization $-\int P_M(x,t)\ln P_M(x,t)dx$ grows proportionally to $-\ln[P_M(0,t)] = -\ln[M_{PE}(t)]$. In words, the Polarization Echo is a measure of the spreading of the polarization density at the echo time. While this is determined only from the diagonal terms in the density matrix, its spreading is a consequence of the correlation loss of non-diagonal terms at the inversion time.

The key point implied by our experiments is that the rate of increase of S_L is controlled by the system dynamics itself. If Σ is a unitary perturbation which we can only know in a statistical sense, then it is necessary to perform the ensemble average and to replace $\rho_{\mathcal{H}+\Sigma} \to \rho' = \sum_\alpha p(\alpha)\rho_{\mathcal{H}+\Sigma_\alpha}$, where α is a parameter to control the perturbation Σ and $p(\alpha)$ is its probability distribution. In this case, for *each* realization of α the dynamical instability gives rise to a rapid destruction of the interferences at the rate (determined by $\mathcal{H}$) at which S_L grows. The entropy meaning of S_L is evidenced if we calculate

$$\begin{aligned} \mathrm{Tr}\rho'^2 &= \mathrm{Tr}\left(\sum_\alpha p(\alpha)\rho_{\mathcal{H}+\Sigma_\alpha}\sum_\beta p(\beta)\rho_{\mathcal{H}+\Sigma_\beta}\right) \\ &= \sum_\alpha p^2(\alpha)\,\mathrm{Tr}\rho_0^2 + 2\sum_{\alpha>\beta} p(\alpha)p(\beta)\,\mathrm{Tr}\left(\rho_{\mathcal{H}+\Sigma_\alpha}\rho_{\mathcal{H}+\Sigma_\beta}\right) \end{aligned} \tag{16}$$

where the second term is essentially the overlap associated with M_{PE}. Assuming that $\mathrm{Tr}\left(\rho_{\mathcal{H}+\Sigma_\alpha}\rho_{\mathcal{H}+\Sigma_\beta}\right) \simeq \mathrm{Tr}\left(\rho_{\mathcal{H}}\rho_{\mathcal{H}+\Sigma}\right)$ where Σ is a typical perturbation, then

$$\begin{aligned} \mathrm{Tr}\rho'^2 &= \mathrm{Tr}\left(\rho_{\mathcal{H}}\rho_{\mathcal{H}+\Sigma}\right) + \sum_\alpha p^2(\alpha)\left(\mathrm{Tr}\rho_0^2 - \mathrm{Tr}\left(\rho_{\mathcal{H}}\rho_{\mathcal{H}+\Sigma}\right)\right) \\ &\simeq \mathrm{Tr}\left(\rho_{\mathcal{H}}\rho_{\mathcal{H}+\Sigma}\right), \end{aligned} \tag{17}$$

for the case $\sum_\alpha p^2(\alpha) \simeq 0$ (e.g., $p(\alpha) = 1/2^N$ with $N \gg 1$).

The presence of non-unitary decoherent processes also manifests the intrinsic instability of the system. In this case, on the basis of typical arguments of irreversible processes[65] described by the quantum master equation[66], it is obtained $M_{PE}(t) = 2\mathrm{Tr}[\rho_{\mathcal{H}+\Sigma}(t)^2]/\mathrm{Tr}[\rho(0)^2]$-1. Again, it is clear that S_L has an entropic meaning[67]. A point to stress is that whatever the origin of the perturbation, the S_L growth is a manifestation of the intrinsic instability of the chaotic system.

4. Discussion

By using NMR, it has been possible to test experimentally several of the arguments involved in the quantum version of the long-standing Boltzmann-Loschmidt polemic. Systems of coupled spins, in conjuntion with carefully designed pulse sequences, were used to "compute" the quantum dynamics for a model many-body system, whose solution is not yet accesible to the most powerfull digital computers.

We first checked the correspondence between a many-body dynamics of a local excitation and a diffusive process. We observed that although the excitation follows a general diffusive trend, it also shows some features of quantum interferences such as *quantum beats* (associated with the discreteness of the lattice) and *mesoscopic echoes* (associated with the Heisenberg time in a finite size cluster). These effects are however washed out by long range interactions and the unbounded nature of the involved set of spins. Quantum coherence of the many-body spin system is also well shown by the high frequency oscillations that are always present on the top of the *Polarization Echoes.* In fact, these echoes appear because the dynamics of these interactions can be time reversed. This is an actual quantum realization of the *gedanken* experiment discussed by Boltzmann and Loschmidt. The efficiency of this Loschmidt's daemon is, however, poorer than expected on naive grounds. Since irreversibility arises from the failure of quantum correlations to rebuild a local excitation, it gives also a decoherence time scale. In some cases, this inefficiency is clearly due to the presence of uncontrolled "environmental" degrees of freedom (represented as a perturbation Σ to the Hamiltonian). What is very remarkable, is that the time reversal is also limited even in the *absence* of appreciable interaction with the environment (experimentally three orders of magnitude beyond the observed times). Since τ_ϕ is always proportional to the spin-spin collision time, it is interpreted as an intrinsic property of the system which evidences the *quantum chaos* produced by the many-body interaction when an infinite number of particles (spins) are involved. The natural definition of quantum chaos in our experimental context is the instability of dynamical quantum correlations with respect to infinitesimal perturbations in the Hamiltonian.

This means that the apparently "diffusive" dynamics of the many-body interaction transforms infinitesimal residual interactions into efficient mechanisms to stabilize an irreversible diffusion[53]. Irreversible interactions (those we can not control) are often amplified even by those interactions we *can* control. This is analogous to the Drude approximation for the electrical resistance of an impure metal at low temperatures. There, the *reversible* elastic scattering with impurities with rate $1/\tau_o$ facilitates the interaction with the thermal bath. This bath acts through the uncontrollable electron-phonon interactions or other dephasing collisions with rate $1/\tau_\phi$. The remarkable consequence is that, in first approximation, the diffusion constant (and also Joule heating) does not depend on τ_ϕ but it is $D \propto v^2\tau_o$, where v is a typical velocity for ballistic propagation of the excitation. The lesson to be learned is that an interaction, which in principle is reversible, can provide an evolution close enough to diffusion (chaos). This relaxes the selection rules present for the electron-phonon interactions in crystals, triggering the energy dissipation. Then, when these processes occur, they stabilize[17] the evolution into a "truly" irreversible diffusion. The observed diffusion constant remains fixed by the reversible interactions. This, in turn, justifies the *stosszahlansatz* or assumption of randomness after successive collisions in which no memory of the previous quantum state is retained. This is equivalent to considering a collision with an impurity as a classical measurement or wave function collapse. In systems of independent electrons, this approximation breaks down close to the localized regime where interferences play a fundamental role[68]. Since localization phenomena might occur also in many-body (spins) systems[69], one can expect a consequent breaking of ergodicity in some regime of parameters, e.g. $\alpha \gg 1$ in Eq. (5).

In formal terms, the experiments and numerical results are consistent with the existence of a nontrivial thermodynamic limit. Thus, $\lim_{\Sigma \to 0} \lim_{N \to \infty} \tau(\Sigma, N) = \tau_\phi$ differs from the non-thermodynamic one $\lim_{N \to \infty} \lim_{\Sigma \to 0} \tau(\Sigma, N) = 0$, where Σ represents a non inverted part in the evolution. In a finite-size numerical calculation this is already insinuated. The first case corresponds to perturbations which are above a critical value as determined from the study of parametric correlations[25]. The non-thermodynamic limit is the other way around. We recall that Σ can be either: 1) unitary $\Sigma = \Delta$, as in the case of an external potential or 2) irreversible $\Sigma = \Delta + \mathrm{i}\Gamma$. This last is usually expressed in a superoperator formulation to describe a thermal bath[66]. The first case is totally described by a reversible Quantum Mechanics and still we obtain decoherence and irreversibility in the thermodynamic limit. This, in turn, justifies the use of the second formulation to describe the interaction with chaotic degrees of freedom of an environment[65].

Studying decoherence time does not only point to philosophical foundations of Physics. There are other immediate applications of our studies. The chaotic origin of decoherence proposed in this work reveals an increasing difficulty to push away the limit of what can be considered decoherence. This, however, should not be discouraging. By learning the mechanisms that produce decoherence, we also learn how to control it. Such an attempt may provide interesting results.

For example, the comparative study of ferrocene and cobaltocene hinted that decoherence can serve to distiguish among two systems that have identical spin networks but different mechanism of interaction with the environment. No difference between them is established until the REPE sequence is applied. This sequence is able to slow down the spin dynamics revealing the underling unreverted interaction which then determines the decoherence time.

The appearance of interference phenomena in the spin dynamics, which in principle can have practical applications ranging from spin editing to quantum computation, is also limited by the decoherence induced by the many-body environment. One must limit the size of the spin system and the instability of the effective spin Hamiltonian. It has been shown that decoherence can be limited by using pulse sequences that restrict the many-body interactions of the Hamiltonian. In Ref. 56 the many-body Heisenberg (isotropic) interaction is reduced to its XY (planar) component, mapable to a one-body system.

As an application beyond NMR, the discussed analogy with electron transport suggests that chaos might be responsible for the saturating decoherence time[70] that limits interference phenomena in mesoscopic conductors at very low temperatures. If sources of environmental decoherence can be ruled out, as it is claimed, the chaotic mechanism would start to manifest.

The described results serve as a landmark to focus new theoretical and experimental work in the field of quantum chaos.

From an experimental point of view, it would be interesting to study decoherence in systems that limit the number of interacting spins to a small number. This would provide an idea of how the localization acts in the many-body system. A control of environmental degrees of freedom could be achieved by gradually changing the concentration of magnetic impurities in crystals. Experiments in liquid crystals could show how the fluctuations of order affect the decoherence times.

On the basis of the previous discussions, a full picture of Quantum Chaos is beginning to emerge. It connects static properties with dynamical ones. It also shows classical thermodynamic behavior as a limit of increasingly complex interference phenomena on large bounded systems. There is still much theoretical work needed to clarify many details in this picture. For example, many people might feel uncomfortable with a possible distinction between one-body and many-body quantum chaos. This distinction was proposed on the basis of the discrepancy between the exponential decoherence predicted for the one-body chaotic system and the Gaussian decoherence found experimentally in our many-body system. This is an issue that needs more analytical and numerical calculations but the task ahead is not easy. The obtention of the right thermodynamic limits would require special tricks. Before, even the simplest one-body chaotic systems should be calculated beyond the semiclassical approximation. The role of different classical Lyapunov exponents in decoherence should be studied. There is also much room to extend on the referred connections between spectral signatures of chaos and dynamical chaos as defined in this work.

Acknowledgements

HMP dedicates this paper to the memory of Rolf Landauer from whom he received inspiration in early stages of his carreer. GU acknowledges support from CONICET. The experiments, supported by grants from ANPCyT, CONICET, CONICOR, SeCyT-UNC and Fundación Antorchas were performed at the LANAIS de RMS (CONICET-UNC) at Córdoba and in Strasbourg at the Laboratoire de RMN et Modelisation Moleculaire UMR7510 associated to CNRS-Bruker S.A.-Université Louis Pasteur.

References

[1] Boltzmann, L. 1896. *Ann. Phys.* (Leipzig) **57**, 773; translated and reprinted in: Brush, S. G. 1966. *Kinetic Theory 2*, Pergamon, Elmsford.

[2] Loschmidt, J. 1876. *Wiener Ber.* **73**, 128.
Loschmidt, J. 1877. *Wiener Ber.* **75**, 67.

[3] Bohr N. 1928. *Nature,* **121**, 580; reprinted in ref: 4.

[4] Wheeler, J. A., W. H. Zurek, (eds.) 1983. *Quantum Theory and Measurements*, Princeton University Press, Princeton.

[5] Einstein, A., B. Podolsky and N. Rosen. 1935. *Phys. Rev.* **47**, 777, reprinted in Ref. 4. Born M. and A. Einstein. 1971. *The Born-Einstein Letters, with comments by M. Born*, Walker, New York.

[6] Kac, M. 1957. *Probability and Related Topics in Physical Sciences*, Interscience. New York.

[7] Lebowitz, J. L. 1993. *Physics Today,* **46**(9), 32.

[8] Zaslavsky, G. M. 1999. *Physics Today,* **52**(8), 39.

[9] Lorenz, E. N. 1993. *The Essence of Chaos*, Univ. of Washington Press, Seattle.

[10] Crutchfield, J. P., J. Doyne Farmer, N. H. Packard and R. S. Shaw. 1986. *Scientific American,* **247**(12), 38.

[11] Binder, P. -M., J. M . Pedraza, and S. Garzón.1999. *Am. J. Phys.,* **67** , 1093.

[12] Bellemans, A. and J.. Orban, 1967. *Phys. Lett.*, **24A**, 260.

[13] Bardeen, J. 1990. *Physics Today,* **43**(12), 25.

[14] Goldstein, S. 1998. *Physics Today,* **51** (3), 42 .

[15] DeWitt, B. 1970. *Physics Today,* **23**(9), 70.

[16] Zurek, W. H. 1991. *Physics Today,* **44**(10), 36; also in ref. 4 p.175.

[17] Pastawski, H. M. 1991. *Phys. Rev. B*, **44**, 6329, see Eqs. (3.9) and (3.10).

[18] Pastawski, H. M. 1992. *Phys. Rev. B*, **46**, 4053.

[19] Chirikov, B. V. 1981. *Phys. Rep.,* **52**, 263.

[20] Fink, M. 1997. *Physics Today,* **50**(3), 34.

[21] F. M. Izrailev, 1990. *Phys. Rep.,* **196**, 299.

[22] Guhr, T., A. Müller-Groeling and H. A. Weidenmüller. 1998. *Phys. Rep.*, **299**, 189.

[23] Bohigas, O., M. J. Giannoni and C. Schmit. 1984. *Phys. Rev. Lett.*, **52**, 1.

[24] Prigodin, V. N., B. L. Altshuler, K. B. Efetov and S. Iida.Guber, J. L., J. Aliaga, H. A. Cerdeira, P. A. Mello and A. N. Proto. 1997. *Phys. Rev. E*, **55**, 6370.

[25] Szafer, A., and B. Altshuler. 1993. *Phys. Rev. Lett.*, **70**, 587.
Altshuler, B. L. and B. D. Simons. 1995. in *Mesoscopic Quantum Physics*, Eds.: E. Akkermans, G. Montambaux, J.-L. Pichard and J Zinn-Justin, Procc. Les Houches session LXI, Elsevier, Amsterdam. p.1.

[26] Heller, E. J. and S. Tomsovic. 1993. *Phys. Today,* **46**(7), 38.

[27] Nakamura, K. 1993. *Quantum Chaos, a new paradigm of nonlinear dynamics,* Cambridge University Press, Cambridge.

[28] Penrose, R. 1997. *The Large the Small and the Human Mind*, Cambridge University Press, Cambridge.

[29] Wheeler, J. A. 1994. in *Physical Origins of Time Asymmetry*, J. J. Halliwell, J. Pérez Mercader and W. H. Zurek eds., Cambridge University Press, p.1

[30] Landauer, R. 1970. *Philos. Mag.*, **21**, 863.
Landauer, R. 1987. *Z. Phys. B.,* **68**, 217.

[31] Peres, A. 1984. *Phys. Rev. A*, **304**, 1610.

[32] Pastawski, H. M. and R. Jalabert. 1999. *unpublished.*

[33] Laughlin, R. B. 1987. *Nucl. Phys. B (Proc. supl.),***3**, 213.

[34] Gutzwiller, M. C. 1991. in: *Chaos and Quantum Physics, Les Houches Session LII.* Eds.: M.-J. Giannoni, A. Voros and J. Zinn-Justin. Elsevier. p.202.

[35] Cucchietti, F. M., H. M. Pastawski and R. Jalabert. 2000. *Physica A*, (at press). also at cond-mat/0002207

[36] Forster, D. 1990. *Hydrodynamic Fluctuations, Broken Symmetry and Correlation Functions,* Addison Wesley, Reading.

[37] Morita, T. 1971. *J. Math. Phys.*, **12**, 2062.
Müller G. 1988. *Phys. Rev. Lett.* **60**, 2785.
Gerling, R. W. and D. P. Landau. 1989. *Phys. Rev. Lett.*, **63**, 812.
Bohm, M., H. Leschke, M. Henneke, V. S. Viswanath, J. Stolze and G. Müller. 1994. *Phys. Rev. B*, **49,** 417.

[38] Blume, M. and J. Hubbard. 1970. *Phys. Rev. B*, **1**, 3815.

[39] Misra, B. and E. C. G. Sudarshan. 1977. *J. Math. Phys.,* **18**, 756.

[40] Levstein, P.R., H.M. Pastawski and J.L. D'Amato. 1990. *J. Phys.: Condens. Matter.*, **2**, 1781.

[41] Cruz, H. B. and L. L. Gonçalves. 1981. *J. Phys. C,* **14**, 2785.

[42] de Gennes, P. G. 1958. *J. Phys. Chem. Solids*, **4**, 223.
Mori, H. and K. Kawasaki. 1962, *Progr. Theoret. Phys.* (Kyoto), **27**, 529.
Bennentt, H. S. and P. C. Martin. 1965. *Phys. Rev.*,**138**, 608.
De Leener M. and P. Resibois. 1966. *Phys. Rev.*, **152**, 318.

[43] Pastawski, H. M. and G. Usaj. 1998. *Phys. Rev. B,* **57**, 5017.

[44] Pastawski, H. M., P. R. Levstein and G. Usaj, 1995. *Phys. Rev. Lett.,* **75**, 4310.

[45] Anderson, P. W. 1954. *J. Phys. Soc. Japan* **9**, 316; see also:
Abragam, A. 1961. *Principles of Nuclear Magnetism*, Clarendon, Oxford. Chap. X.

[46] Levstein, P. R., H. M. Pastawski and R. Calvo. 1991. *J. Phys. Condens. Matt.*, **3**, 1877.
Gennaro, A. M. and P. R. Levstein.1991. *J. Phys. Condens. Matt.*, **3**, 455.
Siqueira, M. L., R. E. Rapp and R. Calvo. 1993. *Phys. Rev. B*, **48**, 3257.

[47] Zhang, W. and D. G. Cory. 1998. *Phys. Rev. Lett.*, **80**, 1324.

[48] Zhang, S., B. H. Meier and R. R. Ernst. 1992. *Phys. Rev. Lett.*, **69**, 2149.

[49] Seiler, P. and J. D. Dunitz. 1979. *Acta Crystallogr. Sect. B,* **35**, 1068.

[50] Slichter, C. P. 1992. *Principles of Magnetic Resonance*, Springer, New York.

[51] Hartmann, S. R. and E. L. Hahn. 1962. *Phys. Rev.*, **128**, 2042.

[52] Müller, L. A. Kumar, T. Baumann and R. R. Ernst. 1974. *Phys. Rev. Lett.,* **32**, 1402

[53] Levstein, P. R., G. Usaj and H. M. Pastawski. 1998. *J. Chem. Phys.*, **108**, 2718.

[54] Pastawski, H. M., G. Usaj and P. R. Levstein. 1996. *Chem. Phys. Lett.*, **261**, 329.

[55] Washburn, S. and R. A. Webb. 1986. *Adv. Phys.*, **35**, 375.
Imry, Y. and R. A. Webb. 1989. *Scientific American*, **255**(4), 56.

[56] Madi,Z.L., B. Brutscher, T. Schulte-Herbrüggen, R. Brüschweiler, R. R.. Ernst. 1997. *Chem. Phys. Lett.*, **268**, 300.

[57] Brewer, R. G. and E. Hahn. 1984. *Scientific American,* **250**(6), 50.

[58] Levitt, M., D. Sutter, and R. R. Ernst. 1986. *J. Chem. Phys.*, **84**, 4243.

[59] Rhim, W. K., A. Pines and J. S. Waugh. 1971. *Phys. Rev. B*, **3**, 684.

[60] Usaj, G., H. M. Pastawski and P. R. Levstein. 1998. *Mol. Phys.*, **95**, 1229. Also available at *cond-mat/0002332.*

[61] Skrevbnev, V. A. and R. N. Zaripov. 1999. *Appl. Magn. Reson.*, **16**, 1.

[62] DiVincenzo, D. P. 1995. *Science,* **270**, 255.

[63] Pastawski, H. M., P. R. Levstein, G. Usaj, J. Raya and J. Hirschinger. 2000. *Physica A,* (at press)

[64] Usaj, G., H. M. Pastawski and P. R. Levstein. (to be published)

[65] Zurek, W. and J. P. Paz. 1994. *Phys. Rev. Lett.*, **72**, 2508; Zurek, W. and J. P. Paz. 1995. *Physica D,* **83**, 300.

[66] Ernst, R. R. G. Bodenhausen and A. Wokaun. 1991. *Principles of Nuclear Magnetic Resonance in One and Two Dimensions*, Oxford University Press. Oxford.

[67] Wehrl, A. 1978. *Rev. of Mod. Physics*, **50**, 221.

[68] Cucchietti, F. M. and H. M. Pastawski. 2000, *Physica A*,(at press); also: cond-mat/0002178.

[69] Altshuler, B., Y. Gefen, A. Kamenev and S. L. Levitov. 1997. *Phys. Rev. Lett.*, **78**, 2803.

[70] Mohanty, P., E. M. Q. Jariwala, and R. A. Webb. 1997. *Phys. Rev. Lett.*, **77**, 3366.

OOPS! I MUST HAVE CONFUSED PROFESSOR GRUYERE'S SEMINAR ON CHEESE HOLES, WITH PROFESSOR GAGGERO'S ON GAS HOLES...

Exchange Energy of a Hole Gas and the Thomas-Fermi-Dirac Approximacion

Luis Manuel Gaggero Sager *

Abstract

We present the Thomas-Fermi-Dirac approximation for an inhomogenous hole gas. We carry out self-consistent calculations considering the particularities the exchange potential has in the Local Density Approximation in the case of a hole gas. We explain the characteristics of this type of calculation and present a general framework for carrying out these calculations more successfully. Using the Thomas-Fermi-Dirac approximation, we present a model for the Hartree-Fock potential. We compare the results obtained from our model with those obtained through self-consistent calculations in two systems, that of Be-delta-doped quantum wells in GaAs, and B-delta-doped quantum wells in Si. As a result of this comparison, we saw that the model presented, in spite of its simplicity, gives very good results. In particular, it practically fixes the energy differences between the occupied levels of the well. In both systems, there is agreement between the results of our two calculations (self-consistent and with our model) with the experimental results.

1. Introduction

The many-body effects have been considered extensively in quantum mechanics in terms of the exchange and correlation potential (see, for instance [1,7]). However, the discussion of the particularities of the effects of many bodies in a hole gas is rather more scarce. The application of the Thomas-Fermi-Dirac theory in semiconductor quantum systems is something we haven't seen as yet. Nor have we seen a framework which considers the particularities of the holes and presents an analytical solution to the Hartree-Fock potential.

The Thomas-Fermi-Dirac approximation is the generalisation of the Thomas-Fermi approximation when the effects of many bodies are considered through an exchange potential. This approximation is known in the literature and has been used in electronic gases[1]. In this case, that of an inhomogeneous electronic gas, it does not lead to a self-consistent solution in which the electron density $n(\mathbf{r})$ either tends to zero at infinity for the neutral atom, nor indeed a solution of finite radius in which the boundary density is zero. Rather, when used for neutral atoms, there is inevitably discontinuity in $n(\mathbf{r})$ at the boundary of the atom which is manifestly unsatisfactory [1]. If we were to use the same approximation, but to describe the conduction band bending of a semiconductor quantum system, caused by an electric charge in some region of the space, exactly the same thing would happen. However, when the charge in question is holes, things are completely different, although

*Universidad Autónoma de Zacatecas Zacatecas, México email:lgaggero@cantera.reduaz.mx

this approximation has never been used to describe hole gases. As we will see further on, in this case, the solution which is obtained is self-consistent and satisfactory.

Delta-doped quantum wells have great application in electronic devices[8-14,18-23]. It is of the utmost importance in the design of these devices to have a self-consistent analytical expression of the potential. Having a self-consistent analytical expression of the potential is also a good tool for more complex theoretical calculations. Also, the application of the Thomas-Fermi-Dirac approximation in this system is interesting.

In any case, even when EFA is directly used, and calculating in a self-consistent manner, (a iterative solution of the Schrödinger and Poisson equation of the system) there are many works which deal with the exchange of hole gas as though it were an exchange of a gas of electrons with positive charges. A couple of works clarify the enormous difference between these. In the analysis of the importance of the phenomena of many bodies for a particular system, it is necessary to bear these differences in mind and it is very important to analyse in each system the importance of the effects of many bodies with charge concentration, as it could turn out that in a single system, the effect of exchange is very important in one range of concentrations and less important in another. In principal, the Hartree potential and the Hartree-Fock potential do not depend on concentration in the same manner.

In epigraph II we present the Thomas-Fermi-Dirac (TFD) approximation for an inhomogenous hole gas, and then we discuss a simple model for p type delta-doped wells (M). In epigraph III we present the self-consistent calculation using the envelope function approximation. In epigraph III (a) we discuss the self-consistent results and using the proposed model for p type delta-doped wells in Si and comparison with the experiments. In epigraph III (b) we do the same but for p type delta-doped wells in GaAs. Finally, in epigraph IV we present the conclusions.

2. The Thomas-Fermi-Dirac approximation

We shall assume to be in the low temperature limit. In an ideal, uniform electron gas, the number of electrons per unit volume is

$$n = \frac{2}{(2\pi)^3}\frac{4\pi}{3}\frac{p_F^3}{\hbar^3} = \frac{1}{3\pi^2}\frac{p_F^3}{\hbar^3} \tag{1}$$

where p_F is the radius of the Fermi sphere in the momentum space. If we consider the gas as locally homogeneous; i.e., that the ideal gas approximation is valid at each point, we have

$$n(\mathbf{r}) = \frac{1}{3\pi^2}\frac{p_F^3\,(\mathbf{r})}{\hbar^3} \tag{2}$$

The probability $I_r(\vec{p})\;d\vec{p}$ that the momentum of the electron has a value between p and $p + dp$

$$I_r(\vec{p})\;d\vec{p} = \frac{4\pi\,p^2\;dp}{4/3\;\pi\;p_F^3}\,\theta(p_F - p) \quad . \tag{3}$$

If the light-hole (lh) and heavy-hole (hh) bands are considered as independent bands, the kinetic energy for a single hole can be written as [16]:

The kinetic energy per unit volume, t,

$$t = \frac{1}{2\pi^2 \left[m_{hh}^{3/2} + m_{lh}^{3/2}\right]^{2/3} \hbar^3} \int_0^{p_F} p^4 dp = \frac{\hbar^2 \left[3\pi^2 n(\mathbf{r})\right]^{5/3}}{10\pi^2 \left[m_{hh}^{3/2} + m_{lh}^{3/2}\right]^{2/3}}. \tag{4}$$

In the above expression $m_{lh}$and m_{hh} are the effective band mass of the light holes and the effective band mass of the heavy holes respectively.

The kinetic energy functional is;

$$T = \int t\, d\mathbf{r} = \frac{3}{10 \left[m_{hh}^{3/2} + m_{lh}^{3/2}\right]^{2/3}} \int n(\mathbf{r}) \left[3\pi^2\, \hbar^3\, n(\mathbf{r})\right]^{2/3} d\mathbf{r} \tag{5}$$

and the energy functional which corresponds to the electron-electron interaction is

$$V_{ee} = \frac{e^2}{2\varepsilon_r} \int\int \frac{n(\mathbf{r}')\, n(\mathbf{r})}{|\mathbf{r} - \mathbf{r}'|} d\mathbf{r}\, d\mathbf{r}' \quad . \tag{6}$$

Here, it is assumed that the electrons are in a homogeneous medium of dielectric constant ε_r (the so-called *hydrogenic model*).

The potential energy from the external potential $(U(\mathbf{r}))$ functional is $\int n(\mathbf{r})\; U(\mathbf{r}) d\mathbf{r}$. Therefore, the TF energy-density functional is given by

$$E_{TF} = \frac{3}{10 \left[m_{hh}^{3/2} + m_{lh}^{3/2}\right]^{2/3}} \int n(\mathbf{r}) \left[3\pi^2\, \hbar^3\, n(\mathbf{r})\right]^{2/3} d\mathbf{r} + \frac{e^2}{2\varepsilon_r} \int\int \frac{n(\mathbf{r}')\, n(\mathbf{r})}{|\mathbf{r} - \mathbf{r}'|} d\mathbf{r}\, d\mathbf{r}' + \int n(\mathbf{r})\, U(\mathbf{r}) d\mathbf{r} \tag{7}$$

It is known that the variational principle leads to $\delta(E_{TF} - \mu N) = 0$ where $N = \int n(\mathbf{r})\; d\mathbf{r}$ and μ is the Lagrange multiplier. Calculating the variational by the standard procedure we find

$$\mu = \frac{1}{2 \left[m_{hh}^{3/2} + m_{lh}^{3/2}\right]^{2/3}} \left[3\pi^2\, \hbar^3 n(\mathbf{r})\right]^{2/3} + \frac{e^2}{2\varepsilon_r} \int \frac{n(\mathbf{r}')}{|\mathbf{r} - \mathbf{r}'|} d\mathbf{r}' + U(\mathbf{r}) \tag{8}$$

but, taking into account that

$$V_H(\mathbf{r}) = \frac{e^2}{2\varepsilon_r} \int \frac{n(\mathbf{r}')}{|\mathbf{r} - \mathbf{r}'|} d\mathbf{r}' \tag{9}$$

The exchange functional is

$$E_x = \int n(\mathbf{r})\; \varepsilon_x(n(\mathbf{r})) d\mathbf{r} \tag{10}$$

The exhange energy for a hole gas with valence band coupling is [6,7]

$$\varepsilon_x(n(\mathbf{r})) = -\varsigma(w) \frac{3a_0^*}{2\pi} \left(3\pi^2 n\left(\mathbf{r}\right)\right)^{1/3} R_y^* \tag{11}$$

where

$$\varsigma(w) = 2^{-1/3} + (1 - w^2) \left[w^2(aw + b) + c(4w^3 + 3w^2 + 2w + 1\right] \tag{12}$$

with $a = 0.679$, $b = -0.0686$, and $c = 1/4 + 3\pi^2/64 - 2^{-1/3} \approx -0.0811$. This factor $\varsigma(w)$, is always smaller when we consider the uncoupled bands, in which case, $\varsigma(w) = (1+w^4)/(1+w^3)^{4/3}$ [7]. In this point, as the problem is not complicated by the consideration of the coupled bands, we use the formula (12) for $\varsigma(w)$.

The functional of energy is now

$$E = \frac{3}{10\left[m_{hh}^{3/2} + m_{lh}^{3/2}\right]^{2/3}} \int n(\mathbf{r}) \left[3\pi^2 \hbar^3 n(\mathbf{r})\right]^{2/3} d\mathbf{r} + \frac{e^2}{2\varepsilon_r} \int\int \frac{n(\mathbf{r}')\, n(\mathbf{r})}{|\mathbf{r}-\mathbf{r}'|} d\mathbf{r}\, d\mathbf{r}' +$$

$$\int n(\mathbf{r})\, U(\mathbf{r}) d\mathbf{r} - \varsigma(w) \frac{3a_0^*}{2\pi} \left(3\pi^2\right)^{1/3} R_y^* \int n(\mathbf{r})^{4/3} d\mathbf{r} \tag{13}$$

It is known that the variational principle leads to $\delta(E_{TF} - \mu N) = 0$where $N = \int n(\mathbf{r})\ d\mathbf{r}$ and μ is the Lagrange multiplier. Calculating the variational by the standard procedure we find

$$\mu = \frac{1}{2}\left[\frac{3\pi^2\, \hbar^3 n(\mathbf{r})}{m_{hh}^{3/2} + m_{lh}^{3/2}}\right]^{2/3} + \frac{e^2}{2\varepsilon_r} \int \frac{n(\mathbf{r}')}{|\mathbf{r}-\mathbf{r}'|} d\mathbf{r}' + U(\mathbf{r}) - \varsigma(w) \frac{2}{\pi} \left(3\pi^2\right)^{1/3} a_0^* n(\mathbf{r})^{1/3} R_y^* \tag{14}$$

If we consider the effective Rydberg $R_y^* = e^2/(2\varepsilon_r a_0^*)$ and $a_0^* = \varepsilon_r \hbar^2/(e^2 m_{hh})$, considering $n_{ua}(\mathbf{r}) = (a_0^*)^3 n(\mathbf{r})$ we have

$$\mu = \left[\frac{3\pi^2 n_{ua}(\mathbf{r})}{1 + \left(\frac{m_{lh}}{m_{hh}}\right)^{3/2}}\right]^{2/3} R_y^* + V_H + U(\mathbf{r}) - \varsigma(w) \frac{2}{\pi} \left(3\pi^2\right)^{1/3} n_{ua}(\mathbf{r})^{1/3} R_y^* \tag{15}$$

Thus,

$$n_{ua}(\mathbf{r}) = \left[1 + \left(\frac{m_{lh}}{m_{hh}}\right)^{3/2}\right]^2 \frac{1}{3\pi^2} \frac{1}{\pi^3} \left[\varsigma(w) - \sqrt{\varsigma^2(w) + \pi^2 \frac{\mu^* - V_H^*(\mathbf{r}) - U^*(\mathbf{r})}{\left[1 + \left(\frac{m_{lh}}{m_{hh}}\right)^{3/2}\right]^{2/3}}}\right]^3 \tag{16}$$

the Poisson equation, remains

$$\frac{d^2 V_H^*(\mathbf{r})}{dz_{ua}^2} = -\left[1 + \left(\frac{m_{lh}}{m_{hh}}\right)^{3/2}\right]^2 \frac{8}{3\pi^4} \left[\varsigma(w) - \sqrt{\varsigma^2(w) + \pi^2 \frac{\mu^* - V_H^*(\mathbf{r}) - U^*(\mathbf{r})}{\left[1 + \left(\frac{m_{lh}}{m_{hh}}\right)^{3/2}\right]^{2/3}}}\right]^3 \tag{17}$$

the solution to this equation would provide a Hartree-Fock self-consistent solution. However, the analytical solution to this equation is not trivial. It must always be carried out numerically, in which case, it should be no more convenient than carrying out the self-consistent calculation (as equation (17) is non linear. In order for it to be advantageous, we shall seek to carry out an additional approximation. The presence of exchange generates a change in the electronic structure of levels in two ways; firstly, the presence of the exchange potential in the Schrödinger equation generates a change in the electronic structure. Then, this change in the electronic structure alters the charge density, thus modifying the Hartree potential. Let us suppose that the Hartree potential remains practically unchanged by the presence of exchange. In this case, using equation (16), we can calculate the exchange potential as a function of the Hartree potential (as in the articles [4,5]). As in this case, [6]

$$V_x(\mathbf{r}) = -\varsigma(w)\frac{2}{\pi}\left(3\pi^2\right)^{1/3} n_{ua}(\mathbf{r})^{1/3} R_y^* \tag{18}$$

substituting (16) in (18), we obtain

$$V_x^*(\mathbf{r}) = -\frac{2\varsigma(w)}{\pi^2}\left[1+\left(\frac{m_{lh}}{m_{hh}}\right)^{3/2}\right]^{2/3}\left[\varsigma(w) - \sqrt{\varsigma^2(w) + \pi^2\frac{\mu^* - V_H^*(\mathbf{r}) - U^*(\mathbf{r})}{\left[1+\left(\frac{m_{lh}}{m_{hh}}\right)^{3/2}\right]^{2/3}}}\right] \tag{19}$$

that is, we are supposing that the presence of exchange alters the level structure and the charge density, but we discard change in the form of the Hartree potential by the presence of exchange, all of this within a quasi-classical approximation.

When we have a quasi-bidimensional gas of holes around one layer of ionised impurities (i.e., a delta-doped quantum well). The Hartree potential (in the Thomas-Fermi approximation) is

$$\frac{d^2V_H^*(z)}{dz_{ua}^2} = -\frac{8}{3\pi}\left[1+\left(\frac{m_{lh}}{m_{hh}}\right)^{3/2}\right]\left[\mu^* - V_H^*(\mathbf{r})\right]^{3/2} + 8\pi p_{2D}^{ua}\delta(z) \tag{20}$$

We propose the solution

$$V_H^*(z) - \mu^* = -\frac{\alpha^2}{(\alpha\left|z\right| + z_0)^4} \tag{21}$$

and the substitution in (14) gives:

$$\alpha = \frac{2}{15\pi}\left[1+\left(\frac{m_{lh}}{m_{hh}}\right)^{3/2}\right] \quad and \quad z_0 = \left(\frac{\alpha^3}{\pi p_{2D}^{ua}}\right)^{1/5} \tag{22}$$

Thus, the charge density will be

$$n_{ua}(\mathbf{r}) = \left[1+\left(\frac{m_{lh}}{m_{hh}}\right)^{3/2}\right]^2 \frac{1}{3\pi^2}\frac{1}{\pi^3}\left[\varsigma(w) - \sqrt{\varsigma^2(w) + \pi^2\frac{1}{\left[1+\left(\frac{m_{lh}}{m_{hh}}\right)^{3/2}\right]^{2/3}}\frac{\alpha^2}{(\alpha\left|z\right| + z_0)^4}}\right]^3 \tag{23}$$

Expressions (15)-(17) summarize the results of the model. With the use of the TF potential $V(z)$obtained above, a Schrödinger wave equation is numerically solved in order to look for the energy levels in the well. In this case

$$V_x^*(\mathbf{r}) = -\frac{2\varsigma(w)}{\pi^2}\left[1+\left(\frac{m_{lh}}{m_{hh}}\right)^{3/2}\right]^{2/3}\left[\varsigma(w) - \sqrt{\varsigma^2(w) + \pi^2\frac{\alpha^2\left(\alpha\left|z\right| + z_0\right)^{-4}}{\left[1+\left(\frac{m_{lh}}{m_{hh}}\right)^{3/2}\right]^{2/3}}}\right] \tag{24}$$

Then the total potential, $V(z) = V_H(z) + V_x(z)$, is

$$V^*(z) = -\frac{\alpha^2}{(\alpha|z| + z_0)^4} - \frac{2\varsigma(w)}{\pi^2}\left[1 + \left(\frac{m_{lh}}{m_{hh}}\right)^{3/2}\right]^{2/3}\left[\varsigma(w) - \sqrt{\varsigma^2(w) + \pi^2 \frac{\alpha^2\,(\alpha|z| + z_0)^{-4}}{\left[1 + \left(\frac{m_{lh}}{m_{hh}}\right)^{3/2}\right]^{2/3}}}\right] \tag{25}$$

Equation (25) summarizes the proposed model for the total potential of p delta-doped quantum wells. The generalisation of this model for the case in which exchange effects are considered is immediate. Its main virtue is that of avoiding carrying out long self-consistent calculations.

3. Self-consistent calculation in Envelope Function Approximation

We use the EFA (Envelope Function Approximation) for the description of hh and lh bands taking them as independent bands. Thus we have two uncoupled Schrödinger-like equations with corresponding ladders of hole subbands.

The spatial band bending is described by a Poisson equation:

$$\frac{d^2V(z)}{dz^2} = \frac{4\pi e}{\varepsilon_r}(\rho_{imp}(z) + \rho_e(z)) \tag{26}$$

where ρ_{imp} is the well impurity density ($\rho_{imp} = p_{2D}\delta(z)$), ρ_e is the free charge density in the well region and ϵ_r is the dielectric constant. We take the energy origin at z=0 (V(0)=0) and the electric field tends to zero when z tends to infinity. The solution of this equation is

$$V(z) = \frac{4\pi e}{\varepsilon_r}\int_0^z (z - \chi)\rho_e(\chi)d\chi + \frac{2\pi e^2}{\varepsilon_r}p_{2D}|z| \ , \tag{27}$$

where the charge density, $\rho_e(z)$, is given by

$$\rho_e(z) = -\sum_{j=1}^{2}\frac{em_j^*}{\pi\hbar^2}\sum_{i=1}^{p_j}[E_F - E_i^j]\theta(E_F - E_i^j)|F_i^j(z)|^2, \tag{28}$$

and E_F is the Fermi level, j labels the bands ($j = 1$ for the hh band and $j = 2$ for the lh band). E_i^j is the i-th energy level of j-th band, and p_j is the occupation number for the j-th band. Charge neutrality requires

$$ep_{2D} = -\int_{-\infty}^{\infty}\rho_e(\chi)d\chi, \tag{29}$$

and the Fermi level is

$$E_F = \frac{\sum_{j=1}^{2} m_j^* \sum_{i=1}^{p_j} E_i^j + \pi\hbar^2 p_{2D}}{\sum_{j=1}^{2} m_j^* p_j}. \tag{30}$$

This is a transcendental equation, because the occupation level p_j is a function of the Fermi level (see Refs. 7,17,18 for details). We will consider the effects of exchange using formula (18).

These calculations, which consist of the simultaneous solution of the Schrödinger and Poisson equations, tend to be rather awkward and furthermore, on occasion present divergence. They are carried out in an iterative manner, that is, using in each step an input potential, the charge density is calculated and with this, what we call output potential. The input potential of the following iteration will be a mixture of the input and output potentials from the step of the previous iteration. One of the key issues in obtaining the quickest convergence, and of course, non-divergence, is this mixture. In this section, we will discuss a method which we have developed for the mixture after some experience in self-consistent calculations[]. The ideal self-consistent calculation is that in which the input and output potentials are identical. Let us imagine that we construct a bidemensional space formed by input potential at point z on the x-axis and the output potential at the same point z on the y-axis, the self-consistent solution would be when the point were exactly above the line with slope 1. If $V^{m\ in}(z)$ ($V^{m\ out}(z)$) is the input (output) potential in step m of self-consistency, in general the point in the space will not be above the line with slope 1, we will construct a straight line with the two points from the previous steps and with the intersection of these points on the said line, we will obtain the following input potential. In what follows, we will suppose that the zero of energies is placed at the origin. The mixture was made in the following manner.

$$V^{(m+1)\ i}(r) = A(r)V^{m\ in}(r) + (1 - A(r))V^{m\ out}(r) \tag{31}$$

where

$$A(r) = \left\{ \begin{array}{ll} \frac{V^{m\ out}(r) + V^{(m-1)\ out}(r)}{V^{m\ out}(r) - V^{(m-1)\ out}(r) - V^{m\ in}(r) + V^{(m-1)\ in}(r)} & if\ \Delta V^{\ m}(r)\ \Delta V^{(m-1)}(r)\ \leq 0 \\ \frac{|V^{m\ in}(r)|}{V^{m\ in}_{\max}(\infty)}\beta & in\ other\ case \end{array} \right\} \tag{32}$$

where $V^{m\ in}_{\max}(\infty)$ is the absolute value of the maximum value of the input potential in step m , $\beta = 1 - \frac{m}{n_{\max}}$ where $n_{\max}$ is the maximum number of iterations proposed to be carried out, changes, but is ususally $n_{\max} \sim 20$ and $\Delta V^{\ m}(r) = V^{m\ in}(r) \ - \ V^{m\ out}(r)(\Delta V^{\ (m-1)}(r) = V^{(m-1)\ in}(r) \ - \ V^{(m-1)\ out}(r))$. The first part of the equation for $A(r)$, demands that both points (that corresponding to step m and that corresponding to step $m-1$) be on both sides of the line in order to avoid divergence. The second part of the equation for $A(r)$at the origin, $V^{(m+1)\ i}(0) = V^{m\ out}(0)$, so that the input potential of step $m+1$ slowly begins to differentiate iteself from the output potential from step m as it distances itself from the origin. On some occasions, sets of parameters in some systems exist which, even with this framework, present difficulties in convergence. An even more complex framework, when the previous one did not work, or worked very slowly, is the following; the idea is the same, but instead of just the two previous steps of convergence here there are three.

Let

$$A_1(r) = \frac{V^{m\ out}(r)\ +\ V^{(m-1)\ out}(r)}{V^{m\ out}(r)\ -\ V^{(m-1)\ out}(r)\ -\ V^{m\ in}(r)\ +\ V^{(m-1)\ in}(r)}\quad , \tag{33}$$

$$A_2(r) = \frac{V^{(m-1)\ out}(r)\ +\ V^{(m-2)\ out}(r)}{V^{(m-1)\ out}(r)\ -\ V^{(m-2)\ out}(r)\ -\ V^{(m-1)\ in}(r)\ +\ V^{(m-2)\ in}(r)} \tag{34}$$

and

$$A_3(r) = \frac{V^{m\ out}(r)\ +\ V^{(m-2)\ out}(r)}{V^{m\ out}(r)\ -\ V^{(m-2)\ out}(r)\ -\ V^{m\ in}(r)\ +\ V^{(m-2)\ in}(r)}\quad . \tag{35}$$

Let us define,

$$V_k^{(m+1)\,in}(r) = \left\{ \begin{array}{ll} 1/2\sum_{j=1}^{2} A_j(r)\, V^{(m-j+1)\,in}(r) + (1 - A_j(r))\, V^{(m-j+1)\,out}(r) & if\ k=1 \\ 1/2\sum_{j=2}^{3} A_j(r)\, V^{(m-j+1)\,in}(r) + (1 - A_j(r))\, V^{(m-j+1)\,out}(r) & if\ k=2 \\ 1/2\sum_{j=1,\, j\neq 2}^{3} A_j(r)\, V^{(m-j+1)\,in}(r) + (1 - A_j(r))\, V^{(m-j+1)\,out}(r) & if\ k=3 \end{array} \right\} \quad (36)$$

then the input potential of step $m+1$ will be given by,

$$V^{(m+1)\,in}(r) = \left\{ \begin{array}{ll} V_1^{(m+1)\,in}(r) & if\ \Delta V^m(r)\,\Delta V^{m-1}(r) \leq 0\ and\ \Delta V^{m-1}(r)\,\Delta V^{m-2}(r) \leq 0 \\ V_2^{(m+1)\,in}(r) & if\ \Delta V^{m-1}(r)\,\Delta V^{m-2}(r) \leq 0\ and\ \Delta V^{m}(r)\,\Delta V^{m-2}(r) \leq 0 \\ V_3^{(m+1)\,in}(r) & if\ \Delta V^m(r)\,\Delta V^{m-1}(r) \leq 0\ and\ \Delta V^{m}(r)\,\Delta V^{m-2}(r) \leq 0 \\ \frac{|V^{(m-1)\,in}(r)|}{V_{\max}^{(m-1)\,in}(\infty)}\beta V^{(m-1)\,in}(r) + (1 - \frac{|V^{(m-1)\,in}(r)|}{V_{\max}^{(m-1)\,in}(\infty)}\beta) V^{(m-1)\,out}(r) & in\ other\ case \end{array} \right. \quad (37)$$

This expression is rather more complex than equation (32); sometimes there are parameters for which convergence becomes extraordinarily difficult, in which case it is better to use (37).

4. Results and discussion

4.1. *B-δ-doped Si quantum well*

We use the following input parameters: $m_{hh}^* = 0.52m_0$, $m_{lh}^* = 0.16m_0$, m_0 being the free electron mass, $\epsilon_r = 11.7$, and 2×10^{12} cm^{-2} $< p_{2D} < 2 \times 10^{13}$ cm^{-2}For greater values of concentration of impurities it is necessary to consider the split off band.

In table 1 we present the self-consistent results with (x) and without exchange (H). In this table we can see that, from the relative point of view, while lesser the concentration, the greater the importance of the exchange effect. For $p_{2D} = 2\times10^{12}$ cm^{-2} the change of levels, taking as the the origin of energies the bottom of the band, is of the order of 80%. However, if we take the bottom of the well as the origin of energies, there is practically no difference, but, there is a change of 25% at the bottom of the well as exchange is considered. Therefore, this is not a good system of energetic reference for comparison. For $p_{2D} = 2 \times 10^{13}$ cm^{-2} the change is of 20 %.

Using equation (25) we calculate the levels of energies in our Thomas-Fermi-Dirac approximation and the Fermi level, with these self-values, using equation (25). In table 2 we present the comparison between the approximation defined in equation (25) (M) and the self-consistent calculation with exchange (X). The medium levels with respect to the bottom of the band, present a difference of approximately 20%. Meanwhile, if we measure them with respect to E_F (Table 3) the difference in the occupied levels is less than 5%. This indicates that the difference between energy levels in our model gives very good results. It is often the level difference that is measured experimentally.

Jiang-hong Zhu et al. [17] made a B-δ-doped Si with a Schottky barrier. They showed the shifts of the conductance peaks in the spectra with the peak doping concentration of $p_{3D} \approx 2 \times 10^{20}$ cm $^{-3}$ and doped thicknesses of 1.2 nm (corresponding to a bidimensional density of impurities of $p_{2D} = 2.4 \times 10^{13}$ cm $^{-2}$). The activation energy reported is 110 ± 20 meV. The difference between the basic level and the top of the valence band, in our calculations (extrapolating), is of ≈ 101 meV for the self-consistent calculation and is of ≈ 110 meV using our Thomas-Fermi-Dirac approximation (ec. (25)) .

4.2. *Be-δ-doped GaAs quantum well*

p_{2D}	hh0(x)	hh0(H)	hh1(x)	hh1(H)	lh0(x)	lh0(H)	E_F(x)	E_F(H)	V_0(x)	V_0(H)
2	18.0	10.9			12.2	6.4	9.6	2.8	38.6	30.1
3	23.1	14.2			15.6	8.6	10.8	2.6	50.0	39.8
4	17.6	17.8			18.6	10.5	11.4	2.0	60.1	48.4
5	31.7	20.8			21.3	12.3	11.6	1.2	69.3	56.4
6	35.5	23.7			23.8	14.1	11.6	0.3	77.9	63.9
7	39.1	28.3	4.9	0.8	26.2	17.0	11.4	0.7	86.0	72.4
8	42.4	31.5	5.2	1.2	28.3	19.5	10.9	0.8	93.6	80.4
9	45.6	35.2	5.6	1.7	30.5	22.1	10.4	1.0	101.0	88.2
10	48.6	38.8	5.8	2.2	32.4	24.6	9.6	1.1	108.0	95.8
20	86.3	71.1	16.1	6.7	47.1	46.7	8.4	0.1	177.7	163.0

Table 1: Energy levels (hh0,hh1,lh0), Fermi Level (E_F) and deep of the well (V_0) in meV measured from the bottom of the band obtained by means of the self-consistent calculation with exhange (X) and without exchange (H) calculations for a p-type B-δ-doped Si quantum well, as functions of the impurity concentration $p_{2D} = (10^{12}\ cm^{-2})$.

p_{2D}	hh0(X)	hh0(M)	hh1(X)	hh1(M)	lh0(X)	lh0(M)	E_F(X)	E_F(M)	V_0(X)	V_0(M)
2	18.0	15.0			12.2	9.7	9.6	6.7	38.6	36.3
3	23.1	20.7			15.6	13.8	10.8	8.5	50.0	48.4
4	27.6	26.1			18.6	17.5	11.4	10.0	60.1	59.5
5	31.7	31.2			21.3	21.1	11.6	11.2	69.3	69.8
6	35.5	36.1			23.8	24.6	11.6	12.3	77.9	79.6
7	39.1	40.8	4.9	6.6	26.2	28.0	11.4	13.1	86.0	88.9
8	42.4	45.4	5.2	7.7	28.3	31.3	10.9	13.9	93.6	98.0
9	45.6	49.9	5.6	8.8	30.5	34.6	10.4	14.6	101.0	106.7
10	48.6	54.2	5.8	9.9	32.4	37.8	9.6	15.2	108.0	115.2
20	86.3	94.6	16.1	21.0	47.1	67.5	8.4	19.2	177.7	191.2

Table 2: Energy levels (hh0,hh1,lh0), Fermi Level (E_F) and deep of the well (V_0) in meV measured from the bottom of the band obtained by means of the self-consistent calculation with exhange (X) and our model (M) calculations for a p-type B-δ-doped Si quantum well, as functions of the impurity concentration $p_{2D} = (10^{12}\ cm^{-2})$.

p_{2D}	hh0(X)	hh0(M)	hh1(X)	hh1(M)	lh0(X)	lh0(M)
2	8.4	8.3			2.6	3.1
3	12.3	12.2			4.8	5.2
4	16.2	16.1			7.2	7.5
5	20.1	20.0			9.7	9.9
6	23.9	23.8			12.2	12.4
7	27.7	27.6	-6.5	-6.5	14.8	14.9
8	31.5	31.5	-5.7	-6.2	28.3	17.4
9	35.2	35.3	-4.8	-5.7	20.1	20.0
10	39.0	39.1	-3.8	-5.2	22.8	22.6
20	77.9	75.5	7.7	1.8	38.7	48.2

Table 3: Energy levels (hh0,hh1,lh0,hh2) in meV measured from the Fermi level(E_F) obtained by means of the self-consistent calculation with exhange (X) and our model (M) calculations for a p-type B-δ-doped Si quantum well, as functions of the impurity concentration $p_{2D} = (10^{12}\ cm^{-2})$.

The following input parameters: $m^*_{hh} = 0.62m_0$; $m^*_{lh} = 0.087m_0 m_0$ being the free electron mass, $\epsilon_r = 12.5$, and 2×10^{12} cm^{-2} $< p_{2D} < 1 \times 10^{14}$ cm^{-2}. This interval for p_{2D} includes its experimentally interesting values. The range of impurity densities studied includes the values of experimental interest.

In table 4 we present the self-consistent results with (x) and without exchange (H). In this table we can see that, from the relative point of view, while lesser the concentration, the greater the importance of the exchange effect. For $p_{2D} = 2 \times 10^{12}$ cm^{-2}the change of levels, taking as the the origin of energies the bottom of the band, is of the order of 70%. For $p_{2D} = 2 \times 10^{13}$ cm^{-2} the change is less than 5 %. As in the case of the p type delta-doped wells in Si, if we take the bottom of the well as the origin of energies, there is practically no difference between considering the exchange or not.

In table 5 we present the comparison between the approximation defined in equation (25) (M) and the self-consistent calculation with exchange (X) (the self-values are measured with respect to the bottom of the well). For $p_{2D} < 1 \times 10^{13}$ cm^{-2} (incidentally the range of concentrations measured experimentally until now) the difference between our model and the self-consistent calculation is of a few meV and is less as the concentration is less. The differences are higher for higher concentrations. While is we measure them with respect to E_F (table 6) the difference is, in the occupied levels, less than 5% (a considerable difference has recently appeared for the third excited level). This indicates that the difference between the energy levels in our model gives very good results; on many occasions, it is the difference of levels which is measured experimentally.

Wagner et al [25] have grown $Al_{1-x}Ga_xAs/GaAs/Al_{1-x}Ga_xAs$ quantum wells in which they place a Be doping spike with an intended dopant density of 8×10^{12} cm^{-2} at the centre of the $GaAs$ layer. The width of the doping spike measured by secondary-ion mass spectroscopy was found to be less than 10 nm. The two emission peaks observed were assigned to recombination involving two different hole subbands. If the energy difference between the two subbands is just given by the peak energy difference, they find a subband separation of 36 meV. Although our calculation refers to a simpler system, namely (i) strict δ -doping without considering the spread of the dopands and (ii) the δ -well in a GaAs matrix,

p_{2D}	hh0(x)	hh0(H)	hh1(x)	hh1(H)	lh0(x)	lh0 (H)	E_F(x)	E_F(H)	V_0 (x)	V_0 (H)
2	16.2	10.5			8.1	4.4	7.1	3.0	35.7	28.7
3	21.0	14.1			10.6	6.0	7.7	2.4	46.5	38.0
4	25.3	17.4			12.8	7.6	7.7	1.6	56.1	46.4
5	29.3	22.2			14.9	10.9	7.5	2.4	65.0	56.0
6	33.2	26.2	4.1	1.9	16.8	12.4	7.0	2.6	73.2	64.6
7	36.5	28.4	4.6	2.5	18.6	14.5		2.8	81.0	72.9
8	39.8	34.0	5.0	3.0	20.3	16.5	5.4	2.8	88.4	80.8
9	43.9	37.7	6.0	3.6	22.7	18.5	5.7	2.8	96.6	88.5
10	46.3	41.4	7.1	4.2	25.3	20.5	6.3	2.7	104.9	96.0
20	86.3	75.7	16.1	10.7	47.1	39.4	8.4	1.9	177.7	164.3
30	117.5	108.8	22.9	19.5	64.7	58.9	6.1	1.1	237.6	225.7
40	150.0	140.0	32.3	28.0	84.3	77.4	6.3	0.1	296.3	282.7
50	182.0	169.4	42.5	36.1	104.2	95.0	7.3	0.1	352.1	336.0
60	212.5	200.3	52.1	46.8	122.9	114.5	7.3	0.3	405.7	389.4
70	241.5	229.7	61.3	56.71	140.8	133.0	6.6	0.1	453.3	440.1
80	269.6	259.4	70.2	66.2	158.1	151.3	5.6	0.4	505.0	491.2
90	296.7	286.6	78.9	76.6	174.9	169.1	4.1	0.4	552.0	537.0

Table 4: Energy levels (hh0,hh1,lh0), Fermi Level (E_F) and deep of the well (V_0) in *meV* measured from the bottom of the band obtained by means of the self-consistent calculation with exhange (X) and without exhange (H) calculations for a p-type Be-δ-doped GaAs quantum well, as functions of the impurity concentration $p_{2D} = (10^{12}\ cm^{-2})$.

p_{2D}	hh0(X)	hh0(M)	hh1(X)	hh1(M)	lh0(X)	lh0(M)	E_F(X)	E_F(M)	V_0(X)	V_0(M)
2	16.2	15.4			8.1	7.8	7.1	6.4	35.7	35.8
3	21.0	21.3			10.6	11.0	7.7	8.0	46.5	47.8
4	25.3	26.7			12.8	14.1	7.7	9.2	56.1	58.7
5	29.3	31.9			14.9	17.1	7.5	10.1	65.0	69.0
6	33.2	36.9	4.1	6.7	16.8	20.0	7.0	10.9	73.2	78.7
7	36.5	41.8	4.6	7.9	18.6	22.8	6.3	11.4	81.0	88.0
8	39.8	46.5	5.0	9.1	20.3	25.6	5.4	11.9	88.4	97.0
9	43.9	51.1	6.0	10.4	22.7	28.3	5.7	12.3	96.6	105.7
10	46.3	55.6	7.1	11.6	25.3	31.0	6.3	12.6	104.9	114.1
20	86.3	97.0	16.1	24.0	47.1	56.2	8.4	17.7	177.7	190.0
30	117.5	134.7	22.9	36.4	64.7	79.7	6.1	21.4	237.6	256.9
40	150.0	170.0	32.3	48.7	84.3	102.2	6.3	24.0	296.3	318.6
50	182.0	204.2	42.5	61.0	104.2	124.0	7.3	25.8	352.1	376.8
60	212.5	237.1	52.1	73.3	122.9	145.2	7.3	27.1	405.7	432.3
70	241.5	269.1	61.3	85.5	140.8	166.1	6.6	29.4	453.3	485.8
80	269.6	300.6	70.2	97.8	158.1	186.5	5.6	31.4	505.0	537.6
90	296.7	330.8	78.9	109.9	174.9	206.7	4.1	33.2	552.0	588.0

Table 5: Energy levels (hh0,hh1,lh0), Fermi Level (E_F) and deep of the well (V_0) in meV measured form the bottom of the band obtained by means of the self-consistent calculation with exhange (X) and our model (M) calculations for a p-type Be-δ-doped GaAs quantum well, as functions of the impurity concentration $p_{2D} = (10^{12}\ cm^{-2})$.

p_{2D}	hh0(X)	hh0(M)	hh1(X)	hh1(M)	hh2(X)	hh2(M)	lh0(X)	lh0(M)
2	9.1	9.0		-4.3			1.0	1.4
3	13.3	13.3		-4.9			2.9	3.1
4	17.6	17.6		-4.9			5.1	5.0
5	21.8	21.8		-4.7			7.4	7.0
6	26.2	26.1	-2.9	-4.2			9.8	9.2
7	30.2	30.3	-1.7	-3.6			12.3	11.4
8	34.4	34.6	-0.4	-2.8			14.9	13.7
9	38.2	38.8	0.3	-2.0			17.0	16.0
10	42.0	43.0	0.8	-1.0			19.0	18.4
20	77.9	79.3	7.7	6.3			38.7	38.5
30	111.4	113.3	16.8	15.0	-1.3	-9.6	58.6	58.3
40	143.7	146.3	26.0	24,8	1.3	-7.1	78.0	78.3
50	174.7	178.5	35.2	35.3	4.1	-3.7	96.9	98.2
60	205.2	210.0	44.8	46.2	6.9	0.3	118.1	115.6
70	234.9	239.7	54.7	56.2	10.3	3.6	134.2	136.7
80	264.0	268.9	64.6	66.4	14.1	7.1	152.5	155.2
90	292.6	297.6	74.8	76.7	18.2	11.0	170.8	173.5

Table 6: Energy levels (hh0,hh1,lh0,hh2) in meV measured from the Fermi level(E_F) obtained by means of the self-consistent calculation with exhange (S) and our model (M) calculations for a p-type p-type Be-δ-doped GaAs quantum well, as functions of the impurity concentration $p_{2D} = (10^{12}\ cm^{-2})$.

not inside a quantum well, we find an energy difference between the first and second hh levels of 35 meV in self-consistent calculation and 37 meV in M, which is gives a plausible approximation.

Dament et al [27]. These authors have grown Be δ-doped GaAs with p-doping density in the range from 6×10^{12} to 2×10^{14} cm^{-2}. They observed that in this system there are two levels for $p_{2D} = 6 \times 10^{12}$ cm^{-2}. The Fermi level is close to the last level. The difference between the Fermi level and the basic level is $\approx$ 22 meV. For densities up to 6×10^{12} cm^{-2}, several more hole levels are claimed to be observed without further specification. In our calculation it is impossible to observe levels higher than the Fermi level, but the difference between the Fermi level and ground level is about 26 meV in both calculations.

Richards et al.[26] have studied the subband structure of a quasi-two-dimensional hole gas formed at a single Be δ-doped layer in GaAs by means of photoluminescence spectroscopy. They also performed self-consistent subband energies calculations. For an 8×10^{12} cm^{-2} acceptor concentration, with a dopant spread of 2 nm, their X result for the difference $hh0 - lh0$ is ≈ 15.1 meV. Our result for that difference is ≈ 20 meV in both calculations. According to the photoluminescence spectroscopy experiments reported in that work, the spacing between both levels is ≈ 19 meV. Thus, our numerical values are in quite good agreement with the experiment, even when, in our case, no dopant spread is considered.

5. CONCLUSIONS

An analysis of tables 1 and 4 clearly shows that the importance of exchange decreases with concentration in both systems. In general, as the exchange potential depends on $n^{1/3}(z)$, while the Hartree potential depends linearly on concentration, exchange must be more important for lower concentration. It would be a little imprudent to affirm that exchange is indispensable in a particular system as for concentrations higher than those considered, the opposite conclusion can be drawn. However, the few works which calculate the effect of exchange in p type delta-doped wells do not consider the particularities of a hole gas nor do they analyse this effect with concentration[28,29]. The conclusions that both works reach in similar systems illustrate this problem. Both works arrive at antagonistic conclusions. In [28] the effect of exchange and correlation for a p type delta-doped well in GaAs is greater than 20% and therefore cannot be ignored and in [29], the effect of exchange and correlation for a p type delta-doped well in Si is less than 10% and may not be appreciated in an experiment with less precision. The fundamental difference between these two calculations is that very different concentrations are presented; while in [28] the concentration is of $p_{2D} = 3 \times 10^{12}$ cm^{-2}, in [29] it is $p_{2D} = 9 \times 10^{13}$ cm^{-2}, and this is why the works arrive at such apparently antagonistic conclusions. The fact is that in both systems it is indispensable to consider the effects of many bodies for low concentrations, and can be obviated for high ones.

The consideration of the particularities of the holes in the moment of considering the many body effects studied in references [6] and [7] make the importance of these effects less than when these considerations are not made.

The comparison of our results with those of experiments, both those from self-consistent calculations nand those obtained using the model proposed in this work for both systems studied, p type delta-doped wells in Si and p type delta-doped wells in GaAs, are very good. The difference between the experiments and our results are always of a few meV.

From the comparison of the proposed model with the self-consistent calculations (Tables 2,3,5 and 6) we can see that the model, on the whole, works very well. However, it is clear that the results are better for the case of Be in GaAs. The advantages of our model are enormous. Carrying out the self-consistent calculations presented here, in the full range of concentrations studied and in both of the systems presented meant two months work, in spite of experience in this type of calculations. Meanwhile, results were obtained with our model within a week. This is the usual relationship of work-time needed between a self-consistent calculation and some approximation to avoid them (which can be quasi-classical, variational or even LDA). In this sense, this framework, at least in these systems studied, shows itself to be extraordinarily useful.

Work with the Thomas-Fermi-Dirac approximation in systems made up of inhomogeneous hole gases has a great future in solid state physics. It would not be difficult to apply these ideas to doped, simple, double or superlattice semiconductor quantum heterostructures, nor would it be so with double or superlattice delta-doped quantum wells,nor for *nipi* type homostructures. It should play an important role in electronic devices such as FETs, in which some of the channels are formed by p type impurities. In spite of the almost inexistent application of this approximation in solid state and its bad reputation aquired in atomic calculations, one can predict a great utility. On the road which leads from the study of the electronic properties of semicondictor quantum systems to the design of electronic devices using these properties, there is no other way than proposing theoretical approximations which, as they are proved with more precise calculations and in a wide range of parameters, give good results and allow the susbtitution of complex calculations by simple calculations in order to play with the parameters and thus obtain the desired result in a specific design.

6. *

References

[1] S. Lundqvist and N. H. March, Theory of the Inhomogeneous Electron Gas (Plenum, 1983).

[2] G. D. Mahan and K. R. Subbaswamy, Local Density Theory of Polarizability (Plenum, 1990).

[3] M.L. Glaser, F. García-Moliner and V.R. Velasco, Physica Scripta,**43**, 512 (1991).

[4] L. M. Gaggero-Sager, Journal of Mathematical Chemistry **4**, 317 (1999).

[5] L. M. Gaggero-Sager and S. J. Vlaev, physica status solidi (b) **215**, 1049 (1999).

[6] P.A. Bobbert, H. Wieldraaijer, R.van der Weide, M. Kemerink, P.M. Koenraad, and J.H. Wolter, Phys. Rev. B **56**, 3664 (1997).

[7] M. Combescot and P. Nozi`eres, J.Phys. C **5**, 2369 (1972).

[8] C. A. C. Mendoca, F. Plentz, J. B. B. Oliveira, E. A. Meneses, L. M. R. Scolfaro, D. Beliaev, S. M. Shibli and J. R. Leite, Phys. Rev. B **48**, 12316 (1993).

[9] G. Tempel, F. Muller, N. Scharz, F. Koch, G. Weiman, H.P. Zeindl, and I. Eisele, Surface Sci. **228**, 247 (1990).

[10] M. Hirai, H. Ohnishi, K. Fujita, P. Vaccaro, and T. Watanabe, Journal of Crystal Growth **150**, 209 (1995).

[11] L. M. R. Scolfaro, D. Beliaev, P. Enderlein and J. R. Leite, Phys. Rev. B. **50**, 8699 (1994).

[12] D. A. Woolf, K. C. Rose, S. J. Morris, D. I. Westwood, J. Rumberg, F. Reinhardt, W. Richer, and R. H. Williams, Journal of Crystal Growth **150**, 197 (1995).

[13] T. Iida, Y. Makita, S. Kimura, Y. Kawasumi, A. Yamada, S. Uekusa, and T. Tsukamoto, Journal of Crystal Growth **150**, 236 (1995).

[14] N. Y. Li, H. K. Dong, C. W. Tu, and M. Geva, Journal of Crystal Growth **150**, 246 (1995).

[15] L. M. Gaggero-Sager, M. E. Mora-Ramos,and A. Contreras-Solorio, Phys. Rev. B **57**, 6286 (1998).

[16] S. J. Vlaev and L. M. Gaggero-Sager, Phys. Rev. B **58**, 1142 (1998).

[17] L. M. Gaggero-Sager, M. E. Mora-Ramos Solid-State Electronics bf 44, 175 (2000).

[18] L.M. Gaggero Sager and R. Pérez Alvarez, Journal of Applied Physics **74** :12, 4566 (1995).

[19] L.M. Gaggero Sager and R. Pérez Alvarez, Journal of Applied Physics **79**:5, 3351 (1996).

[20] L. M. Gaggero Sager and R. Pérez Alvarez physica status solidi (b) **197**, 105 (1996).

[21] L.M. Gaggero Sager and R. Pérez Alvarez. Applied Physics Letter **70** 212 (1997).

[22] L. M. Gaggero Sager and M. E. Mora Ramos, Materials Science and Engineering B. **47** 279 (1997).

[23] L.M. Gaggero Sager and M. E. Mora-Ramos. physica status solidi (b) **210**, 81 (1998).

[24] J-h Zhu, D-w Gong, B. Zhang, F. Lu, C. Sheng, H-h Sun, and X. Wang, Phys. Rev. B. **52**, 8959 (1995).

[25] J. Wagner, A. Ruiz, and K. Ploog, Phys. Rev. B. **43**, 12134 (1991).

[26] D. Richards, J. Wagner, H. Schneider, G. Hendorfer, M. Maier, A. Fischer, and K. Ploog, Phys. Rev. B. **47**, 9629 (1993).

[27] T. C. Damen, M. Fritze, A. Kastalsky, J. E. Cunningham, R. N. Pathak, H. Wang, and J. Shah, Appl. Phys. Lett. **67**, 515 (1995).

[28] G. M. Sipahi, P. Enderlein, L. M. R. Scolfaro and J. R. Leite, Phys. Rev. B. **53**, 9930 (1996).

[29] G. Karunasiri, Jpn. J. Appl. Phys. **33**, 2401 (1994).

Index

R

S

T

V